Springer Tracts in Modern Physics
Volume 221

Springer Tracts in Modern Physics

Springer Tracts in Modern Physics provides comprehensive and critical reviews of topics of current interest in physics. The following fields are emphasized: elementary particle physics, solid-state physics, complex systems, and fundamental astrophysics.
Suitable reviews of other fields can also be accepted. The editors encourage prospective authors to correspond with them in advance of submitting an article. For reviews of topics belonging to the above mentioned fields, they should address the responsible editor, otherwise the managing editor.
See also springer.com

Michael E. Dreyer

Free Surface Flows under Compensated Gravity Conditions

With 128 Figures

Springer

Michael E. Dreyer
University of Bremen
ZARM
Center of Applied Space Technology
and Microgravity
Am Fallturm
28359 Bremen
E-mail: dreyer@zarm.uni-bremen.de

Library of Congress Control Number: 2006932585

Physics and Astronomy Classification Scheme (PACS):
47.55.N- Interfacial flows, 47.55.nb Capillary and thermocapillary flows, 47.35.Pq Capillary waves

ISSN print edition: 0081-3869
ISSN electronic edition: 1615-0430
ISBN-10 3-540-44626-5 Springer Berlin Heidelberg New York
ISBN-13 978-3-540-44626-2 Springer Berlin Heidelberg New York

Springer is a part of Springer Science+Business Media
springer.com

Typesetting: by the author and techbooks using a Springer LaTeX macro package
Cover production: WMXDesign GmbH, Heidelberg

Printed on acid-free paper SPIN: 11743286 56/techbooks 5 4 3 2 1 0

Preface

The work described in this book was performed during my employment at the Center of Applied Space Technology and Microgravity (ZARM) of the University of Bremen in the years 1994 to 2006. Some of the results were published in papers and reports by (in alphabetical order) Jens Gerstmann, Mark Michaelis, Antje Ohlhoff, Uwe Rosendahl, Michael Stange, Gerrit Wölk, and myself. The contents of these papers are partly included in this work to make it a thorough and comprehensive document. I am very grateful to my colleagues since this book would have never been possible without their cooperation.

First of all I would like to express my gratitude to the head of ZARM, the chair for mechanics and fluid mechanics in the faculty of production engineering and technology of the University of Bremen, Prof. Dr.-Ing. Hans J. Rath, who gave me the freedom and the possibilities to perform this work. Furthermore, I acknowledge with great pleasure the cooperation with my colleagues (in alphabetical order) Dr. Volker Baumbach, Dr. Christian Berg, Frank Ciecior, Dennis Haake, Dr. Armin de Lazzer, Holger Faust, Carsten Fechtmann, Dr. Jens Gerstmann, Dr. Aleksander Grah, Dennis Haake, Prof. Dieter Langbein, Ronald Mairose, Dr. Mark Michaelis, Dr. Karine Odic, Dr. Antje Ohlhoff, Peter Prengel, Uwe Rosendahl, Dr. Uwe Schmid, Markus Stadtlander, Dr. Michael Stange, Malte Stief, and Dr. Gerrit Wölk.

The funding for the research projects from the Federal Ministry for Education and Research through the German Aerospace Center e.V. (DLR) with grant numbers 50 WM 0535, 50 WM 0241, 50 WM 9901, 50 JR 0011, 50 TT 9737, and for the sounding rocket flights TEXUS 41 and TEXUS 42, the funding from the Deutsche Forschungsgemeinschaft DFG under grant number RA 352/11-1, the funding from the European Space Agency ESA for drop tower flights and the sounding rocket flight TEXUS 37 as well as for the topical team on *Liquid Management in Space* under ESTEC Contract 14160/00/NL/JS and the support by NASA John H. Glenn Research Center, Cleveland, Ohio and the National Center for Microgravity Research in Fluids and Combustion in Cleveland, Ohio, to prepare the science concept document for our space experiment on *Critical Velocities in Open Capillary Channel Flows* (CCF) is gratefully acknowledged. Many thanks are due to the EADS ST company in Bremen for their numerous contracts to investigate microgravity liquid management problems.

I personally would like to thank the following people outside ZARM for their cooperation and their contributions to this work, namely from EADS ST Dr. Phillip Behruzi, Burckhard Franke, Dieter Grothe, Dr. Irmin Meier, Dr. Gaston Netter, and Dr. Andreas Rittweger, from CEA Dr. Olivier Lebaigue, from CNES Dr. Bruno Vielle, from the Collège de France Dr. David Quéré, from the German Aerospace Center DLR Dr. Horst Binnenbruck, Dr. Rainer Forke, Karl-Otto Jung, Dr. Rainer Kuhl, Dr. Thilo Kranz, Dr. Claus Lippert, Hans Meusemann, and Dr. Peter Preu, from ESA Dr. Wolfgang Herfs, Dr. Ewald Kufner, Dr. Olivier Minster, and Dr. Martin Zell, from LEGI Prof. Emil Hopfinger, from Lockheed Martin Jim Tegart, from NCMR Dr. Enrique Ramé, from NASA Glenn Research Center in Cleveland Dr. Fred Kohl and Brian Motil, from Portland State University Prof. Mark Weislogel, from the Technical University of Berlin Prof. Günther Frohberg, from the Technical University of München Prof. Johannes Straub, from ONERA the director of the fluid mechanics and energetics branch Paul Kuentzmann, from PMD Technology Don Jaekle Jr., from Purdue university Prof. Steven Collicott, from the University of Groningen Prof. Arthur E.P. Veldman, and from the Friedrich-Alexander University of Erlangen-Nürnberg Prof. Eberhard Bänsch. I like to thank the following persons for valuable comments to the manuscript: S. Collicott, A. Grah, K. Odic, M. Stief, and M. Weislogel.

In particular I am grateful to Prof. Antonio Delgado, the chair for fluid mechanics at the Friedrich-Alexander University of Erlangen-Nürnberg and Prof. Cam Tropea, the chair for fluid mechanics and aerodynamics at the Technical University of Darmstadt for their cooperation in my postdoctoral lecture qualification process and their expert opinion on this professorial dissertation. Finally, I would like to mention the friendly pressure Antonio Delgado put on me to finish this work in due time. Thank you very much for this guidance.

Finalmente quiero agradecerle a mi esposa Tomasa Méndez-Dreyer por su ayuda, paciencia y amor. Yo nunca hubiera terminado esta obra sin su apoyo. A ella le dedico este libro[1].

Bremen — *Michael E. Dreyer*
July 2006

[1] Finally I would like to express my deep gratitude to my wife for her help, patience and love. I would never have finished this work without her support. I dedicate this book to her.

Contents

Part III Capillary Rise in Cylindrical Tubes

Part IV Critical Velocities in Open Capillary Channel Flows

1

Introduction

In the compensated gravity environment of a spacecraft the hydrostatic pressure decreases to very low values depending on the residual acceleration, and surface tension forces become dominant. The term compensated gravity is used through the text synonymous to the terms reduced gravity, microgravity, low gravity, and zero gravity. The physical meaning is the fact that the body force due to gravity (which always exists in low earth and geostationary orbits) is compensated by another body force due to the acceleration of the system under consideration. The difference between these two accelerations is the residual acceleration.

Surface tension forces can be used to transport and position liquids if the residual acceleration results in a hydrostatic pressure that is small compared to the capillary pressure. One prominent application is the use of propellant management devices (PMD) in surface tension satellite tanks. The task for a PMD is to keep the tank outlet covered with liquid whenever outflow is demanded. Furthermore, PMDs are used to ensure expulsion and refilling of tanks for liquids and gases for life support, reactants, and experiment supplies. In the following we consider the behavior of fluids in a low gravity environment with special emphasis on applications in PMD systems. Fundamental knowledge of the fluid mechanics of capillary problems is very important since most of the PMD designs are not testable on ground and thus completely rely on analytical or numerical concepts.

In this work three different flow problems are treated with analytical, numerical and experimental means: the transient contour change between two static surface configurations (surface oscillation), the capillary rise in tubes (capillary rise), and the flow through open capillary channels (critical velocities). These problems are linked together by the same set of equations and boundary conditions which are necessary to model the fluid behavior, and by the same set of characteristic numbers.

The book is divided into four parts. The first part introduces the basic equations, the dimensionless parameters and time scales for all problems under consideration. Some important values of the dimensionless parameters are

Michael E. Dreyer: *Free Surface Flows under Compensated Gravity Conditions*
STMP **221**, 1–3 (2007)
DOI 10.1007/3-540-44628-1_1

given. It is followed by a literature survey of the most relevant papers for this field.

The second part deals with the transient contour change which occurs in right circular cylinders with a confined volume of liquid as well as during the initial stages of the capillary rise into a cylindrical tube from an infinite reservoir, if a step reduction of the body force occurs and the system becomes surface tension dominated. The free surface moves through a damped oscillation from the initially flat surface under gravity to a surface with constant curvature under compensated gravity. The problem is treated experimentally and numerically with special emphasis on the boundary condition at the wall. The results are important for the designer of PMDs since changes in direction and magnitude of the residual acceleration occur during attitude control maneuvers of the spacecraft. The resonance frequencies may influence the performance of the PMD system and the satellite attitude control system as well and the settling time determines the damping of the system until the new surface configuration is established.

In the third part we consider the capillary rise in tubes if the reservoir is infinite and a wetting liquid is allowed to spread into the system without limits. The tube is a basic model for capillary vanes which are used in surface tension satellite tanks to connect the bulk liquid with the reservoir if the residual accelerations due to spacecraft maneuvers have orientated the bulk far away from the outlet, and to allow the filling of the reservoir after the maneuver by capillary forces. We present experimental results for a wide range of the dimensionless parameters and a one-dimensional theory which is corroborated by the experiments within a few percent. The respective importance of the major terms of the equation in the different stages of the capillary rise is shown and enables the PMD designer to build simplified models for their practical needs.

The fourth part addresses the flow through an open capillary channel which can be considered as a propellant pipeline whose cross-sectional area is partly confined by free liquid surfaces. To prevent the collapse of these free surfaces and thus gas ingestion at the outlet, the limits of the flow velocity have to be known. Capillary channels (or vanes) used in PMDs have to guarantee bubble free supply of propellant to the engine. Similarities of this flow problem exist to compressible ducts flows and open channel flows under terrestrial conditions. The flow rate of these flows is limited due to choking. The theory of choked flows predicts a limiting velocity corresponding to a characteristic signal velocity. The signal velocity in open capillary channels is determined by the speed of longitudinal waves. This velocity is derived from the capillary pressure and requires a mathematical description as well as a precise measurement of the free surface. The theory of choking in capillary channels is validated by our experimental data, but necessitates further investigations. The author would be glad to encourage additional theoretical, numerical and experimental research on that topic.

Finally the appendix gives some additional information about the inlet and the entrance pressure losses of the considered flow problems. The inlet flow takes place before the closed cross section of the test channel and is characterized by strong convective accelerations and weak molecular momentum transport. In the entrance region of the test section a boundary layer with strong molecular momentum transport develops from the wall towards the center and the increasing displacement thickness leads to a weak convective acceleration. Depending on the REYNOLDS number of the flow problem and the length of the test section, the laminar flow profile may not be developed and the boundary layer may characterize the velocity distribution in the test section. The two different sections of the flow, the inlet and the entrance flow in the above defined sense, are modeled in the appendix by analytical and numerical means emphasizing empirical relations which can be implemented in the complex flow models.

Part I

State of the Art

The first part of this book introduces the basic equations, here the equation of mass conservation and the equation of linear momentum, and the respective boundary conditions for the isothermal case. The scaling with the characteristic values of the problems under consideration yields a set of dimensionless numbers. The orders of magnitude of these dimensionless numbers are derived from the properties and lengths of the spacecraft propellants and their vessels. The choice of the test liquids aims at a maximum similitude between experiments and application. These applications are the management of storable and cryogenic propellants in satellite and rocket tanks. The literature review focuses first on the heritage of liquid management in space and in particular on surface tension tank technology. The modern design of tanks is mainly based on computational fluid dynamic tools such as FLOW-3D, FLUENT or others. As shown in several benchmarks, the bulk behavior of the liquid is described quite well by such tools, but the analysis requires boundary conditions such as the contact angle. Contact angle hysteresis and a dynamic contact angle behavior cannot be computed on the large mesh sizes required for a full tank scale. Thus, validated empirical correlations with experimental evidence are needed as input for the computations. A large portion of the literature review deals with the scales and the different physical mechanisms of the contact angle. Even though non-isothermal effects are not considered in this book, some thermal effects on the contact angle are described.

Some experimental studies on surface oscillations upon step reduction of the acceleration level (using a drop tower) are available and have been used to design the experiments described in Part II in this study. Rise time, frequencies and settling characteristics in dependence of the contact angle have been reported. The main trends have been corroborated by recent numerical studies showing the influence of the contact angle condition on frequency and damping. This book fills in some gaps left in the space of the non-dimensional numbers for this problem and provides more accurate data of the surface shape using modern high-speed camera systems.

A large amount of papers is available on the capillary rise of liquid in tubes, both in terrestrial and microgravity environments. They have been reviewed in retrospect of a thorough theoretical description given in Part III. Washburn's approach from 1920 is still used today to describe the wicking action of porous media, a capillary driven flow in pores of small size (more precisely at low Bond numbers). But a comprehensive theory corroborated by experiments in all time regimes is still lacking. This gap has been closed with this book.

Capillary dominated flows in channels with open cross section have been treated by very few authors only. The existence of a universal limitation similar to choking in compressible duct flows has been mentioned only once. Therefore, the findings in Part IV describe a totally new flow phenomena.

2

Basic Equations and Dimensionless Numbers

In this chapter we discuss the equation of mass conservation and the equation of linear momentum, and their respective boundary conditions. We assume that material continuum properties of the liquid phases such as density ρ, dynamic viscosity μ, and surface tension σ are constant. For the surface tension we consider the static value, thus the effect of a dynamic surface tension is neglected. This effect may occur at the very first beginning of the creation of a new surface and is related to orientations of the molecules at the free surface and contamination of the free surface. The effect of curvature on surface tension [1, 2, 3] will not be taken into account. Likewise, molecular influences arising in thin liquid films are supposed not to influence the global behavior of the meniscus, thus interfaces are assumed to behave always, as if bulk masses are incorporated.

The equations of mass conservation and of linear momentum read (see, for example, [4]):

$$\nabla \cdot \boldsymbol{u} = 0 \ , \tag{2.1}$$

$$\underbrace{\frac{\partial \boldsymbol{u}}{\partial t}}_{u} + \underbrace{(\boldsymbol{u} \cdot \nabla)\,\boldsymbol{u}}_{c} = \underbrace{-\frac{1}{\rho}\nabla p}_{p} + \underbrace{\nu \nabla^2 \boldsymbol{u}}_{v} + \underbrace{\boldsymbol{k}}_{g} \ . \tag{2.2}$$

For the following discussion the respective terms in the NAVIER-STOKES-Equation (2.2) are named following their physical meaning (see [5]): *unsteady* (change of momentum with time), *convective* momentum transport (change of momentum in space), static *pressure* gradient, *viscous* or molecular momentum transport and body force due to *gravitational* or residual accelerations.

The boundary conditions are:

1. No slip at solid walls (The stress singularity, which occurs due to the no-slip condition if the contact line moves over the solid, is considered in Sect. 4.3)
2. The kinematic boundary condition for the free surface (The velocity at the free boundary is equal to the velocity of the surface itself)

Michael E. Dreyer: *Free Surface Flows under Compensated Gravity Conditions*
STMP **221**, 7–12 (2007)
DOI 10.1007/3-540-44628-1_2

3. The stress boundary condition at the free surface

$$\underbrace{-\boldsymbol{n}\,(p-\widehat{p})}_{\text{I}}+\underbrace{\boldsymbol{n}\cdot\tau-\boldsymbol{n}\cdot\widehat{\tau}}_{\text{II}}=\underbrace{\sigma\,(\nabla_S\cdot\boldsymbol{n})\,\boldsymbol{n}}_{\text{III}}-\underbrace{\nabla_S\,\sigma}_{\text{IV}}\ . \tag{2.3}$$

with the normal vector $\boldsymbol{n}$ of the interface, the viscous stress tensor τ of the liquid and $\hat{\tau}$ of the gas. The operator ∇_S yields the gradient of the normal vector (giving the curvature) and the gradient of the surface tension at the interface. The terms I - IV can be described as follow:

Term I: pressure difference across the interface
Term II: viscous stress difference across the interface
Term III: normal stress difference due to the curvature of the interface
Term IV: tangential stress difference due surface tension variation

In the following terms II and IV are neglected (surface tension is constant, passive overlaying gas phase, no tangential forces act along the free surface) and the pressure difference across the free surface reduces to the GAUSS-LAPLACE equation

$$p-p_a=\Delta p_c=-2\sigma H=-\sigma\left(\frac{1}{R_1}+\frac{1}{R_2}\right)\ , \tag{2.4}$$

with the main radii of curvature R_1 and R_2 of the free surface[1]. The gas pressure above the liquid surface is always the ambient pressure for the problems discussed in this work.

4. The contact angle γ of the free surface at the contact line with the solid wall. The contact angle depends on the velocity of the contact line as well as on the physical (i.e. roughness) and chemical nature of the solid surface. The particular boundary conditions are specified with more detail in each section.

2.1 Dimensionless Numbers and Time Scales

Scaling (2.1) and (2.2) with the characteristic velocity V, length L, pressure σ/L and acceleration k (gravity acceleration on earth, or any other acceleration), thus:

$$\boldsymbol{u}^*=\frac{\boldsymbol{u}}{V},\quad \nabla^*=L\nabla,\quad {\nabla^*}^2=L^2\nabla^2,\quad t^*=\frac{tV}{L},\quad p^*=\frac{pL}{\sigma},\quad \boldsymbol{k}^*=\frac{\boldsymbol{k}}{k}\ ,$$

yields the dimensionless equations of mass and momentum conservation:

$$\nabla^*\cdot\boldsymbol{u}^*=0\ , \tag{2.5}$$

[1] In this work we follow the sign convention that the radius of curvature is positive if it lies on the concave side of the curve. Or in other words, the capillary pressure is higher in the fluid with the convex surface shape.

$$\frac{\partial \boldsymbol{u}^*}{\partial t^*} + (\boldsymbol{u}^* \cdot \nabla^*)\, \boldsymbol{u}^* = -\frac{1}{\mathrm{We}} \nabla^* p^* + \frac{1}{\mathrm{Re}} {\nabla^*}^2 \boldsymbol{u}^* + \frac{1}{\mathrm{Fr}} \boldsymbol{k}^* \tag{2.6}$$

with the boundary condition:

$$\Delta p_c^* = -2H^* = -\left(\frac{1}{R_1^*} + \frac{1}{R_2^*}\right) . \tag{2.7}$$

The dimensionless numbers occurring in (2.6) are the WEBER number We (ratio of the convective momentum flux to the capillary pressure or ratio of system velocity to capillary dominated disturbance speed)

$$\mathrm{We} = \frac{\rho L V^2}{\sigma} , \tag{2.8}$$

the REYNOLDS number Re (ratio of convective momentum flux to viscous momentum flux)

$$\mathrm{Re} = \frac{LV}{\nu} \tag{2.9}$$

($\nu = \mu/\rho$), which can also be written as

$$\mathrm{Re} = \frac{\mathrm{We}}{\mathrm{Ca}}$$

with the capillary number Ca (ratio of viscous momentum flux to capillary pressure)

$$\mathrm{Ca} = \frac{\mu V}{\sigma} , \tag{2.10}$$

and the FROUDE number Fr (ratio of system velocity to gravity dominated disturbance speed)

$$\mathrm{Fr} = \frac{V^2}{kL} , \tag{2.11}$$

which can be written as

$$\mathrm{Fr} = \frac{\mathrm{We}}{\mathrm{Bo}}$$

with the BOND number Bo (ratio of hydrostatic pressure to capillary pressure)

$$\mathrm{Bo} = \frac{\rho k L^2}{\sigma} . \tag{2.12}$$

2.1.1 Global Time Domain

In most cases, as considered in this work, a characteristic velocity is not known a priori in capillary driven problems. An appropriate velocity scale is therefore $V = \sqrt{\sigma/(\rho L)}$, well know from the dispersion relation for capillary waves (see [6]). Inserting this velocity into (2.6) yields

$$\frac{\partial \boldsymbol{u}^*}{\partial t^*} + (\boldsymbol{u}^* \cdot \nabla^*)\, \boldsymbol{u}^* = -\nabla^* p^* + \mathrm{Oh}\ \Delta^* \boldsymbol{u}^* + \mathrm{Bo}\ \boldsymbol{k}^* . \tag{2.13}$$

The OHNESORGE number as a measure for the molecular momentum transport in capillary dominated regimes is introduced here:

$$\mathrm{Oh} = \frac{\mathrm{Ca}}{\sqrt{\mathrm{We}}} = \left(\frac{\rho\nu^2}{\sigma L}\right)^{1/2} . \tag{2.14}$$

The OHNESORGE number is dedicated to VON OHNESORGE who was one of the first researchers observing the different types of liquid jet disintegration [7]. On occasion in the Anglo-American literature the SURATMAN number is used instead of the Ohnesorge number with the definition $\mathrm{Su} = 1/\mathrm{Oh}^2$.

2.1.2 Initial Time Domain

If additionally the characteristic length is not known, the capillary length

$$L_c = \left(\frac{\sigma}{\rho k}\right)^{1/2} \tag{2.15}$$

may be used, i.e. for the very initial response of a free surface upon step reduction of the BOND number. Then the momentum transport equation reads:

$$\frac{\partial \boldsymbol{u}^*}{\partial t^*} + (\boldsymbol{u}^* \cdot \nabla^*)\, \boldsymbol{u}^* = -\nabla^* p^* + \mathrm{Mo}\, {\nabla^*}^2 \boldsymbol{u}^* + \boldsymbol{k}^* . \tag{2.16}$$

The new dimensionless number here is the MORTON number Mo

$$\mathrm{Mo} = \left(\frac{\nu^4 \rho^3}{\sigma^3} k\right)^{1/4} , \tag{2.17}$$

which is well known in two-phase flows. The MORTON number can be seen as an OHNESORGE number computed with the capillary length. When a step reduction of the BOND number is considered the initial value of the acceleration k_i has to be used.

If only two terms of the NAVIER-STOKES-Equation (2.6) are dominating the flow, the comparison of these terms yields different time and length scales. If only the pressure and the unsteady term are dominant, i.e. in the beginning of a capillary driven flow, the time scale reads

$$t_{pu} = \sqrt{\frac{\rho L^3}{\sigma}} , \tag{2.18}$$

known as the characteristic time scale for the oscillation of drops and bubbles. The same time scale results from comparing the pressure and the convective term. The corresponding velocity is

$$u_{pu} = \left(\frac{\sigma}{\rho L}\right)^{1/2} . \tag{2.19}$$

In the case of $L = L_c$, this time reads

$$t_{puLc} = \left(\frac{\sigma}{\rho k^3}\right)^{1/4} \tag{2.20}$$

and the velocity becomes

$$u_{puLc} = \left(\frac{k\sigma}{\rho}\right)^{1/4} . \tag{2.21}$$

If only pressure and viscous forces are dominating the flow, the respective time scale

$$t_{pv} = \frac{\mu L}{\sigma} , \tag{2.22}$$

is an important measure for the time needed for a disturbance to settle to the final static surface configuration. Comparing the viscous and the unsteady terms yields the well known viscous time scale

$$t_{vu} = \frac{L^2}{\nu} . \tag{2.23}$$

The corresponding time with $L = L_c$ reads

$$t_{vuLc} = \frac{\sigma}{\mu k} . \tag{2.24}$$

If gravity or residual accelerations play a dominant role in the fluid flow, the following time and length scales are useful: pressure $\sim$ gravity $\rightarrow L_c = \sqrt{\sigma/\rho k}$ (capillary length), gravity $\sim$ unsteady $\rightarrow t_{gu} = \sqrt{L/k}$, viscous $\sim$ gravity $\rightarrow$ $t_{vg} = \nu/(kL)$. The convective time scale is defined as $t_{cu} \sim L/V$ (comparison of the unsteady and the convective terms).

As a result of these considerations the dimensionless numbers can be expressed as ratios of characteristic times or lengths. To simplify the notation of the dimensionless number the kinematic viscosity $\nu = \mu/\rho$ and the kinematic surface tension $\beta = \sigma/\rho$ are used in the following.

$$\mathrm{Bo} = \frac{1}{\beta} k L^2 = \frac{t_{pu}^2}{t_{gu}^2} = \frac{L^2}{L_c^2}$$

$$\mathrm{We} = \frac{1}{\beta} L V^2 = \frac{t_{pu}^2}{t_{cu}^2}$$

$$\mathrm{Ca} = \frac{\nu}{\beta} V = \frac{t_{pv}}{t_{cu}}$$

$$\mathrm{Oh} = \sqrt{\frac{\nu^2}{\beta}\frac{1}{L}} = \frac{t_{pu}}{t_{vu}}$$

$$\mathrm{Mo} = \left(\frac{\nu^4}{\beta^3}k\right)^{1/4} = \frac{t_{puLc}}{t_{vuLc}}$$

$$\mathrm{Re} = \frac{1}{\nu}LV = \frac{t_{vu}}{t_{cu}}$$

$$\mathrm{Fr} = \frac{V^2}{Lk} = \frac{t_{gu}^2}{t_{cu}^2}$$

$$\mathrm{Ga} = \frac{kL^3}{\nu^2} = \frac{t_{gu}^2}{t_{vg}^2}$$

References

1. R.C. Tolman: J. Chem. Phys. **17**, 333 (1949)
2. S. Ono, S. Kondo: Molecular Theory of Surface Tension in Liquids. In: *Handbuch der Physik, Bd. 10: Struktur der Flüssigkeiten*, ed by S. Flügge (Springer, Berlin Göttingen, Heidelberg 1960) p. 134
3. H.T. Davis: NASA CR-168848, 129 (1982)
4. F.M. White: *Fluid Mechanics*, (McGraw Hill, New York 1986) p. 209
5. R.B. Bird, W.E. Stewart, E.N. Lightfoot: *Transport Phenomena*, (John Wiley, New York 1960) p. 76
6. L.D. Landau, E.M. Lifschitz: *Lehrbuch der Theoretischen Physik, Band 6, Hydrodynamik*, (Akademie Verlag, Berlin 1991) p. 275
7. H. Brauer: *Grundlagen der Einphasen- und Mehrphasenströmungen*, (Verlag Sauerländer, Aarau (Schweiz) 1971)

3

Orders of Magnitude of the Dimensionless Numbers

Liquids used in spacecrafts are among others storable propellants (monomethylhydrazine MMH, nitrogen tetroxide N_2O_4, hydrogen peroxide H_2O_2) and cryogenic propellants (liquid hydrogen, liquid oxygen, liquid methane). The relevant properties for this study (density ρ, dynamic viscosity μ, surface tension σ, and refractive index n) are given in Table 3.1 together with some test liquids to compute the order of magnitude of the dimensionless numbers. The data for the cryogenic liquids have been taken from Barron [1].

Table 3.1. Properties of liquid propellants and test liquids at $p = 101.3$ kPa

Liquid	T (°C)	T (K)	ρ (kg/m³)	μ (kg/m s) 10^{-3}	σ (kg/s²) 10^{-3}	n (-)
MMH	45	318	851	0.576	31	
N_2O_4	45	318	1385	0.309	21	
H_2O_2	20	293	1450	1.250	80	
Methane	−161	112	424	0.118	14	
Oxygen	−183	90	1141	0.189	13	
Hydrogen	−253	20	71	0.013	2	
SF 0.65	25	298	757	0.494	15	1.377
SF 1.0	25	298	814	0.816	17	1.384
SF 10.0	25	298	931	9.350	20	1.401
FC-77	25	298	1760	1.422	14	1.281
Butylbenzene	25	298	857	1.260	25	1.488
Detra	25	298	880	2.205	30	1.491
Heptane	25	298	679	0.390	20	1.385
Argon	−186	87	1394	0.252	11	
Nitrogen	−196	77	807	0.165	9	1.205
Neon	−246	27	1206	0.130	5	
Helium-4	−269	4	125	0.004	0.09	

Michael E. Dreyer: *Free Surface Flows under Compensated Gravity Conditions*
STMP **221**, 13–15 (2007)
DOI 10.1007/3-540-44628-1_3

Table 3.2. Combinations of liquid properties

Liquid	T (°C)	T (K)	$\mu/\rho = \nu$ (m^2/s) 10^{-6}	$\sigma/\rho = \beta$ (m^3/s^2) 10^{-6}	σ/μ (m/s)
MMH	45	318	0.677	36	54
N_2O_4	45	318	0.223	15	69
H_2O_2	20	293	0.862	55	64
Methane	−161	112	0.278	33	119
Oxygen	−183	90	0.167	12	69
Hydrogen	−253	20	0.188	28	147
SF 0.65	25	298	0.65	20	31
SF 1.0	25	298	1.00	20	20
SF 10.0	25	298	10.0	21	2
FC-77	25	298	0.81	8	10
Butylbenzene	25	298	1.47	29	20
Detra	25	298	2.51	34	14
Heptane	25	298	0.57	29	51
Argon	−186	87	0.18	8	44
Nitrogen	−196	77	0.20	11	55
Neon	−246	27	0.11	4	37
Helium-4	−269	4	0.03	0.7	26

Since the properties ρ, μ, and σ enter the dimensionless numbers always in a combination, these combinations are given in Table 3.2.

To compare the applicability of the experimental data on real spacecraft tank systems, the dimensionless numbers for propellants and some characteristic lengths are computed in Table 3.3. For the computation of the MORTON number and the capillary length the initial level of the acceleration k_i should be used because the initial configuration of the free surface plays an important role in the pressure distribution. The BOND number is composed using the final acceleration k_f of the spacecraft because this defines the final, steady

Table 3.3. Dimensionless numbers for propellants. The characteristic length is taken as $L = 1$ m

Liquid	T (K)	Mo (-) $k_i = 1\ g_0$	Oh (-)	Bo (-) $k_f = 10^{-5} g_0$	L_c (m) $k_i = 1\ g_0$
MMH	318	$3 \cdot 10^{-3}$	$1 \cdot 10^{-4}$	3	$2 \cdot 10^{-3}$
N_2O_4	318	$2 \cdot 10^{-3}$	$6 \cdot 10^{-5}$	6	$1 \cdot 10^{-3}$
H_2O_2	293	$2 \cdot 10^{-3}$	$1 \cdot 10^{-4}$	2	$2 \cdot 10^{-3}$
Methane	112	$1 \cdot 10^{-3}$	$5 \cdot 10^{-5}$	3	$2 \cdot 10^{-3}$
Oxygen	90	$1 \cdot 10^{-3}$	$5 \cdot 10^{-5}$	8	$1 \cdot 10^{-3}$
Hydrogen	20	$9 \cdot 10^{-4}$	$4 \cdot 10^{-5}$	4	$2 \cdot 10^{-3}$

Table 3.4. Values of the characteristic times and velocities for propellants. The characteristic length is taken as $L = 1$ m, the characteristic initial acceleration as $k_i = 1\ g_0$

Liquid	T (K)	t_{puLc} (s)	u_{puLc} (m/s)	t_{pu} (s)	u_{pu} (m/s)	t_{vu} (s)
MMH	318	$1 \cdot 10^{-2}$	$1 \cdot 10^{-1}$	$2 \cdot 10^{2}$	$6 \cdot 10^{-3}$	$1 \cdot 10^{6}$
N_2O_4	318	$1 \cdot 10^{-2}$	$1 \cdot 10^{-1}$	$3 \cdot 10^{2}$	$4 \cdot 10^{-3}$	$4 \cdot 10^{6}$
H_2O_2	293	$2 \cdot 10^{-2}$	$1 \cdot 10^{-1}$	$1 \cdot 10^{2}$	$7 \cdot 10^{-3}$	$1 \cdot 10^{6}$
Methane	112	$1 \cdot 10^{-2}$	$1 \cdot 10^{-1}$	$2 \cdot 10^{2}$	$6 \cdot 10^{-3}$	$4 \cdot 10^{6}$
Oxygen	90	$1 \cdot 10^{-2}$	$1 \cdot 10^{-1}$	$3 \cdot 10^{2}$	$3 \cdot 10^{-3}$	$6 \cdot 10^{6}$
Hydrogen	20	$1 \cdot 10^{-2}$	$1 \cdot 10^{-1}$	$2 \cdot 10^{2}$	$5 \cdot 10^{-3}$	$5 \cdot 10^{6}$

shape of the free surface. The length $L = 1$ m has been chosen with respect to applications in rocket upper stages (characteristic diameters between 4 and 6 m).

The following discussion will show that the MORTON and the BOND number may have the same order of magnitude in the experiments (in the drop tower or a sounding rocket) and the original configuration. Thus, empirical relations may be applied without extrapolation. Contrary to that the OHNESORGE number is always up to 2 orders of magnitude larger in the experiment than in the original application. No equivalent liquids are available to be comparable to propellants in terms of their properties density, viscosity and surface tension. Thus the original damping behavior cannot be achieved in experiments.

The initial response time t_{puLc} for all propellants is in the order of 10^{-2} s yielding initial velocities of 10^{-1} m/s. The times t_{pu} scaling the eigenfrequency of the system are in the order of 10^2 s and the corresponding damping is in the order of 10^6 s (see Table 3.4).

These estimations illustrate about the orders of magnitude of the dimensionless numbers involved in liquid management in space. A comparison between experiment and original situation should always include a more thorough analysis of the scales involved. In particular the specific length scales should be chosen with care. The BOND number reads

$$\mathrm{Bo} = \frac{\rho k L_1 L_2}{\sigma} , \qquad (3.1)$$

with L_1 scaling the hydrostatic height and L_2 scaling the curvature of the free surface, which are not necessarily the same.

References

1. R.F. Barron: *Cryogenic Heat Transfer*, (Taylor & Francis, Philadelphia 1999) pp. 357

4

Literature Review

The literature review is divided into the following parts: liquid management and surface tension tank technology, contact angle, surface oscillations and reorientation, capillary rise, and open capillary channel flows. The first part introduces the technology base from which the fluid mechanic problems were extracted. The reorientation takes place upon step reduction of the BOND number and is related to the onset of the ballistic phase of a spacecraft. The contact angle is the most important boundary condition for this damped oscillation and the relevant literature is reviewed in detail. The capillary rise takes place in PMD's if the BOND number decreases after a maneuver. The review focuses on the capillary rise in capillary tubes and summarizes many of the papers written on that subject. Finally, the literature on flows through open capillary channels is discussed in the context with the application in PMD technology. Similarities to choked flows in ducts with compressible fluids and open channel flows on earth are obvious but no papers are known that give theoretical or experimental evidences. Thus the topic of Part IV is effectively a new flow phenomena.

4.1 Liquid Management in Space

An intensive microgravity fluid dynamics program was initiated by NASA in the late 1950s because of the need to manage liquid propellants in space vehicles. Therefore, some of the earliest experimental results on microgravity fluid behavior were obtained by NASA engineers and scientists. Practical problems, such as the prediction of the liquid-vapor interface, reorientation of the free surface, propellant withdrawal and refilling of tanks and slosh dynamics were the driving considerations. The following papers are part of the result of this effort.

Petrash et al. [1, 2, 3, 4] have studied equilibrium surfaces in spherical tanks with different contact angles, reorientation times (see also [5]), capillary rise of liquids in cylindrical tubes in drop tower tests and the effect of the acceleration

Michael E. Dreyer: *Free Surface Flows under Compensated Gravity Conditions*
STMP **221**, 17–48 (2007)
DOI 10.1007/3-540-44628-1_4

disturbance on the liquid interface. Otto [6], Reynolds et al. [7, 8] and Habip [9] give a review of the state of the art until 1964 with particular emphasis on scaling laws. Masica et al. [10, 11] investigated the hydrostatic stability of a free surface under low gravity conditions in a drop tower with residual acceleration. They found a criterion for different orientations of the tube axis to the gravity or residual acceleration vector. The free surface is stable against axial acceleration up to Bo = 0.84, and Bo = 3.33 if the meniscus is fixed at a solid edge. Later, Masica [12] found the hydrostatic stability limit Bo = 1.25 for circular tubes for lateral acceleration perpendicular to the tube axis in drop tower tests. Paynter and Barksdale [13] reported on drop tower tests with residual accelerations for perforated plates and screens. They found the maximum stable BOND number Bo = 0.84 for perforated plates (based on pore radius) and Bo = 0.45 for screens (based on one-half the open width). Nussle et al. [14] and Symons et al. [15] performed drop tower tests for liquid outflow and liquid inflow and detected a maximum WEBER number for both cases.

4.2 Surface Tension Tank Technology

Since this work is dedicated to the use of capillary forces in spacecraft tank design the reader may refer to the literature cited in the following to gain more insight into this technology. It is not the scope of this book to describe all the problems which the designer encounters during the layout process for a spacecraft tank. But it is necessary to identify the main problems and the fluid physics behind to contribute to the knowledge of spacecraft tank design.

Special work dedicated to the development of PMDs was done in drop towers and aircrafts by Clodfelter [16], de Brock [17], Bizell & Crane [18], Yeh [19] and Boulay & Larigaldie [20] together with numerical calculations for static liquid free surfaces in a surface tension tank. Dowdy et al. [21] prescribed the surface tension propellant control for the Viking 75 orbiter with central baffle, communication channel, pressurization and vent tubes. Chapter & Rider [22] give a flow analysis in a surface tension tank propellant management system of the space shuttle reaction control system (RCS). Kim & Bennett [23] investigated a screen PMD (total communication) for cryogenic liquid with analytical modeling. Purohit & Loudenback [24] and Baralle [25] prescribe an industrial application using capillary forces for propellant management devices. Lee [26] gives a calculation of the flow including the pressure drops in gallery type PMD with screens for cryogenic liquid He I and II. Bicknell & Czysz [27] report on low-g venting of a surface tension tank and tank resupply while venting in drop tower and parabolic flights.

Tegart [28] calculated static fluid interfaces with residual acceleration using the *Surface Evolver* program and corroborated the stability limits found by [10, 11]. Jaekle [29, 30] presents the engineering analysis for the design of PMD components such as vanes and sponges. The analysis for vanes will be discussed in more detail in Chap. IV. Two space shuttle mid-deck experiments were performed by Dominick & Driscoll [31] and Dominick & Tegart [32] with a screen channel PMD for a total communication surface tension tank. The aim was to verify filling, venting and expulsion under worst case conditions (low filling rate and accelerations) and to determine the critical Weber number for the expulsion of liquid from the tank.

4.3 Contact Angle – Isothermal Conditions

The contact angle is a boundary condition for the free surface of the liquid and thus needed for the computation of the mass and momentum conservation equations. In the case of a moving contact line the contact angle is not always treated as constant. Viscous forces deform the free surface near the contact line causing a dynamic contact angle γ_d. The results of the numerical simulations reveal a significant influence of the dynamic contact angle boundary condition imposed on the moving contact line on the surface oscillations, i.e. on the resonance frequencies and the damping. It is important to use a suitable boundary condition for the dynamic contact angle which reproduces the observed behavior.

In principle one can distinguish between complete wetting ($\gamma_s = 0$) and partial wetting (Fig. 4.1). In the case of complete wetting a film is formed which thickness depends on the interaction between cohesion (of the liquid) and adhesion (between liquid and solid) forces. The thickness of this film is governed by the disjoining pressure (see [34]). The static contact angle γ_s is defined as the angle between the tangent to the liquid-vapor interface at the contact line and the solid surface. In case of partial wetting we observe a certain contact line with a finite contact angle γ_s (Fig. 4.1). The measured value of the contact angle depends on the closeness of the observation [35]. Following Shikmurzaev [36] three different notations are introduced, the *microscopic* contact angle, the *macroscopic* contact angle and the *apparent* contact angle (Fig. 4.2). The three notations are described in the following.

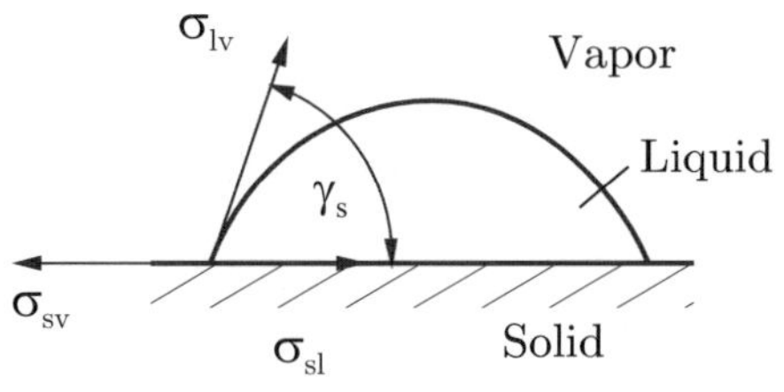

Fig. 4.1. Partial wetting with static contact angle γ_s. The surface energies are denoted by σ_{sv} for the solid-vapor interface, σ_{sl} for the solid-liquid interface, and σ_{lv} for the liquid-vapor interface. The latter is commonly known as the surface tension of the liquid

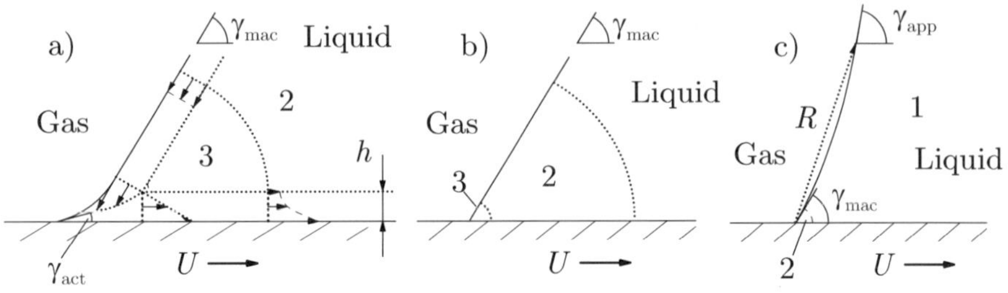

Fig. 4.2. Different contact angle definitions based on different characteristic flow domains: (**a**) microscopic or actual contact angle, (**b**) macroscopic contact angle and (**c**) apparent contact angle. 1,2, and 3 characterize the different flow domains

Microscopic Contact Angle

If the characteristic length scale L is in the order of the thickness of the interfacial layer ($h \approx 10^{-8} - 10^{-9}$ m, flow domain 3 in Fig. 4.2 a), where the properties strongly differ from those of the bulk, the microscopic (or actual) contact angle γ_{act} is defined. If $L \sim h$, the interface is no longer a geometrical surface with a surface tension. Density changes from liquid to vapor along h and forces of non-hydrodynamic origin should be considered.

Macroscopic Contact Angle

The usual hydrodynamic approach, γ_{mac}, is applicable for $L \gg h$ (flow domain 2 Fig. 4.2 b). The interfaces are modeled as geometrical surfaces of zero thickness with intrinsic surfaces properties like the surface tension. The three phase interaction zone becomes a structureless contact line. The angle γ_{mac} is composed from the tangent plane to the free surface, the contact line and the solid boundary. γ_{mac} is used as a geometric boundary condition for the macroscopic hydrodynamic equations determining the free surface shape. When the fluid is at rest γ_{mac} is considered in the Young equation (4.1).

Three phases are in contact at the contact line, the solid s, the liquid l and the corresponding gas or vapor v. Each interface, solid-liquid, solid-vapor and liquid-vapor has a certain free energy per unit area, σ_{sl}, σ_{sv} and σ_{lv}, respectively (Fig. 4.1). Young [37] related these free energies for a drop to the static contact angle γ_s and gave the following equation,

$$\sigma_{sv} - \sigma_{sl} = \sigma_{lv} \cos \gamma_s \ . \tag{4.1}$$

Apparent Contact Angle

The apparent contact angle γ_{app} is the angle formed by the solid wall and the tangent plane to the free surface at a certain distance from the contact line (flow domain 1 in Fig. 4.2 c). In the case of moving contact lines γ_{app} is used to consider bending of the free surface due to viscous stresses and relates the angle between the tangent plane to the free surface at a distance R ($R \gtrapprox 10^{-5}$ m) from the contact line and the solid boundary with the macroscopic contact angle γ_{mac}. The apparent contact angle γ_{app} is not a characteristic of a model applied to describe the flow associated with moving contact lines, but this model is often used as an auxiliary concept in interpreting the results of the experimental studies. In the case of the moving contact line, the Young equation (4.1) is no longer applicable. The surface near the contact line will be deformed by viscous forces.

The Moving Contact Line Problem

In the case of the moving contact line two problems must be solved. To describe the flow field at the moving contact line mathematically a *dynamic contact angle* must be defined as boundary condition. In addition a further boundary condition must be defined at the surface of the solid near the moving contact line. The assumption of the no-slip boundary condition at the surface of the solid yields a non-integrable stress (singularity) at the contact line, which is not physical. The resulting shear stress singularity must be removed. Moffat [38] describes the two-dimensional flow if a flat plate is drawn with the velocity U into a viscous fluid with a free surface (Fig. 4.3). The plate and the free surface form the angle α, the surface tension is neglected. It is supposed that the gravity keeps the free surface horizontal. He assumed that the Stokes approximation, where inertia forces are negligible compared with the viscous forces, is valid near the contact line. The planar flow is described with the stream function $\Psi(r, \Theta)$, where r and Θ are the polar coordinates. The solution of the Stokes equation for the stream function

$$\nabla^4 \Psi = 0 \,, \tag{4.2}$$

is in the form of

$$\Psi = Urf(\Theta) \,. \tag{4.3}$$

The corresponding velocity components are

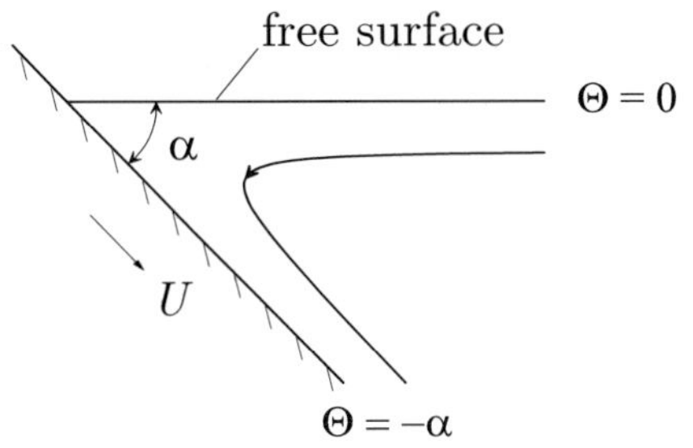

Fig. 4.3. Flow in a corner

$$u = \frac{1}{r}\frac{\partial \Psi}{\partial \Theta} = U f'(\Theta), \qquad v = -\frac{\partial \Psi}{\partial r} = -U f(\Theta) \, . \tag{4.4}$$

The function $f(\Theta)$ is of the following form for the problem under consideration:

$$f(\Theta) = (\sin\alpha \cos\alpha)^{-1} (\Theta \cos\Theta \sin\alpha - \alpha \cos\alpha \sin\Theta) \ . \tag{4.5}$$

The velocity of particles on the free surface is

$$u_s = U f'(\Theta = 0) = -U \left\{ 1 - \frac{(\alpha - \sin\alpha)(1 + \cos\alpha)}{\alpha - \sin\alpha \cos\alpha} \right\} \ . \tag{4.6}$$

Equation (4.6) shows that a particle on the free surface moves towards the wall ($u_s < 0$) with a speed independent of r and less than the velocity of the solid wall. When this particle meets the wall the speed suddenly increases to the value of U and the particle turns around the corner. This infinite acceleration is caused by an infinite stress and pressure, both of order r^{-1} at the corner. The solution breaks down very near the corner, this behavior is named the *moving contact line problem*.

One possibility to remove the stress singularity is to allow the contact line to move (*slip*) along the solid surface. Several slip models exist in the literature derived from different basic models. The most common model is the Navier slip condition formulated by Lamb [39]. This model is based on the assumption that the tangential traction on the inner surface must ultimately balance the force exerted on the outer surface by the solid. The slipping is resisted by a tangential force proportional to the relative velocity w with regard to the solid,

$$-\mu \frac{\partial w}{\partial r} = \beta w \ , \tag{4.7}$$

or

$$w = -l_s \frac{\partial w}{\partial r} \ , \tag{4.8}$$

with $l_s = \mu/\beta$. μ is the viscosity and the constant β expresses the ratio of the tangential force to the relative velocity and is called *slip coefficient.* The slip coefficient l_s is often named to as the *slip length.* However the slip length is not the length over which slip is enabled. A more restrictive expression would be *extrapolation length* [40]. The slip coefficient measures the distance from the solid surface to a fictitious point within the solid at which the extrapolated liquid velocity would be the same as that of the solid.

If we consider a moving plate and allow slip, the relative velocity between fluid and solid is the difference between the velocity of the fluid u and the plate velocity U. In general (4.7) can be formulated in the following form,

$$\mathbf{n} \cdot \mathbf{P} \cdot (\mathbf{I} - \mathbf{nn}) = \beta(\mathbf{u} - \mathbf{U}) \ , \tag{4.9}$$

where $\mathbf{n}$ is the outward normal vector to the solid surface, $\mathbf{I}$ is the metric tensor and $\mathbf{P}$ is the strain tensor. The left side of (4.9) denotes the shear stress exerted on the liquid-solid interface. The right side expresses the product of the slip coefficient and the relative velocity. The coefficient of sliding β was used with different definitions:

- the simplest definition is to set $\beta = \text{const}$ [41, 42],
- the slip coefficient is a function of the local film thickness h, $\beta = \beta(h)$ [43, 44],
- $\beta = 0$ within a preset distance L_s, where L_s is the slip region. The slip region L_s is defined as
$$L_s = U\tau \; , \tag{4.10}$$
where τ is the relaxation time. For $r > L_s$ is $\beta = \infty$, where r is the distance from the contact line [45].
- Imposing a yield criterion σ_c $\beta = \infty$ if $|\mathbf{n} \cdot \mathbf{P} \cdot (\mathbf{I} - \mathbf{nn})| < \sigma_c$, where σ_c is the yield stress. If the wall shear exceeds the critical value then $\beta = \sigma_c / |u - U|$ [46].

Another possibility to remove the stress singularity at the contact line is to prescribe a slip velocity profile as a function of the distance from the contact line r:

- $u = Ur/(l_s + r)$ [47],
- $u = Ur^2/(l^2 + r^2)$ [47],
- $u = Ur^{1/2}/(l^{1/2} + r^{1/2})$ [47],
- $u = U(1 - \exp(-r/l_s))$ [55, 49],

where r is the distance from the contact line. Dussan V. [47] describes the sensitivity of the overall flow field in dependence of the slip boundary condition. She used three different slip boundary conditions and found that the slip boundary condition affects the flow field only at the magnitude of the slip length but has no effect on the macroscopic behavior.

4.3.1 Dynamic Contact Angle

In the case of a moving contact line the contact angle is no longer constant. Viscous forces deform the free surface near the contact line to a dynamic contact angle γ_d. For a steady movement without inertia effects, the dynamic contact angle depends on the direction of the motion. One has to distinguish between γ_a for the advancing contact line and γ_r for the receding contact line.

General Observations

Figure 4.4 shows a typical experimental result for the dependence of the apparent dynamic contact angle γ on the contact line velocity [50], where U_c denotes the slowest speed at which this experiment was performed. Dussan

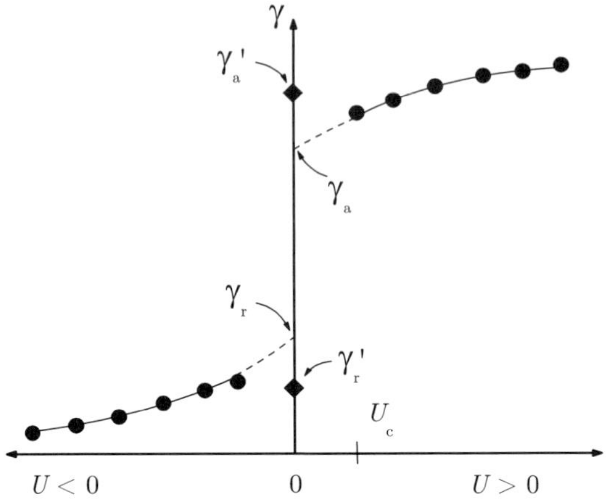

Fig. 4.4. Dependence of the dynamic contact angle on the contact line velocity [50]

V. [50] considers the case of partial wetting with a finite static contact angle. For the initially static contact line, $U = 0$, the contact angle varied between $\gamma_a^{'}$ and $\gamma_r^{'}$ until the contact line starts to move. The interval $\gamma_a^{'} - \gamma_r^{'}$ is referred to as the contact angle hysteresis. De Gennes [51] invokes three major causes which could explain the source of the hysteresis: the surface roughness, chemical contaminations or inhomogeneities in the solid surface and solutes in the liquid, like surfactants or polymers.

Hoffman [53] investigated the effect of the flow on the apparent contact angle of an advancing liquid-air interface in a glass capillary over the range in which viscous and interfacial forces are the dominant forces controlling the system. He plotted the apparent contact angle versus the capillary number Ca plus a shift factor (Fig. 4.5). The capillary number is expressed here with the velocity of the advancing meniscus U_m,

$$\mathrm{Ca} = \frac{\mu U_m}{\sigma} \; . \tag{4.11}$$

The shift factor is determined solely by the static contact angle. He found one curve which correlates the data from different fluids with different static contact angles. Jiang [52] correlated the experimental data from [53] and determined the following mathematical relationship,

$$\frac{\cos\gamma_s - \cos\gamma_{app}}{\cos_s + 1} = \tanh\left(4.96\,\mathrm{Ca}^{0.702}\right) \; . \tag{4.12}$$

Semi Empirical Model

Ehrhard and Davis [42] used a semi empirical model for the apparent advancing dynamic contact angle based on the experimental results from [50]. They

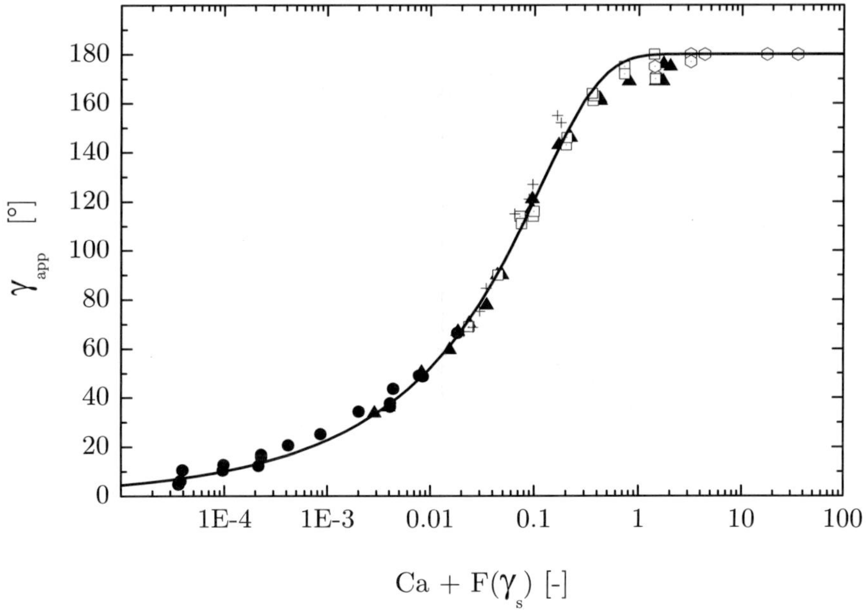

Fig. 4.5. Effect of flow on the apparent contact angle of an advancing liquid-air interface [53]

studied the isothermal and the non-isothermal spreading of a viscous-liquid drop on a smooth horizontal surface. They generalized the angle versus the contact line velocity to

$$U = \kappa(\gamma - \gamma_a)^m \,, \tag{4.13}$$

where γ_a is the static advancing contact angle, κ is an empirical coefficient, and m is called mobility exponent. The form with $m = 1$ was used by Greenspan [43], while the case $m = 3$ is suggested for the apparent contact angle by Hoffman [53] and Tanner [54]. The case of constant contact angle, $m \to \infty$, has been considered by [56] and others. Ehrhard & Davis [42] considered the case $m = 1$ and $m = 3$ and found good agreement between theory and isothermal experiments for the mobility exponent $m = 3$.

Expansion Model

Another type of model cuts off the neighborhood of the contact line together with the shear-stress singularity, which is attributed to the breakdown of the continuum mechanics modeling, and prescribes the value of the contact angle at the cut-off distance from the contact line [55, 56, 57, 58, 59, 60, 61, 62, 63]. The models employ matched asymptotic expansions of creeping flow with three regions of expansion. The inner, intermediate and outer region are characterized by dimensionless lengths r/l_s, $\mathrm{Ca}\ln(r/L)$ and r/L, where r is the distance from the contact line, l_s is the slip length and L a geometry dependent macroscopic length scale.

In these models, the detailed nature of the inner-region hydrodynamics, where the classical hydrodynamics fail, needs not to be specified. Only its asymptotic behavior must be established. In the intermediate region near the contact line surface tension and viscous forces dominate. The interface deformation in the intermediate region is independent of the system geometry and gravity, usual hydrodynamic controls the fluid motion. The outer region depends on the macroscopic geometry. In the case of low capillary number viscous forces are negligible and the shape of the fluid interface is primarily controlled by the balance between gravitational and surface tension forces.

The analysis of the problem yields a solution for the interface shape in the intermediate region of the form

$$\theta(r) = g^{-1}\left[g(\theta_R) + \mathrm{Ca}\ \ln(r/R)\right] , \tag{4.14}$$

where

$$g(\theta) = \int_0^\theta \frac{\phi - \cos(\phi)\sin(\phi)}{2\sin(\phi)} d\phi . \tag{4.15}$$

The angle between the moving solid and the tangent to the liquid free surface at a distance r from the dynamic contact line is denoted with $\theta(r)$. In addition, θ_R represents this angle at some distance R located within the intermediate region. As seen in (4.14), the slope of the interface within the intermediate region is independent of the macroscopic geometry L. Although the exact magnitude of θ_R depends upon the specific inner model, the form of (4.14) is independent of it. If θ_R is known, for example from experiments, θ_R could be used as a material boundary condition for the outer problem, replacing the true dynamic contact angle boundary condition. Cutting off the domain at $r = R$ and providing this slope characterization would permit a solution for the free surface shape without resorting to specify any of the details of the fluid physics in the inner region. The asymptotic hydrodynamic models are restricted for $\mathrm{Ca} < 0.1$. For $\mathrm{Ca} > 0.1$ the curvature of the interface in the intermediate region differs from the curvature seen in the experiments [64].

Theoretical Approach

Friz [65] made a theoretical hydrodynamic analysis of a slug of liquid being pushed through a tube (Fig. 4.6), and derived an equation for the shape of the liquid front and the apparent contact angle. He assumed that the wall of the tube was already wetted with a thin layer of liquid of thickness δ_p, and depicted the dynamic contact angle as an asymptotic limit to the liquid/vapor interface near the wall. His equation for the dynamic contact angle is

$$\tan\gamma_d = 3.4\ \mathrm{Ca}^{1/3} . \tag{4.16}$$

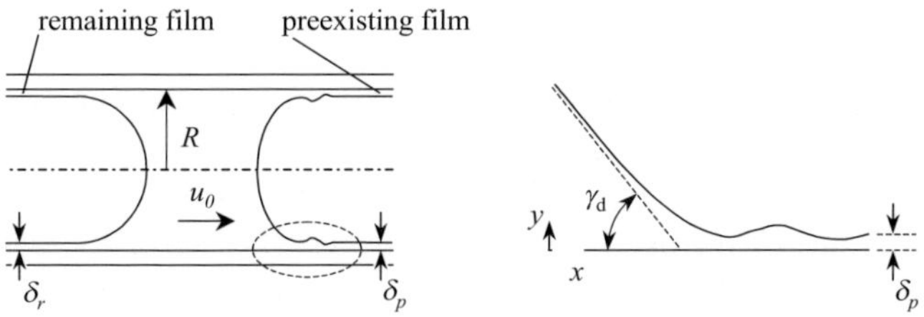

Fig. 4.6. Moving slug in a pipe

The apparent dynamic contact angle is not depending on the film thickness δ_p. The same problem was considered experimentally by Rose & Heinz [66]. The comparison of the analytical data with the experimental data shows a great gap in the curves of the dynamic contact angle versus velocity. The experimental data are two orders of magnitude higher than the analytical data.

Molecular Kinetic Theory

Besides the hydrodynamic approach the molecular kinetic theory is used to model the spreading process. The principal hypothesis is that the motion of the three-phase line is ultimately determined by the statistical kinetics of molecular events occurring within the three-phase zone. Blake & Haynes [67] derived a theoretical treatment of dynamic contact angles for a liquid/liquid displacement (Fig. 4.7). Using the theory of absolute reaction rates, they picture the essential motion to be a sliding of the molecules along the solid surface from the liquid to the vapor side of the three phase contact line. Blake & Haynes assume the driving force to be the out-of-balance surface tension

$$\sigma \cos(\gamma_s - \gamma_d) \ . \tag{4.17}$$

This work is used to raise or lower the activation energy for forward or reverse molecular migration along the solid surface. Blake & Haynes found the following relationship between the dynamic contact angle γ_d and the forward velocity V,

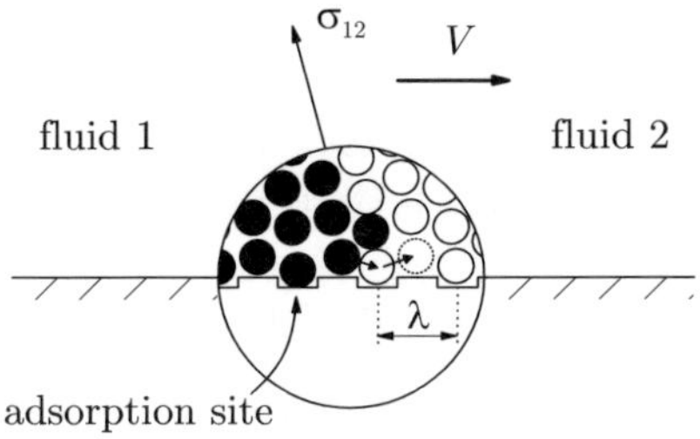

Fig. 4.7. Adsorption/desorption model of molecular displacement within the three-phase zone

$$V = 2K\lambda \sinh\left[\frac{\sigma}{\Delta nkT}\cos\left(\gamma_s - \gamma_d\right)\right] , \tag{4.18}$$

where K is the number of molecular displacements occurring per unit time per unit length of the contact line, λ is the average distance between centers of adsorption sites, Δn is the number of sites affected at the solid/liquid and the solid/vapor interface, per unit area, k is the Boltzmann constant and T is the temperature. In reality the molecules of the three-phase zone will be influenced by interactions with the solid surface and also viscous interactions between the fluid molecules themselves. This behavior is taken into account in the *combined* theory in which the molecular kinetics are linked to the hydrodynamics [68]. The molecular kinetic theory was successfully applied by van Mourik et al. [69] to the test case of a free surface oscillation [70].

Surface Layer Concept

If the fluid is at rest, the Young equation is valid and the force balance at the contact line yields the well known relation (4.1). With an advancing contact line motion, the liquid particles which initially form an element of the gas-liquid interface arrive on the solid surface in a finite time and form an element of the solid-liquid interface (Fig. 4.8). This behavior was observed in experiments by Dussan V. & Davis [71] and is called the rolling motion. During this process the surface properties of this element change from the equilibrium surface properties of the free surface to those of the liquid-solid interface. The change in the surface properties is not instantaneous: it is a result of some transitional process with its specific kinetics depending on the properties of the contacting materials. The surface properties relax within a finite relaxation time τ to the new equilibrium properties. The finiteness of the surface tension relaxation time gives rise to the formation of regions with large surface tension gradients, which can considerably influence the whole motion of the liquid in the vicinity of the contact line. Shikhmurzaev [36, 72, 73, 74, 75, 76] developed a model which includes this relaxation process in a thin layer of liquid adjacent to the solid and the gas. The flow field is subdivided into three regions (Fig. 4.9) with the following characteristic length scales.

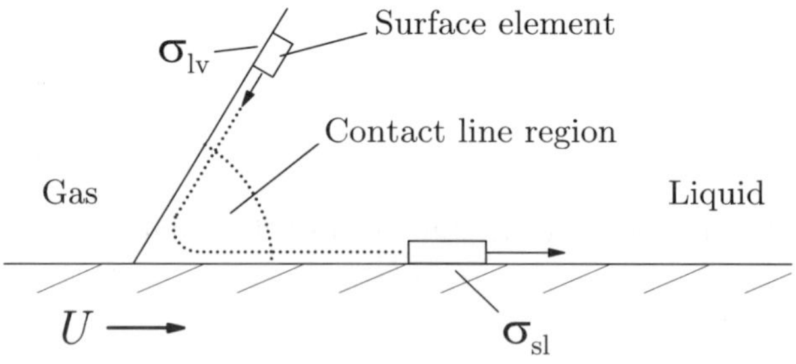

Fig. 4.8. Transition of the properties of an element at the free surface to those of the liquid-solid interface

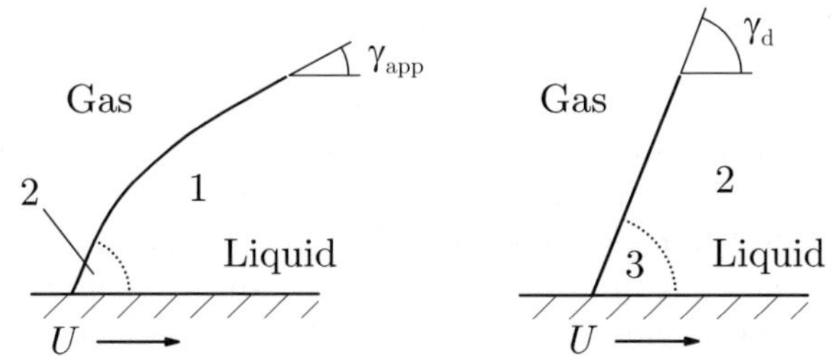

Fig. 4.9. Asymptotic regions in the moving contact-line analysis: 1-'outer' region (characteristic length scale L); 2-'intermediate' region (length scale ϵL); 3-'inner' region (length scale $\epsilon \mathrm{Ca} L$); γ_{app} apparent contact angle measured far from the contact line; γ_d macroscopic contact angle

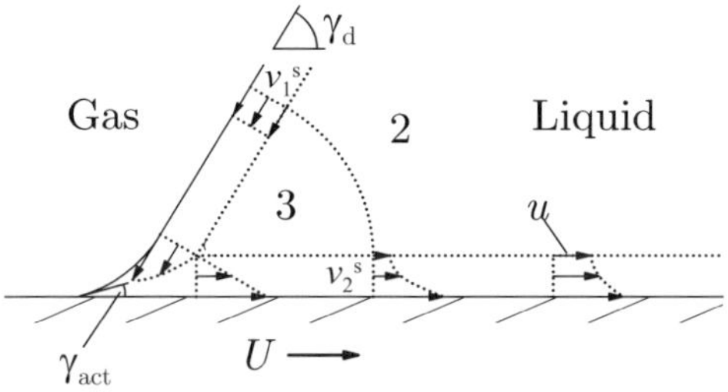

Fig. 4.10. A detailed sketch for the flow in immediate vicinity of the contact line. The interfaces are depicted as layers of finite thickness in order to show schematically the velocity distribution across each interface. The velocity on the side of the interfacial layer facing the liquid tends toward that of the solid as the distance from the contact lines increases. γ_{act} is the actual contact angle, i.e., the angle measured on the microscopic length scale

- L, a characteristic length scale of the outer region, where the classical solution [38] with the no slip boundary condition on the solid and zero tangential stress at the free surface is valid.
- $l = \varepsilon L$, the intermediate or slip region, where $\varepsilon = U\tau/L$ is the ratio of the characteristic relaxation length $U\tau$ to the characteristic length scale of the flow domain L.
- $\varepsilon \mathrm{Ca} L$, the inner region.

The solutions obtained in these regions must be asymptotically matched. To describe the relaxation of the surface properties in the intermediate region Shikhmurzaev [75] introduced the thin layer of the liquid adjacent to the solid or gas (Fig. 4.10). Within the intermediate region the surface tension changes and therefore apparent[1] slip takes place.

[1] On the solid-side interface the velocity is equal to U of the solid (no-slip condition). On the liquid-side interface, the boundary condition of the bulk, the velocity u is less than U and therefore apparent slip takes place.

4.3.2 Oscillating Contact Line

Ting & Perlin [77] investigated the dynamic contact angle behavior for an oscillating contact line experimentally. They considered an upright plate which oscillates vertically with sinusoidal motion in dye-laden water with an air interface. The definition of the coordinate system is shown in Fig. 4.11. The angle γ between the tangent on the free surface and the x -coordinate is defined as $\gamma = \tan^{-1}(\partial\eta/\partial x)$. η describes the y-coordinate of the free surface position at the distance x from the wall. The contact angle γ_c which is measured between the tangent of the free surface and the solid at the contact point is consequently

$$\gamma_c = \gamma + \frac{1}{2}\pi \, , \tag{4.19}$$

$$\gamma_c = \tan^{-1}(\partial\eta/\partial x) + \frac{1}{2}\pi \, . \tag{4.20}$$

The frequencies f varied in the 1-20 Hz range, and the stroke amplitude A in the range from 0.5 mm to 6 mm. The static contact angles at the beginning of the experiments varied in the range $27.8^\circ < \gamma_s < 52.1^\circ$.

As the plate moves downward, the contact angle begins to increase and reaches its maximum value near the minimum plate position. Then the contact angle decreases and the contact point undergoes a free slip motion ($U_r/U_{p\,\max} = 1$). The contact line motion is more stick-like ($U_r/U_{p\,\max} = 0$) when the plate position is near the minimum plate position.

The link between the dynamic angle and the relative contact line velocity is realized by the non-dimensional capillary coefficient λ

$$\frac{U_r}{U_{p\,\max}} = \left(\frac{\partial\eta}{\partial t} - U_p\right)\frac{1}{U_{p\,\max}} = \lambda\left[\tan^{-1}(\partial\eta/\partial x) + \frac{1}{2}\pi\right] = \lambda\gamma_c \, . \tag{4.21}$$

The capillary coefficient λ depends on the time t, the stroke amplitude s and the angular frequency of oscillation ω. A typical run of the curve λ versus time

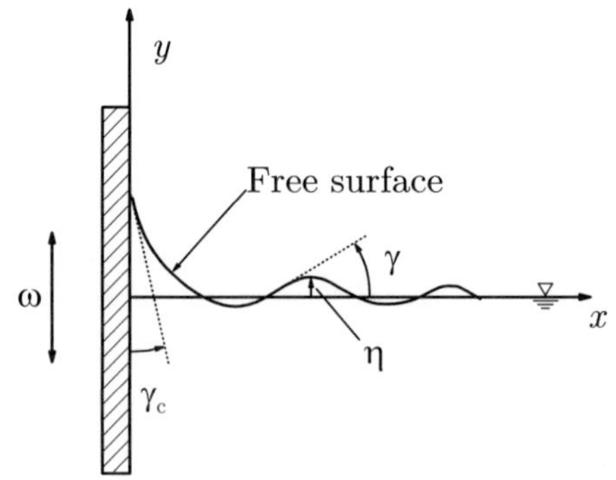

Fig. 4.11. Free surface evaluation for a moving contact line

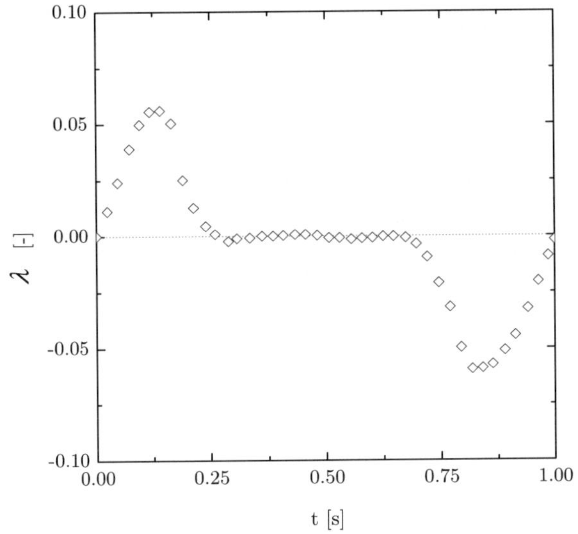

Fig. 4.12. Time variation of the relative contact line velocity divided by the contact angle for a 2 Hz oscillation, stroke amplitude 3 mm

is shown in Fig. 4.12. The results show that the frequency effect is unimportant relative to the amplitude effect. Therefore the frequency effect is neglected in the model for the boundary condition, $\lambda = \lambda(t, s)$.

Hocking [78] and Young & Davis [79] use an oscillating contact-line boundary condition in order to calculate the waves generated by a vertically oscillating plate. The plate is partially immersed in a non-wetting fluid. Hocking neglects the viscosity and supposes a linear wave motion. Two different contact angle conditions have been investigated, with and without contact angle hysteresis. The form of the edge condition is consistent with the known behavior of the contact angle [50] and has the same form as that used by [79] for low speeds and include the approximation of a the linear relationship between the slope and the speed.

Satterlee & Reynolds [80] used a linear condition for the dynamic contact angle in a lateral sloshing problem. They investigated the special case of a cylindrical tank partially filled with a nonwetting liquid, the static contact angle is equal to 90 degree (Fig. 4.13). The contact angle condition imposed at the contact line depends on the z coordinate of the free surface at the wall h and is proportional to a free parameter Γ:

$$\frac{\partial h}{\partial r} = \Gamma \, (h - h_0) \ , \tag{4.22}$$

where h_0 is the static height. The case $\Gamma = 0$ corresponds to the fixed contact angle condition whereas the contact line is free to move. When $\Gamma = \infty$ the contact angle is allowed to vary and the contact line is fixed. Kamotani et al.

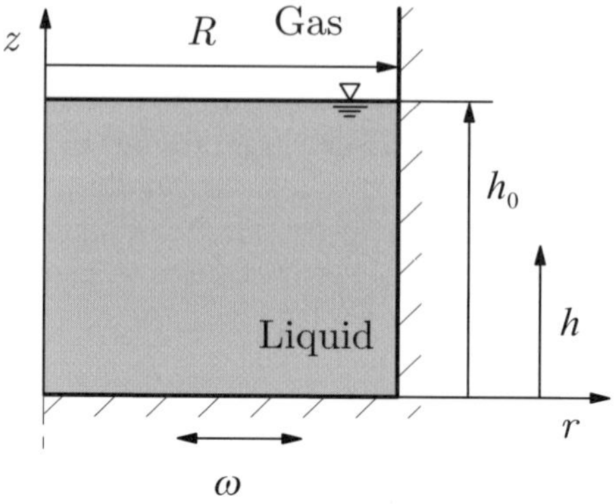

Fig. 4.13. Lateral sloshing in a cylindrical tank

[81] used the same condition in the numerical simulation of g-jitter effects on the free surface deformation for liquids in an open container.

4.4 Contact Angle – Non-isothermal Conditions

Different phenomena appear in non-isothermal conditions which influence the flow in the contact line region and therefore the dynamic contact angle. These effects are the temperature dependence of the liquid properties and the static contact angle, evaporation and condensation effects, thermocapillary effects (Marangoni), and intermolecular interaction between liquid and solid, if a thin liquid film is present in front of the contact line. In the following the different phenomena are discussed individually.

Temperature Dependence of the Static Contact Angle

The Young equation (4.23) relates the free energies at the three phase interaction line with the static contact angle γ_s

$$\sigma_{sv} - \sigma_{sl} = \sigma_{lv} \cos \gamma_s \ . \tag{4.23}$$

With increasing temperature the surface tension of the liquid-vapor interface decreases. If we assume that the temperature dependence of the liquid-vapor surface tension is larger than that of the solid-vapor and the solid-liquid surface tension, the static contact angle decreases with increasing temperature. This is usually seen in liquid-solid systems.

Marangoni Effects

Thermocapillary or Marangoni effects result from forces on the liquid-vapor interface caused by temperature-induced variations of the liquid-vapor surface tension. Almost all single component liquids of low molecular weight have surface tensions that decrease with temperature. The liquid is pulled by the surface tension from the region near the contact line towards the region of higher surface tension. Ehrhard & Davis [42], Ehrhard [82] and Anderson & Davis [83] describe the influence of the Marangoni flow on the dynamic contact angle in the case of a liquid droplet spreading over a heated surface. They show that the dynamic contact angle increases with increasing temperature difference.

Vapor Recoil

Since a fluid particle conserves its mass flux upon phase change, a slowly moving liquid particle at the interface accelerates greatly when it becomes vaporized; the vapor particle has a considerably lower density than the liquid particle. The back reaction on the interface is called vapor recoil. Vapor recoil effectively causes an increased pressure on the interface due to the momentum transfer from the escaping particles. The evaporation rate is highest near the contact line caused by the highest temperature difference. Since the mass flux is larger near the contact line, the pressure is correspondingly stronger near the contact line. Thus, the nonuniform pressure causes an increase of the contact angle. This effect was observed by Anderson and Davis [83] who considered a volatile liquid droplet on a uniformly heated horizontal surface.

Evaporation

The evaporation withdraws mass from the fluid to the vapor phase. The evaporation rate increases with increasing temperature differences. Thus, the nonuniform temperature distribution will cause a higher evaporation rate near the contact line. Renk & Wayner [84, 85] studied experimentally and analytically a steady evaporating meniscus in which capillary-induced flow towards the meniscus balances the liquid loss due to evaporation. They found that the evaporating meniscus profile was a function of the evaporated heat flux. Moosman & Homsy [86] followed with a more detailed analysis of the same configuration in which they included the effect of the meniscus extending to an adsorbed layer. They reported an increase in the apparent contact angle when evaporation is present. Wayner [87] considered the motion of a meniscus due to an evaporation/condensation process wherein liquid evaporates from the thicker portion of the layer and condenses on a microscopic film ahead of the meniscus. He correlated an increase in the macroscopic contact angle with the contact line motion due to this evaporation/condensation process. Ripple [88] analyzed the motion of a meniscus on a heated surface on microscopic scales, where the contact line region extends to a monolayer away from the bulk fluid. He described the effects of liquid evaporation, thermocapillarity, and intermolecular forces on the meniscus. He found that the microscopic contact angle could be calculated and that it increases with evaporation rate.

4.5 Surface Oscillations

Experimental Investigations

The first experimental study concerning the reorientation behavior of a liquid interface which is suddenly exposed to a step reduction in gravity was performed in 1964 by Siegert et al. [5]. They used containers with cylindrical, spherical and annular shapes (diameter 22 mm $< D <$ 156 mm) partly filled with different liquids having a static contact angle of $\gamma_s = 0$ with the vessel wall. The experiments were carried out in a drop tower. By reason of rather low OHNESORGE numbers (and thus a low damping behavior of the flow) and a limited experiment time, the total reorientation time of the liquid interface could not be observed. As a result of their examination, they presented a correlation for a characteristic time t_{cr} until the surface center point $h\,(r = 0, t)$ passes the first time the equilibrium position after entering microgravity condition (for the definition of the time t_{cr} compare also Fig. 7.12). The correlation for t_{cr} in a cylindrical container is a function of the pressure unsteady time scale and reads

$$t_{cr} = 0.146\sqrt{\frac{8\rho R^3}{\sigma}} = 0.413\ t_{pu}\ . \tag{4.24}$$

Kaukler [89] investigated the oscillation behavior of a liquid/liquid interface in a cylindrical container (D = 25 mm) suddenly exposed to a step reduction in gravity (by the use of a drop tower). He investigated the effect of critical point wetting. Critical point wetting occurs in systems of two immiscible liquids which are isothermally exposed to different temperatures. Above a wetting temperature T_W one of the two liquids totally wets the cylinder wall with a static contact angle of $\gamma_s = 0$. For temperatures below T_W partial wetting with $\gamma_s > 0$ occurs. Thus, by varying the ambient temperature T above and below T_W, the effect of partial ($T < T_W$) and complete wetting ($T > T_W$) on the reorientation behavior of the liquid interface was investigated. An evaluation of the frequency of the surface center point shows that for partial wetting ($T < T_W$) the frequency decreases with decreasing static contact angle (increasing temperatures for isothermal conditions). For temperatures above T_W (complete wetting) the frequency remains constant.

In an extensive series of drop tower experiments Weislogel & Ross [90] investigated the behavior of the free liquid/gas interface in a circular cylinder by varying the cylinder radius (9.52 mm $< D <$ 30 mm), the test liquids ($1.3 \cdot 10^{-3} <$ Oh < 1.1) and the static contact angle ($0 < \gamma_s < 74°$), where no differentiation between the initial and final static contact angle was given. They focussed their examination on the determination of the total reorientation time t_R of the liquid interface by plotting the location of the surface center point as a function of time, where they defined the reorientation time to be reached when regular oscillations are no longer detectable or the meniscus locations becomes steady in time for experiments with creeping flow. Besides the evaluation of the reorientation time they also determined the frequency of the center point oscillation and counted the number of nodes. They reported an increasing frequency if the static contact angle is varied to higher values. As a further result of their study they presented an empirical correlation for the settling time t_R, based on a scale analysis. The correlations for t_R which is scaled with the time scale $t_{vu} = R^2/\nu$ ((2.23) with the length scale R) reads

$$t_R = t_{vu}\left(1+\alpha^2\right)^{-1}\left(10^B\zeta^A + 0.01\alpha^2\right) \tag{4.25}$$

and is basically a function of the OHNESORGE number and the static contact angle. The parameters are

$$\zeta = \frac{\sqrt{\cos\gamma_s}}{1-\sin\gamma_s}\mathrm{Oh}\ , \tag{4.26}$$

$$A = -1.2\alpha^2 + 2.2\alpha + 0.28\ , \tag{4.27}$$

$$B = 3.9A - 3.32 \tag{4.28}$$

and α from (5.7). Figure 4.14 shows the settling time t_R from (4.25) scaled with t_{vu} for different test liquids if the contact angle is varied in a range $1° < \gamma_s < 87°$ for a cylinder radius $R =$ 10 mm. The maximum time for the surface to reach equilibrium is required for test liquids with intermediate contact angles. The effect of the OHNESORGE number is more definite, where higher OHNESORGE numbers cause a stronger damping behavior of the flow

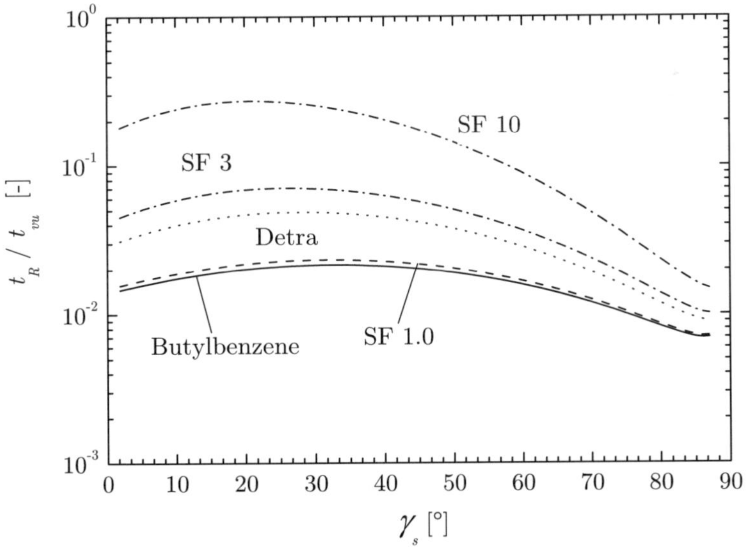

Fig. 4.14. Settling time t_R given by Equation (4.25) presented by [91] scaled with t_{vu}. The different lines are plotted for the test liquids used in this investigation with $R =$ 10 mm

with higher dimensionless times t_R/t_{vu}. The authors claim an accuracy of (4.25) better than ±50% for over 85% of the data and ±20% for experiments with test liquids with a static contact angle of $\gamma_s = 0$.

Numerical Investigations

In a theoretical investigation Bauer & Eidel [92] examined the natural frequencies ω_0 of a free liquid/gas interface in a circular cylinder assuming potential flow for microgravity conditions. The investigated flow field did not have a moving contact line, thus it was considered to be pinned. By varying the static contact angle in a range $0 < \gamma_s < 180°$ it was found that the natural frequency, scaled with the time t_{pu} (2.18), increases with rising static contact angle γ_s. When γ_s exceeds the value of $\gamma_s = 90°$ the frequency decreases again. Moreover they investigated the sloshing mass of the liquid depending on the fill height, which is considerably larger for liquids with small static contact angles than for liquids with a non-wetting character.

In a numerical study Wölk et al. [93] modeled the entire process of the free surface reorientation upon step reduction in gravity with a commercial finite element code (FIDAP). To model the reorientation process, they introduced a boundary condition for the moving contact line which is a modification of (4.22) from [80]

$$\frac{\partial h}{\partial r} = -\Gamma\,(h - h_0) + \cot(\gamma_s) \ , \tag{4.29}$$

to be able to regard static contact angles $\gamma_s < 90°$. They were able to validate experimental results from [91] by variation of the free parameter Γ ($\Gamma = 0$ corresponds to a fixed contact line condition while $\Gamma \longrightarrow \infty$ represents a free contact line condition). Although they did not consider the velocity dependent behavior of the dynamic contact angle as well as contact angle hysteresis, they pointed out the limits of systems with $\Gamma = 0$ and $\Gamma \longrightarrow \infty$.

In a recent numerical work from Gerstmann et al. [94] the influence of the Ohnesorge number and γ_s on the settling time and the natural frequency of the free surface in the center line were investigated in more detail. Using also a finite element code, they confirmed the behavior of a rising frequency for higher static contact angles in good agreement with experimental data from [91]. However, a strong influence of γ_s on the settling time was not observed in the investigated range of contact angles.

4.6 Capillary Rise in Tubes and Vanes

In this section several publications on capillary-driven flows will be discussed in chronological order to demonstrate the present advances regarding the theoretical and experimental methods applied on this phenomenon. The characteristic scales of the experimental investigations discussed here are listed in Table 4.1. The scales are calculated using (10.1)-(10.4) with $t_{pu} = \sqrt{\rho R^3/\sigma}$

Table 4.1. Characteristic scales of experiments on capillary rise

Author	t_{pu} (s)	t_{vu} (s)	Oh (-)	Bo (-)
[95]	$< 6.0 \cdot 10^{-3}$	$0.66 \ldots 1.36$	$2.6 \cdot 10^{-3} \ldots 6.4 \cdot 10^{-3}$	< 0.35
[96]	$< 1.2 \cdot 10^{-3}$	$0.02 \ldots 1.21$	$7.3 \cdot 10^{-4} \ldots 6.9 \cdot 10^{-3}$	< 0.04
[97]	$< 1.4 \cdot 10^{-3}$	$0.04 \ldots 0.30$	$3.3 \cdot 10^{-3} \ldots 2.5 \cdot 10^{-2}$	< 0.05
[98]	$< 1.2 \cdot 10^{-3}$	$0.04 \ldots 0.12$	$4.4 \cdot 10^{-3} \ldots 1.3 \cdot 10^{-2}$	< 0.04
[99]	$< 2.5 \cdot 10^{-1}$	$1.12 \ldots 337$	$5.3 \cdot 10^{-4} \ldots 2.2 \cdot 10^{-3}$	$\ll 1$
[2]	$< 2.2 \cdot 10^{-1}$	$3.79 \ldots 79.6$	$1.9 \cdot 10^{-3} \ldots 4.1 \cdot 10^{-3}$	$\rightarrow 0$
[100]	$< 1.7 \cdot 10^{-3}$	$0.02 \ldots 0.35$	$3.4 \cdot 10^{-3} \ldots 7.2 \cdot 10^{-3}$	< 0.05
[101]	$< 9.1 \cdot 10^{-6}$	$< 2.0 \cdot 10^{-4}$	$2.0 \cdot 10^{-2} \ldots 3.5 \cdot 10^{-1}$	$< 5.8 \cdot 10^{-5}$
[102]	$< 6.9 \cdot 10^{-3}$	$< 2.9 \cdot 10^{-3}$	$1.7 \ldots 5.8 \cdot 10^{2}$	< 0.46
[103]	$< 2.0 \cdot 10^{-2}$	$0.06 \ldots 5.29$	$1.9 \cdot 10^{-3} \ldots 6.3 \cdot 10^{-3}$	< 1.38
[104]	$< 1.4 \cdot 10^{-1}$	$0.45 \ldots 61.3$	$8.6 \cdot 10^{-4} \ldots 5.6 \cdot 10^{-2}$	$\rightarrow 0$
[105, 106]	$< 4.9 \cdot 10^{-1}$	$0.25 \ldots 238$	$9.2 \cdot 10^{-4} \ldots 8.7 \cdot 10^{-3}$	$\rightarrow 0$
[107]	$< 3.7 \cdot 10^{-3}$	$3.7 \cdot 10^{-4} \ldots 1.12$	$2.4 \cdot 10^{-3} \ldots 3.8$	< 0.20
[108]	$< 4.9 \cdot 10^{-3}$	$0.56 \ldots 1.46$	$2.4 \cdot 10^{-3} \ldots 3.0 \cdot 10^{-3}$	< 0.20
[109]	$< 2.4 \cdot 10^{-3}$	$4.4 \cdot 10^{-3} \ldots 0.81$	$2.1 \cdot 10^{-3} \ldots 2.8 \cdot 10^{-2}$	< 0.11
This work [127]	$0.02 \ldots 2.26$	$3.64 \ldots 1770$	$5.8 \cdot 10^{-4} \ldots 4.7 \cdot 10^{-3}$	$\rightarrow 0$

and $t_{vu} = R^2/\nu$, thus the proportionality constants in (10.1) and (10.2) are set to unity.

We start with the studies performed 1906 by Bell & Cameron [95], 1918 by Lucas [111] and 1921 by Washburn [96]. These authors considered only the capillary force at the meniscus and the friction force inside a horizontal tube assuming the parabolic POISEUILLE velocity distribution. Balancing the two forces yields the meniscus velocity

$$\dot{h} = \frac{\sigma \cos \gamma_s R}{4 \mu h} \tag{4.30}$$

with the static contact angle γ_s between liquid and tube wall. Integration with respect to time leads to the commonly known Lucas-Washburn equation

$$h^2 = \frac{\sigma \cos \gamma_s R}{2 \mu} t \, . \tag{4.31}$$

Equation (4.31) yields that the distance h traveled by the meniscus is proportional to $\sqrt{t}$. However, since (4.30) predicts an infinite meniscus velocity at the initial stage of the flow, the limited validity of (4.31) becomes obvious. Equation (4.31) was also found empirically by [95] prior to the mathematical derivation by Lucas and Washburn.

Washburn reported good agreement of (4.31) with experiments performed using water and mercury and capillaries with internal radii of $R = 0.1454$ mm and 0.369 mm. The tubes were mounted vertically and, to eliminate the influence of the hydrostatic pressure, horizontally. For the wider tube Washburn noticed deviations between the experimental data and (4.31) in the vicinity of $h = 0$ and explained this with sticking friction that interrupted the liquid motion. The Lucas-Washburn equation (4.31) is an approximate model for large times when all influences apart from the capillary force and the friction force have ceased. It is still applied to predict the capillary penetration of liquids into porous structures [112, 113, 114].

Other researchers introduced several improvements to the Lucas-Washburn equation. Rideal [97] added the inertia force of the moving liquid mass inside the tube. Experiments with horizontal tubes were conducted but due to the small time scales only the behavior $h \sim \sqrt{t}$ was observed and the influence of inertia could not be proved. Bosanquet [115] added the pressure at the tube entrance to the theoretical considerations of the capillary rise. For the solution of the equation he introduced a constant $t_c = R^2/(8\nu)$ (with $t_c \sim t_{vu}$). For small t_c (small R) his equation reduces to the Lucas-Washburn equation (4.31).

LeGrand & Rense [98] studied the capillary rise in vertical tubes experimentally and found Lucas-Washburn behavior. They remarked that for the theoretical treatment of the phenomenon, a mass of liquid below the tube entrance, a non-constant contact angle, and deviations from laminar flow at the tube entrance and in the vicinity of the meniscus should be considered.

Siegel [99] was the first to perform experiments under reduced gravity (Bo $\ll 1$) using a free falling platform with 0.74 s experiment duration. Due to the limited experiment time Lucas-Washburn behavior was not reached. A linear dependence $h \sim t$ was observed instead. The experimental data were compared with a theoretical model considering the capillary force, the inertia force, and the friction force. The theoretically predicted meniscus heights were found to be much larger than the experimental values. Thus, additional terms for the entrance pressure loss and a turbulent drag were added to the model but no general improvement was reached. As another source of error Siegel discussed that in contrast to his theoretical model the contact angle may not have been constant during the experiments but a function of the meniscus speed $\dot{h}$. To consider this effect he multiplied the capillary force with $1 - m\dot{h}$, where m was chosen to fit the experimental data. One must admit that the first applicable equations for the velocity dependence of the contact angle were published some years after Siegel's study (see [40]). Siegel improved the accuracy of his theoretical predictions by considering the inertia force of a certain amount of liquid below the tube entrance due to the acceleration from rest to the velocity that prevails inside the tube.

Petrash et al. [2] performed the first drop tower experiments (Bo $\rightarrow 0$) with 2.25 s microgravity duration. Their experimental data indicated a linear $h \sim t$ dependence. Additionally to the above mentioned forces, the retarding capillary pressure of the liquid surface inside the reservoir was considered for their theoretical model. In experiments under normal gravity the liquid surface in the reservoir remains flat, but under reduced gravity the liquid surface of the reservoir becomes curved and the arising capillary force counteracts the driving capillary force of the meniscus inside the tube. The calculated $h \sim t$ curves exhibit an initial $h \sim t^2$ behavior which changes rapidly into a linear dependence. A satisfactory agreement between the calculated capillary rise and the experimental data was reported.

Szekeley et al. [116] studied the capillary flow into a tube by applying a macroscopic energy balance. The consideration of inertia removed the anomaly of the initial infinite meniscus velocity predicted by Washburn (4.30). The time $t_c = R^2/(8\nu)$ from [115] was introduced and for small t_c Szekeley et al. [116] found that their equation reduces to the Lucas-Washburn equation (4.31).

Levine et al. [117] presented the most detailed theory for the capillary rise in tubes up to now. They applied the Navier-Stokes equations on the liquid volume inside the tube and an additional momentum balance on the sink flow from the reservoir towards the tube entrance. Hereby they were able to improve the calculation of pressure losses inside the reservoir. A perfectly wetting liquid with a constant contact angle independent from the meniscus velocity ($\gamma_d = \gamma_s = 0$) was assumed and the hydrostatic pressure was considered yielding

$$\ddot{h} = \frac{1}{h + 1.03R} \left\{ \frac{2\sigma}{\rho R} - \frac{8\nu}{R^2} h\dot{h} - \left[\frac{7}{3} + \frac{8}{\mathrm{Re}_d} \right] \frac{1}{2} \dot{h}^2 - g_0 h \right\} \tag{4.32}$$

with $\mathrm{Re}_d = d\dot{h}/\nu$. They formulated their equation for the meniscus rise in dimensionless form introducing the dimensionless time t_* and the dimensionless meniscus height h_* according to

$$t_* = \frac{t}{t_c} \quad \text{with} \quad t_c = \frac{R^2}{8\nu} \quad \text{and} \tag{4.33}$$

$$h_* = \frac{h}{t_c h_c} \quad \text{with} \quad v_c = \sqrt{\frac{2\sigma}{\rho R}} \, . \tag{4.34}$$

The viscous time scale t_c is proportional to t_{vu} (10.2) and was used before by [115] and [116]. The constant v_c has the dimension of a velocity. Levine et al. did not compare the results of their analysis with experimental data.

Jeje [100] discussed the importance of inertia forces for the capillary rise, but due to the parameter range of his laboratory experiments only Lucas-Washburn behavior ($h \sim \sqrt{t}$) was found. Again, Fisher et al. [101] confirmed the Lucas-Washburn equation (4.31) with their experiments.

Letelier [118] presented another mathematical analysis and found the Lucas-Washburn equation (4.31) to be an asymptotic form of their equation for large times. Although they emphasized that the contact angle depends on the flow velocity, a constant value was used in their study. In the year 1980 Levine et al. [119] presented a theory of capillary rise in cylindrical tubes and parallel-plate channels. A constant contact angle was assumed, inertia and entrance effects were neglected, and a generalized Lucas-Washburn equation was found. Batten [120] derived an equation for the meniscus rise based on the analysis of [116]. He added a term for the dissipation of kinetic energy beneath the meniscus and calculated the time for the reorientation of the meniscus from the initial flat shape to the final spherical shape. This approach is questionable here because the proposed viscous time is too large for the reorientation of the surface. Good agreement with experimental data from [98] was found but these data are not adequate to verify additional dissipative effects since the data are well predicted by the simple Lucas-Washburn equation (4.31).

Joos et al. [102] considered an empirical relation for the velocity dependence of the contact angle but neglected inertia and entrance effects. Due to the high-viscous liquids used in their experiments, the viscous time t_{vu} is smaller than the response time t_{pu}. Good agreement between theoretical and experimental data was found. They concluded that the Lucas-Washburn equation (4.31) describes the meniscus movement the better the narrower the capillary is.

Ishikawa & Satoda [103] investigated horizontal tubes under terrestrial conditions. Some of their experimental data indicate an initial $h \sim t^2$ and afterwards a $h \sim t$ behavior. The experiments were reported to be sensitive to the surface conditions of the tubes. Only for prewetted tubes good agreement between experimental and theoretical data was obtained. The deviation between theory and experiment in the case of a dry tube surface was explained with the dynamic contact angle and a dynamic surface tension. Nevertheless both effects were not considered in their theoretical model. Their dimensionless equation for the liquid rise contains the characteristic time for viscous effects $t_c = R^2/(8\nu)$ and a dimensionless number $\mathrm{Me} = 1/\mathrm{Oh}^2$. Their study was later expanded [104] with microgravity experiments ($\mathrm{Bo} \to 0$) in a 10 s drop shaft and, to study the influence of gravity-jitter, parabolic flight experiments in an airplane.

Dreyer et al. [105] examined the capillary-driven flow under microgravity in a 4.7 s drop tower using parallel-plates channels. This geometry is closely related to surface tension tank technology, but the results are also comparable to the tube geometry. In their theoretical model inertia effects, the dynamic contact angle, the friction force, the entrance pressure loss, and the curvature of the free surface in the liquid reservoir were considered. For the dimensionless representation a characteristic time $t_{cpp} = a^2/(12\nu)$ and a characteristic velocity $v_{cpp} = \sqrt{2\sigma/(\rho a)}$, with the plate distance a, were used. The dimensionless equation for the meniscus rise was found to depend on the ratio of plate distance over plate width a/b and the Ohnesorge number $\mathrm{Oh}_{pp} = \sqrt{\nu^2\rho/(2\sigma a)}$ for the parallel-plates geometry. Dreyer et al. [105] describe that the capillary rise process is divided into three successive stages. At the beginning the meniscus position h is proportional to the time squared ($h \sim t^2$) and the flow is determined by inertia forces. After that the meniscus height is proportional to the time ($h \sim t$), with the convective pressure loss at the channel entrance dominating the flow. Finally the distance traveled by the meniscus is proportional to the square root of time $\left(h \sim \sqrt{t}\right)$, the above mentioned viscous dominated Lucas-Washburn behavior. Their theoretical results were confirmed by drop tower experiments with characteristic scales as listed in Table 4.1 (here $R = a/2$ was used for the calculation of the scales).

In contrast to [105] Quéré [107] found that the meniscus height h is proportional to the time t at the beginning of the liquid rise and derived the constant velocity $v_c = \sqrt{2\sigma/(\rho R)}$ for that domain. This theoretical velocity was not confirmed by his experiments where lower velocities were measured. Several reasons for this deviation were proposed by Quéré: the existence of a dynamic contact angle, an inertia effect of the reservoir, and a pressure loss at the sharp edge of the tube entrance. Two time scales were proposed. The inertia regime which was exclusively related to the meniscus shape development was given as $t_{pu} = \sqrt{\rho R^3/\sigma}$ (10.1). This time scale was reported to be in the order of 10^{-3} s and thus too short to be verified by the experiments which were observed with a video camera that took pictures every 0.005 s. The time scale for the transition to the Lucas-Washburn regime was found to be $R^2/(4\nu)$ which equals $2t_c$.

Trutschel & Schellenberger [108] investigated the transient capillary rise in vertical tubes under terrestrial conditions numerically using a commercial three-dimensional finite element method (FIDAP 7.52) instead of the one-dimensional integral approaches of their predecessors. Only the calculated finite meniscus height was compared with their experiments and good agreement was found. Their work focused on specific numerical aspects such as node density, mesh refinement, and numerical accuracy.

Weislogel & Lichter [121] studied the capillary driven flow along interior corners of a partly filled container in microgravity using a 2.2 s drop tower. They reported that after the initial formation of the capillary surface, the tip location h of the liquid column at first moves linearly with time $(h \sim t)$ followed by a $(h \sim t^{3/5})$ dependence and finally shows the $(h \sim \sqrt{t})$ behavior.

Zhmud et al. [109] once more examined the Lucas-Washburn equation (4.31) in comparison with their ground experiments. They extended the Lucas-Washburn equation with a term for inertia effects and discussed the influence of dissipation effects due to the velocity field change in the vicinity of the meniscus. Other effects like the dynamic contact angle were not considered. They reported an initial $(h \sim t^2)$ behavior followed by the final $(h \sim \sqrt{t})$ dependence.

So far, a wetting liquid displacing air has been considered. A number of authors focused on liquid/liquid displacement due to its importance in enhanced oil recovery [122, 123, 124, 125, 126]. Mumley et al. [122, 123] studied the liquid/liquid displacement in glass capillaries and observed an intermittent flow (pseudo-slip-stick). Their theoretical model was based on the analysis of [116]. An excess force was introduced to explain additional retarding forces that arise from hydrodynamic and surface forces at the liquid/liquid interface. The excess force was found to be in the same order of magnitude as the driving capillary force. Mumley et al. suggested that the capillary rise phenomenon must be understood for the simple case of prewetted capillaries and continuous movement before progress in liquid/liquid displacement can be made. An interfacial resistance was also reported by Remoortere & Joos [125]. They found the resistance to be one order of magnitude larger than the interfacial force. Reference [126] reported in their study on oil/water displacement that even at the small pore size of porous media inertial forces are important and thus the pure Lucas-Washburn equation (4.31) cannot be applied.

Considering the previous approaches discussed in this section, it can be summarized that a comprehensive theory for the capillary rise in cylindrical tubes is still missing. This gap has been partly closed in a recent paper by Stange et al. [127][2]. The use of the characteristic scales t_{pu}, t_{vu}, and Oh (10.1)–(10.3) for the discussion of data found in literature indicates why different $h(t)$ dependencies were found by different authors. For ground experiments the time scales t_{pu} and t_{vu} are small, thus high Ohnesorge numbers occurred and only Lucas-Washburn behavior was observed. If wider tubes were used, which became possible by microgravity experiments, first t_{vu} became larger, the Ohnesorge number was reduced, and the $h \sim t$ behavior preceded the Lucas-Washburn flow. If the Ohnesorge number was reduced further, also t_{pu} was increased to considerable values and an initial $h \sim t^2$ dependence was observed. In this case t_{vu} may become larger than the experiment duration, thus the Lucas-Washburn flow was not reached within the experiment time.

4.7 Critical Velocities in Open Capillary Channel Flows

If the Bond number is sufficiently low, the flow path does not necessarily need a closed cross section as the tubes discussed above. Free liquid surfaces along parts of the circumference may serve as a boundary for the liquid. But these capillary channels are sensitive against pressures lower than ambient pressure. The limitation of the flow rate in open capillary channels has not been fundamentally discussed in literature. As far as the literature inquiry shows theoretical work is rare and just a few publication dealing with numerical model computations are available. Obviously experimental investigations have not been performed, except for the experiments presented in this work [128].

The only publication dealing with flow rate limitations in open capillary channel is the paper of Jaekle [29]. He discusses the use of open capillary channels, so-called vanes, as propellant management devices (PMDs) in surface tension tanks. The simplest form of a vane consists of a thin metal sheet mounted perpendicularly to the tank wall (T-shaped geometry). In addition to these sheets at the tank wall some PMDs have a vane structure along the

[2] We acknowledge the permission by the American Institute of Physics to use parts of the publication by M. Stange, M. Dreyer, H. Rath, Capillary driven flow in circular cylindrical tubes, 2003, Physics of Fluids, for this book to provide a comprehensive description of the capillary rise.

longitudinal tank axis (center post). Since the primary aim of the author is to provide tools for the estimation of the flow rate provided by PMDs, the investigated channel consists of a typical configuration of two vanes. One flow path at the tank wall and the second along the longitudinal axis are considered. Both vane geometries are assumed to be V-shaped. For the theoretical prediction of the flow rate a steady and a non-steady model are presented. The flow is assumed to be one-dimensional represented by the mean velocity. Both models consider only the main radius of curvature of the liquid surface in the cross section plane neglecting the curvature in flow direction. The viscous flow losses are estimated by a mean friction factor. Jaekle mentions that the friction can be five times above the used approximation due to the varying cross section. Thus the predicted volume flux might be too high. The steady flow analysis yields a differential equation for the radius of curvature of the liquid surface along the vane. For a given flow rate and downstream fillet radius at the outlet the numerical integration yields the radii of the liquid volume along the flow path. The model computations show that at a certain downstream fillet radius the flow rate in the vane at the tank wall is limited. Jaekle attributes the reason for this limitation to the effect of choking. Choking occurs when the flow velocity approaches a certain limiting velocity, in this case defined by the area wave speed of the vane. Without going into the details the author mentions an analogy of the vane flow to flows in flexible tubes, terrestrial open channel flows and compressible gas flows.

The steady flow analysis is a very useful tool to estimate the characteristics and to evaluate the impact of the vane cross section. But mostly the conditions in a surface tension tank are transient caused by the ignition of the engine, thruster pulsing or depletion of refillable reservoirs. For this reason Jaekle [29] solved the time-dependent momentum and continuity equations using an explicit finite difference algorithm. The model computations show the transient behavior of the flow after a sudden change of the outlet boundary condition. After a certain time the tank fill fraction becomes critical and thus, the flow in the vane along the tank walls becomes choked, while the center post vane still supplies the outlet.

Der [130] analysed the propagation of small disturbances in simple vanes theoretically. Motivated by PMDs this work investigates at which speed and in which direction waves propagate along the vane, if the fluid is at rest. Neglecting the radius of curvature in flow direction as well as the viscosity a linear wave equation analogous to the acoustic wave equation is derived. Thus, the considered waves are of longitudinal type. Der shows that the wave speed depends on the liquid volume attached to the vane. In the case that the corresponding radius of curvature is smaller than the vane height the wave speed depends on the fluid properties and the radius of curvature. This speed can also be derived from the general form of (16.78). In the case of a radius of curvature larger than the vane height the speed additionally depends on the vane height itself.

Aside from the above mentioned papers no other papers dealing with the limitation of flow rate or wave motion in open capillary channel were found. As related research subjects a few papers dealing with capillary channels shall be mentioned. In their work Romero & Yost [131] consider the spreading of a liquid wetting front in a V-shaped surface groove (wedge). The position of the wetting front is proportional to $\sqrt{Dt}$. D is called a diffusion coefficient which depends on the fluid properties, the contact angle and the groove angle. Weislogel & Lichter [121] investigated the flow in an interior corner of an experiment container after a sudden reduction of the gravitational forces. For the theoretical prediction of the liquid flow a one-dimensional diffusion equation is set up. The theoretical results are compared with the experimental data collected using a 2.2 s drop tower. The concept of the flow resistance applied by Weislogel & Lichter is based on the work of [132] and [133]. Both papers investigate the laminar capillary flow at constant flow rate in an V-shaped groove numerically. A two-dimensional solution for the flow is presented and the laminar friction factor depending on the contact angle and half the wedge angle is derived.

The theory of choking in compressible gas flow is discussed in several textbooks, e.g. Faber [134], Landau & Lifschitz [135], White [136], and Schapiro [137]. The publication of Shapiro is emphasized since it gives a substantial explanation of the choking phenomena and detailed derivation of all relevant equations. An explanation and theoretical derivation of choking in an open channel flow under normal gravity is given by Smits [138]. Based on a general approach and using experimental results from in Part IV Rosendahl et al. [129][3] presented the proof that choking in capillary channels exists and that it can be explained in the same way as in other systems.

References

1. D.A. Petrash, R.F. Zappa, E.W. Otto: NASA TN D-1197 (1962)
2. D.A. Petrash, T.M. Nelson, E.W. Otto: NASA TN D-1582 (1963)
3. D.A. Petrash, R.C. Nussle, E.W. Otto: NASA TN D-1577 (1963)
4. D.A. Petrash, R.C. Nussle, E.W. Otto: NASA TN D-2075 (1963)
5. C.E. Siegert, D.A. Petrash, E.W. Otto: NASA TN D-2458 (1964)
6. E.W. Otto: NASA TM X-52016 (1964)
7. W.C. Reynolds, M.A. Saad, H.M. Satterlee: Report No. LG-3, Mechanical Engineering Department, Stanford University (1964)
8. W.C. Reynolds, H.M. Satterlee: Liquid Propellant Behavior at Low and Zero Gravity. In: *The Dynamics Behavior of Liquids in Moving Containers*, ed by H.N. Abramson (NASA SP-106 1966) pp. 387-439

[3] We acknowledge the permission by Cambridge University Press to use parts of the publication by U. Rosendahl, A. Ohlhoff, M. Dreyer, Choked flows in open capillary channels: theory, experiment and computation, 2004, Journal of Fluid Mechanics, for this book to enable a comprehensive description of the phenomenon.

9. L.M. Habip: Astronautica Acta **11.6**, 401 (1965)
10. W.J. Masica, D.A. Petrash, E.W. Otto: NASA TN D-2267 (1964)
11. W.J. Masica, J.D. Derdul, D.A. Petrash: NASA TN D-2444 (1964)
12. W.J. Masica: NASA TN D-4066 (1967)
13. H.L. Paynter, T.R. Barksdale: J. Spacecr. Rockets **7**, 702 (1970)
14. R.C. Nussle, J.D. Derdul, D.A. Petrash: NASA E-2646 (1965)
15. E.P. Symons, R.C. Nussle, K.L. Abdalla: NASA TN D-4628 (1968)
16. R.G. Clodfelter: Aviation Safety Digest, ASD-TDR-63-506, 1 (1963)
17. S.C. DeBrock: AIAA 73-1228, 1 (1973)
18. G.D. Bizzell, G.E. Crane: CR 135004, 1 (1976)
19. T.P. Yeh, G.F. Orton: AIAA 87-1895, 1 (1987)
20. J. Boulay, S. Larigaldie: ONERA-TP-1975-116E, 1 (1975)
21. M.W. Dowdy, R.E. Hise, R.G. Peterson et al: AIAA 76-596, 1 (1976)
22. J.J. Chapter, S.B. Rider: AIAA 80-1098, 1 (1980)
23. I.K. Kim, F.O. Bennett: AIAA 87-1763, 1 (1987)
24. G.P. Purohit, L.D. Loudenback: AIAA 88-2919, 1 (1988)
25. D. Baralle, J.P. Fournier: AIAA 89-2761, 1 (1989)
26. J.M. Lee: Cryogenics **29.5**, 523 (1989)
27. B.A. Bicknell, P.M. Czysz: AIAA 90-2379, (1990)
28. J. Tegart: AIAA 91-2174, 1 (1991)
29. D.E. Jaekle, Jr.: AIAA 91-2172, 1 (1991)
30. D.E. Jaekle, Jr.: AIAA 93-1970, 1 (1993)
31. S. Dominick, S. Driscoll: AIAA 93-2424, 1 (1993)
32. S. Dominick, J. Tegart: AIAA 94-3027, 1 (1994)
33. B.N. Antar, V.S. Nuoti-Antar: *Fundamentals of Low Gravity Fluid Dynamics and Heat Transfer*, (CRC Press, Boca Raton Florida 1993)
34. J. Israelachvili: *Intermolecular and surface forces*, (Academic press, Amsterdam 1992) p. 192
35. C.G. Ngan, E.B. Dussan V.: J. Fluid Mech. **118**, 27 (1982)
36. Y.D. Shikhmurzaev: J. Fluid Mech. **334**, 211 (1997)
37. T. Young: Philos. Trans. Roy. Soc. London **95**, 185 (1805)
38. H.K. Moffatt: J. Fluid Mech. **18**, 1 (1963)
39. H. Lamb: *Hydrodynamics*, (Dover Publication, New York 1932) p. 586
40. S.F. Kistler: Hydrodynamics of Wetting. In: *Wettability*, ed by J.C. Berg (Marcel Dekker, New York 1993) pp. 311-429
41. L.M. Hocking: J. Fluid Mech. **79**, 209 (1977)
42. P. Ehrhard, S.H. Davis: J. Fluid Mech. **228**, 365 (1991)
43. H.P. Greenspan: J. Fluid Mech. **84.1**, 125 (1978)
44. P.J. Haley, M.J. Miskis: J. Fluid Mech. **223**, 223 (1991)
45. C. Huh, S.G. Mason: J. Fluid Mech. **81.3**, 401 (1977)
46. P.A. Durbin: J. Fluid Mech. **197**, 157 (1988)
47. E.B. Dussan V.: J. Fluid Mech. **77**, 665 (1976)
48. D.E. Finlow, P.R. Kota, A. Bose: Phys. Fluids **8**, 302 (1996)
49. M.Y. Zhou, P. Sheng: Phys. Rev. Letters **64**, 882 (1990)
50. E.B. Dussan V.: Ann. Rev. Fluid Mech. **11**, 371 (1979)
51. P.G. de Gennes: Rev. Mod. Phys. **57.3**, 827 (1985)
52. T.S. Jiang, S.O. Oh, J.C. Slattery: J. Colloid Interface Sci. **69.1**, 74 (1979)
53. R.L. Hoffman: J. Colloid Interface Sci. **50.2**, 228 (1975)
54. L.H. Tanner: J. Phys. D: Appl. Phys. **12**, 1473 (1979)

55. D.E. Finlow, P.R. Kota, A. Bose: Phys. Fluids **8**, 302 (1996)
56. L.M. Hocking, A.D. Rivers: J. Fluid Mech. **121**, 425 (1982)
57. C. Huh, L.E. Scriven: J. Colloid Interface Sci. **35.1**, 85 (1971)
58. O.V. Voinov: Fluid. Dyn. **11**, 714 (1976)
59. L.M. Pismen, A. Nir: Phys. Fluids **25**, 3 (1982)
60. R.G. Cox: J. Fluid Mech. **168**, 169 (1986)
61. C.G. Ngan, E.B. Dussan V.: J. Fluid Mech. **209**, 191 (1989)
62. E.B. Dussan V., E. Ramé, S. Garoff: J. Fluid Mech. **230**, 97 (1991)
63. W. Boender, A.K. Chesters, A.J.J. van der Zanden: Intern. J. Multiphase Flow **17.5**, 661 (1991)
64. Q. Chen, E. Ramé, S. Garoff: Phys. Fluids **7**, 2631 (1995)
65. G. Friz: Z. angew. Phys. **19.4**, 374 (1965)
66. W. Rose, R.W. Heins: Colloid Sci. **17**, 39 (1962)
67. T.D. Blake, J.M. Haynes: J. Colloid Interface Sci. **30.3**, 421 (1969)
68. J.G. Petrov, B.P. Radoev: Colloid & Polymer Science **259**, 753 (1981)
69. S. van Mourik, A.E.P. Veldman, M.E. Dreyer: Microgravity Sci. Technology **17.3**, 87 (2005)
70. M. Michaelis, M.E. Dreyer: Multiphase Science and Technology **16.1-3**, 219 (2004)
71. E.B. Dussan V., S.H. Davis: J. Fluid Mech. **65**, 71 (1974)
72. Y.D. Shikhmurzaev: Int. J. Multiphase Flow **19.4**, 589 (1993)
73. Y.D. Shikhmurzaev: Fluid Dynamics Research **13**, 45 (1994)
74. Y.D. Shikhmurzaev: AIChe Journal **42**, 601 (1996)
75. Y.D. Shikhmurzaev: J. Fluid Mech. **313**, 359 (1998)
76. T.D. Blake, M. Bracke, Y. Shikhmurzaev: Phys. Fluids **11.8**, 1995 (1999)
77. C.L. Ting, M. Perlin: J. Fluid Mech. **295**, 263 (1995)
78. L.M. Hocking: J. Fluid Mech. **179**, 253 (1987)
79. G.W. Young, S.H. Davis: J. Fluid Mech. **174**, 327 (1987)
80. H.M. Satterlee, W.C. Reynolds: NASA TR LG-2, 1 (1964)
81. Y. Kamotani, L. Chao, S. Ostrach et al: J. Spacecr. Rockets **32.1**, 177 (1995)
82. P. Ehrhard: J. Fluid Mech. **257**, 463 (1993)
83. D.M. Anderson, S.H. Davis: Phys. Fluids **7.2**, 248 (1995)
84. F.J. Renk, P.C. Wayner, Jr.: J. Heat Transfer **101**, 55 (1979)
85. F.J. Renk, P.C. Wayner, Jr.: J. Heat Transfer **101**, 59 (1979)
86. S. Moosman, G.M. Homsy: J. Colloid Interface Sci **73.1**, 212 (1979)
87. P.C. Wayner, Jr.: Langmuir **9**, 294 (1993)
88. D. Ripple: NISTIR 6351, 1 (1999)
89. W.F. Kaukler: Metallurgical Transactions AIME **19a**, 2625 (1988)
90. M.M. Weislogel, H.D. Ross: NASA TM 103641, (1990)
91. M.M. Weislogel, H.D. Ross: Microgravity Sci. Technol. **3**, 24 (1990)
92. H.F. Bauer, W. Eidel: Appl. microgravity tech. **2**, 212 (1990)
93. G. Wölk, M.E. Dreyer, M. M. Weislogel: J. Spacecr. Rockets **34**, 110 (1997)
94. J. Gerstmann, M. Michaelis, M.E. Dreyer et al: PAMM **3**, 354 (2003)
95. J.M. Bell, F.K. Cameron: J. Phys. Chem. **10**, 658 (1906)
96. E.W. Washburn: Phys. Rev. **17.3**, 273 (1921)
97. E.K. Rideal: Philos. Mag. Ser. 6 **44**, 1152 (1922)
98. E.J. LeGrand, W.A. Rense: J. Appl. Phys. **16**, 843 (1945)
99. R. Siegel: J. Appl. Mech. **83**, 165 (1961)
100. A.A. Jeje: J. Colloid Interface Sci. **69.3**, 420 (1979)

101. L.R. Fisher, P.D. Lark: J. Colloid Interface Sci. **69.3**, 486 (1979)
102. P. Joos, P. van Remoortere, M. Bracke: J. Colloid Interface Sci. **136.1**, 189 (1990)
103. N. Ichikawa, Y. Satoda: J. Colloid Interface Sci. **162**, 350 (1994)
104. N. Ichikawa, M. Misawa, K. Kawasaki et al: AIChe Journal **92**, 186 (1996)
105. M.E. Dreyer, A. Delgado, H.J. Rath: J. Colloid Interface Sci. **163**, 158 (1994)
106. M.E. Dreyer: *Kapillarer Flüssigkeitsanstieg zwischen parallelen Platten unter kompensierter Gravitation*, (VDI-Verlag, Düsseldorf 1994)
107. D. Quéré: Europhys. Lett. **39**, 533 (1997)
108. R. Trutschel, U. Schellenberger: Int. J. Numer. Methods Fluids **26**, 485 (1998)
109. B.V. Zhmud, F. Tiberg, K. Hallstensson: J. Colloid Interface Sci. **228**, 263 (2000)
110. M. Stange, M.E. Dreyer, H.J. Rath: Physics of Fluids **15**, 2587 (2003)
111. R. Lucas: Kolloid-Zeitschrift **23**, 15 (1918)
112. A. Marmur, R.D. Cohen: J. Colloid Interface Sci. **189**, 299 (1997)
113. A. Borhan, K.K. Rungta: J. Colloid Interface Sci. **155**, 438 (1993)
114. A. Borhan, K.K. Rungta: J. Colloid Interface Sci. **158**, 403 (1993)
115. C.H. Bosanquet: Philos. Mag. Ser. 6 **45**, 525 (1923)
116. J. Szekely, A.W. Neumann, Y.K. Chuang: J. Colloid Interface Sci. **35.2**, 273 (1971)
117. S. Levine, P. Reed, E.J. Watson et al: A Theory of the Rate of Rise of a Liquid in a Capillary. In: *Colloid and Interface Science*, ed by M. Kerker (Academic press, Amsterdam 1976) pp. 403-419
118. M.F. Letelier, H.J. Leutheusser, C. Rosas: J. Colloid Interface Sci. **72.3**, 465 (1979)
119. S. Levine, J. Lowndes, E.J. Watson et al: J. Colloid Interface Sci. **73.1**, 136 (1980)
120. G.L Batten, Jr.: J. Colloid Interface Sci. **102**, 513 (1984)
121. M.M. Weislogel, S. Lichter: J. Fluid Mech. **373**, 349 (1998)
122. T.E. Mumley, C.J. Radke, M.C. Williams: J. Colloid Interface Sci. **109.2**, 398 (1986)
123. T.E. Mumley, C.J. Radke, M.C. Williams: J. Colloid Interface Sci. **109.2**, 413 (1986)
124. A. Calvo, I. Paterson, R. Chertcoff et al: J. Colloid Interface Sci. **141**, 384 (1991)
125. P. van Remoortere, P. Joos: J. Colloid Interface Sci. **160**, 397 (1993)
126. K.S. Sorbie, Y.Z. Wu, S.R. McDougall: J. Colloid Interface Sci. **174**, 289 (1995)
127. M. Stange, M.E. Dreyer, H.J. Rath: Phys. Fluids **15.9**, 2587 (2003)
128. M.E. Dreyer, U. Rosendahl, H.J. Rath: AIAA 98-3165 (1998)
129. U. Rosendahl, A. Ohlhoff, M.E. Dreyer: J. Fluid Mech. **518**, 187 (2004)
130. J. Der: AIAA 91-2175 (1991)
131. L.A. Romero, F.G. Yost: J. Fluid Mech. **322**, 109 (1996)
132. T.C. Ransohoff, C.J. Radke: J. Colloid Interface Sci. **121**, 392 (1988)
133. P.S. Ayyaswamy, I. Catton, D.K. Edwards: J. App. Mech. **41**, 332 (1974)
134. T.E. Faber: *Fluid Dynamics for Physicists*, (Cambridge university press, Cambridge 1995)
135. L.D. Landau, E.M. Lifschitz: *Lehrbuch der Theoretischen Physik, Band VI, Hydrodynamik*, (Akademie Verlag, Berlin 1991)
136. F.M. White: *Fluid Mechanics*, (McGraw Hill, New York 1986) p. 511

137. A.H. Shapiro: *The Dynamics and Thermodynamics of Compressible Fluid Flow*, (The Ronald Press Company, New York 1953)
138. A.J. Smits: *A Physical Introduction to Fluid Mechanics*, (John Wiley & Sons, Inc., 2000)

Part II

Surface Oscillations upon Step Reduction of the Bond Number

In this part we consider the effect of a sudden change in the acceleration level on the free surface of a right circular cylinder. In cases where $\mathrm{Bo} \ll 1$, the effects of gravity are small compared to surface tension effects. A reduction of gravity reduces the body force and thus the hydrostatic pressure gradient in the fluid. With vanishing hydrostatic pressure, interface oscillations are started and interfacial forces are no longer masked by earth gravity. The fluid moves from an essentially flat surface configuration towards the curved low gravity equilibrium shape. Depending on the damping of the system, an overshoot may result and oscillations about the equilibrium surface shape occur. Only a small number of experimental studies have been performed concerning surface reorientation and oscillation in low gravity environments. In the following, numerical and experimental investigations of the reorientation and settling (damped oscillation) of a fluid meniscus in a right circular cylinder upon a step reduction in gravity level are presented.

The frequency of the damped oscillation towards the low Bond number equilibrium condition depends on the value of the static contact angle, the contact line condition (free or fixed) and the Ohnesorge number. Experiments with high static contact angles obey a fixed contact line behavior, and the frequency is mainly independent of the contact angle. Experiments with a low static contact angle value follow a free contact line behavior, and the frequency depends slightly on the static contact angle. The same holds for the dependence on the Ohnesorge number: The frequency of high contact angle liquids is constant, whereas the frequency of low contact angle liquids with a free contact line behavior decreases with decreasing Ohnesorge number. The experimental findings could be corroborated by numerical computations using a finite element code.

Liquids with high static contact angles show a short nonlinear period with a following linear oscillation after the contact line is fixed. The damping increases with increasing Ohnesorge number. Due to the fixed contact line the damping is mainly caused by bulk dissipation and a linear dependence of the damping ratio on the Ohnesorge number is observed. A clear trend for liquids with small contact angle could not be found. If the Ohnesorge number is small, say below 2×10^{-3}, a layer is formed following the first rise on the wall and the contact line shows a free contact line behavior. The damping is caused by the oscillating wall layer and in the bulk and the damping ratio is one order of magnitude larger than for large contact angles. If the Ohnesorge number is higher, the same behavior as for high contact angle liquid was observed, but a clear trend in the data could not be found. It is expected that with increasing Ohnesorge number the wall effect diminishes due to increasing pinning of the contact line, and that the damping ratio increases linearly with the Ohnesorge number. Nevertheless the damping ratio remains still one order of magnitude smaller compared to liquids with large contact angle. Thus the damping of the same liquid but with a different contact angle (due to changes the surface energy of the wall) may differ one order of magnitude.

5

Static Surface Shapes

The surface orientation begins with a quiescent free surface at a high Bond number (Bo $\gg$ 1) and ends with a stationary surface shape at a low Bond number (Bo $\ll$ 1). Before starting the investigation of the transient behavior, the initial and final conditions must be known. Figure 5.1 shows the geometry and nomenclature of a partly filled right circular cylinder with radius R. The free liquid/gas interface is denoted with $z = h\,(r,t)$ for the axis-symmetric coordinate system. In the initial situation a gravitational acceleration k_{zi}, aligned with the cylinder axis, is acting on the system. In this case the free surface shape $h(r, t = 0)$ is dominated by hydrostatic forces and capillary forces can be neglected except at the cylinder wall, where the liquid meets the solid with the initial static contact angle γ_{s0}. In the final situation compensated gravity conditions with the residual acceleration k_z are considered and a free surface with constant curvature meets the cylinder wall with the final static contact angle γ_s. Due to the initial filling procedure of the vessel the static contact angle must be distinguished between the initial advancing static contact angle γ_{s0} at the time $t = 0$ and the static contact angle γ_s for $t \to \infty$ after reorientation is completed.

5.1 Initial Condition with Residual Acceleration

To describe the static liquid interface $h\,(r, t_0)$ for the initial condition with the gravitational acceleration k_{zi}, the GAUSS-LAPLACE equation

$$p - p_a = \Delta p_c = -2\sigma H = -\sigma \left(\frac{1}{R_1} + \frac{1}{R_2} \right) \tag{5.1}$$

with the mean curvature $2H$ at the liquid interface $z = h$ in an axis-symmetric system [1]

Michael E. Dreyer: *Free Surface Flows under Compensated Gravity Conditions*
STMP **221**, 51–55 (2007)
DOI 10.1007/3-540-44628-1_5

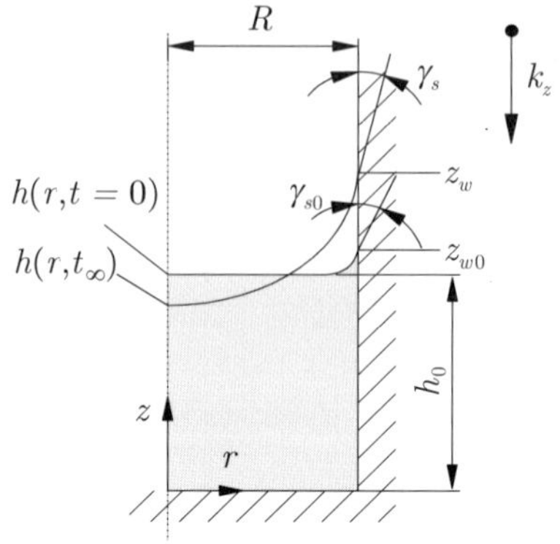

Fig. 5.1. Geometry and nomenclature for the cylindrical container partly filled with liquid. The flow is considered to be axis-symmetric. The coordinate system is located at the bottom of the cylinder. The fill height h_0 denotes the height of the center point $h(0,0)$

$$2H = \frac{1}{r}\frac{\mathrm{d}}{\mathrm{d}r}\left\{\frac{r\dfrac{\mathrm{d}h}{\mathrm{d}r}}{\left[1+\left(\dfrac{\mathrm{d}h}{\mathrm{d}r}\right)^2\right]^{1/2}}\right\} \tag{5.2}$$

and the hydrostatic pressure difference $\Delta p = p_1 - p_2 = C_1 - C_2 + \rho k_{zi} h(r)$ are applied. Nondimensionalization with $r^* = r/R$, $h^* = h/R$, $C = (C_1 - C_2)R/\sigma$ and the BOND number

$$\mathrm{Bo} = \frac{\rho k_{zi} R^2}{\sigma}, \tag{5.3}$$

leads to a nonlinear differential equation for the initial static liquid/gas interface

$$\frac{1}{r^*}\frac{\dfrac{\mathrm{d}h^*}{\mathrm{d}r^*}}{\left[1+\left(\dfrac{\mathrm{d}h^*}{\mathrm{d}r^*}\right)^2\right]^{1/2}} + \frac{\dfrac{\mathrm{d}^2h^*}{\mathrm{d}r^{*2}}}{\left[1+\left(\dfrac{\mathrm{d}h^*}{\mathrm{d}r^*}\right)^2\right]^{3/2}} = C + \mathrm{Bo}h^*. \tag{5.4}$$

This equation was solved numerically with the boundary condition at the cylinder wall

$$\left.\frac{\mathrm{d}h^*}{\mathrm{d}r^*}\right|_{r^*=1} = \cot\gamma_{s0} \tag{5.5}$$

and the fill height in the cylinder axis $h^*(r=0, t=0) = h_0^*$. Thus the static liquid interface in a right cylindrical container with an acceleration k_{zi} acting along the cylinder axis is only a function of the BOND number and the static contact angle γ_{s0}. The BOND number establishes a dimensionless relation between hydrostatic and capillary forces. High BOND numbers lead to a flat looking shape of the initial free surface. At the wall, where hydrostatic and capillary forces are balanced, the wall coordinate $z\,(r = R, t = 0) = z_{w0}$ is on the order of $z_{w0} \sim L_c\sqrt{2\,(1 - \sin\gamma_{s0})}$ (solution for the surface height at a plane wall [2]) with the capillary length

$$L_c = \sqrt{\frac{\sigma}{\rho k_{zi}}}\,. \tag{5.6}$$

5.2 Surface Shape in Reduced Gravity

After step reduction in gravity from k_{zi} to a residual acceleration $k_z \approx 10^{-5} g_0$ a BOND number Bo $\ll 1$ is obtained (for characteristic lengths in the order of 0.1 m and the test liquid given in Chap. 7). The static liquid interface for this BOND number has a spherical shape, which can be described geometrically. As schematically shown in Fig. 5.2, the dimensionless horizontal difference (here the superscript $(^+)$ for dimensionless variables is based on the coordinate system defined in Fig. 5.2) between the coordinate of the contact point $z_w^+ = z_w/R = h^+\,(r^* = 1)$ and the center point $z_c^+ = z_c/R = h^+\,(r^* = 0)$ may be expressed for this configuration with [3]

$$\alpha = h^+\,(r^* = 1) - h^+\,(r^* = 0) = \frac{1 - \sin\gamma_s}{\cos\gamma_s}\,, \tag{5.7}$$

while the dimensionless radius of the spherical surface is described geometrically by

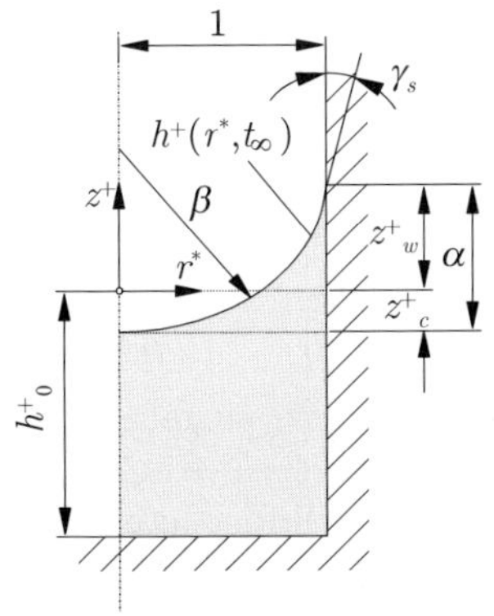

Fig. 5.2. Equilibrium configuration of the liquid interface for reduced gravity k_z in dimensionless notation. The coordinate system (z^+, r^*) is located at the height h_0^+ for an equivalent totally flat free surface, where the superscript $(^+)$ denotes the dimensionless form (scaled with R) for this coordinate system

$$\beta = \frac{1}{\cos \gamma_s} , \tag{5.8}$$

or by

$$\beta = \frac{1+\alpha^2}{2\alpha} , \tag{5.9}$$

if γ_s is substituted by α. The static contact angle γ_s can then be described by

$$\gamma_s = \arccos \frac{2\alpha}{1+\alpha^2} . \tag{5.10}$$

Thus, if the segment α can be determined experimentally from the final equilibrium surface configuration, the final static contact angle γ_s can be evaluated with (5.10), which is done for the determination of the static contact angle γ_s (compare Sect. 7.2.3).

On the other hand, if the static contact angle γ_s is known, the final deflection of the contact point z_w^+ and the center point z_c^+ can be described if the coordinate system is transformed to the height h_0^+. This height is equivalent to the height of a totally flat free surface (this analytical based coordinate system nondimensionalized with R is denoted with the superscript ($^+$) in contrast to the experimental based coordinate system introduced in Fig. 5.1). Comparing the liquid volume above h_0^* with the gas volume below h_0^+ one obtains analytical solutions for the final equilibrium ($t \to t_\infty$) contact point coordinate [4]

$$h^+ (r^* = 1, t_\infty) = z_{we}^+ = \frac{1}{3} \frac{(2 - 3\sin\gamma_s + \sin^3\gamma_s)}{\cos^3\gamma_s} , \tag{5.11}$$

and the coordinate of the center point

$$h^+ (r^* = 0, t_\infty) = z_{ce}^+ = -\frac{1}{\cos\gamma_s} \left[1 - \frac{2(1-\sin^3\gamma_s)}{3\cos^2\gamma_s} \right] . \tag{5.12}$$

Thus the contour of the spherical free surface may be described by

$$h^+ (r^*) = \frac{2(1-\sin^3\gamma_s)}{3\cos^3\gamma_s} - \frac{1}{\cos\gamma_s} \left[1 - (r^* \cos\gamma_s)^2 \right]^{1/2} . \tag{5.13}$$

However, as shown in Fig. 5.3, the volume above the plane $z^* = 0$, denoted with $V_{Lc}^* = V_{Lc}/R^3$, yields to an offset of the coordinate systems with

$$\Delta z_{Lc}^* = \frac{\Delta z_{Lc}}{R} = \frac{V_{Lc}}{\pi R^3} = \frac{1}{\pi} V_{Lc}^* \tag{5.14}$$

due to the capillary rise of the free surface at the cylinder wall. Thus the relation between the coordinate system from experiments and the analytical coordinate system reads for the cylinder axis

$$z_c^* = z_c^+ - \Delta z_{Lc}^* \tag{5.15}$$

and for the contact point

$$z_w^* = z_w^+ - z_{w0}^* + \Delta z_{Lc}^* . \tag{5.16}$$

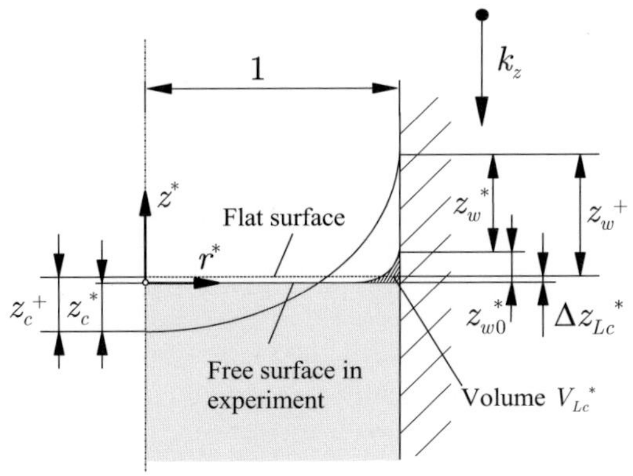

Fig. 5.3. Comparison between the analytical free surface position z^+ and the surface position z^* defined in experiments

References

1. B.N. Antar, V.S. Nuoti-Antar: *Fundamentals of Low Gravity Fluid Dynamics and Heat Transfer*, (CRC Press, Boca Raton Florida 1993) p. 41
2. L.D. Landau, E.M. Lifschitz: *Lehrbuch der Theoretischen Physik, Band VI, Hydrodynamik*, (Akademie Verlag, Berlin 1991) p. 271
3. M.M. Weislogel, H.D. Ross: Microgravity Sci. Technol. **3**, 24 (1990)
4. H.F. Bauer, W. Eidel: Appl. microgravity tech. **2**, 212 (1990)

6

Scaling of the Dynamic Behavior

The following chapter describes the governing equations of the free surface reorientation and gives a scaling to derive the determining parameters of the dynamic flow behavior. The initial and the final conditions of the surface reorientation are given by a static surface described above.

For the description of the dynamic behavior of the surface reorientation the following assumptions are made:

1. the flow is isothermal,
2. the flow is incompressible,
3. the flow is axis-symmetric,
4. no influence of the gas phase above the liquid on the flow.

6.1 Governing Equations

We recall the continuity equation (2.1) and the NAVIER-STOKES equation (2.2) from Part I in vector notation:

$$\nabla \cdot \boldsymbol{u} = 0 \ , \tag{6.1}$$

$$\rho \frac{\partial \boldsymbol{u}}{\partial t} + \rho \left(\boldsymbol{u} \cdot \nabla \right) \boldsymbol{u} = -\nabla p + \mu \nabla^2 \boldsymbol{u} + \rho \boldsymbol{k}. \tag{6.2}$$

The boundary conditions are given by the normal and tangential stress balance at the liquid interface which reduces to (2.7)

$$p - p_a = -2\sigma H \ ,$$

where p_a is the ambient pressure in the gas phase above the liquid interface. The mean curvature $2H$ may be expressed with (5.2). The kinematic boundary condition at the liquid interface reads

$$\boldsymbol{u} \cdot \boldsymbol{n} = 0 \ . \tag{6.3}$$

Michael E. Dreyer: *Free Surface Flows under Compensated Gravity Conditions*
STMP **221**, 57–62 (2007)
DOI 10.1007/3-540-44628-1_6

Furthermore, the boundary conditions in the cylinder axis ($r = 0$)

$$\left.\frac{\partial h}{\partial r}\right|_{r=0} = 0 \tag{6.4}$$

and at the cylinder wall ($r = R$)

$$\left.\frac{\partial h}{\partial r}\right|_{r=R} = \cot \gamma \tag{6.5}$$

have to be satisfied, where γ is the contact angle in general.

The comparison of the different terms leads to the time scales t_{pu} (2.18), t_{pv} (2.22), and t_{vu} (2.23), as discussed in Chap. 2. To characterize the different flow regimes and the determining parameters of the dynamic flow behavior, the governing equations will be nondimensionalized with characteristic length, time, velocity, pressure and acceleration scales, which is carried out in the following sections.

6.2 Scaling of the Initial Capillary Rise (Initial Time Domain)

As described above, the capillary length L_c defines a length scale near the cylinder wall for the initial situation when capillary and hydrostatic forces are balanced. Therefore, L_c should be the appropriate length scale for the very initial capillary rise since other longitudes cannot determine the flow in this region (compare Fig. 6.1). Assuming that capillary and inertia forces dominate the process leads to the pressure unsteady time t_{pu} (2.18) to be the relevant time scale. With the capillary length L_c (5.6) as the appropriate length scale, this time scale becomes

$$t_{puLc} = \left(\frac{\rho L_c^3}{\sigma}\right)^{1/2} = \left(\frac{\sigma}{\rho k_{zi}^3}\right)^{1/4} . \tag{6.6}$$

Thus the characteristic velocity yields

$$u_{puLc} = \frac{L_c}{t_{puLc}} = \left(\frac{k_{zi}\sigma}{\rho}\right)^{1/4} . \tag{6.7}$$

Using (6.6, 6.7), the initial acceleration k_{zi} and the capillary pressure to be the appropriate pressure scale

$$p_c = \frac{\sigma}{L_c} = (\sigma \rho k_{zi})^{1/2} \tag{6.8}$$

for scaling of the mass balance (6.1) and the momentum balance (6.2) leads to the scaled equations in cylinder coordinates (all variables are dimensionless, a superscript is neglected for clarity reasons, the subscript denotes the coordinate direction r and z)

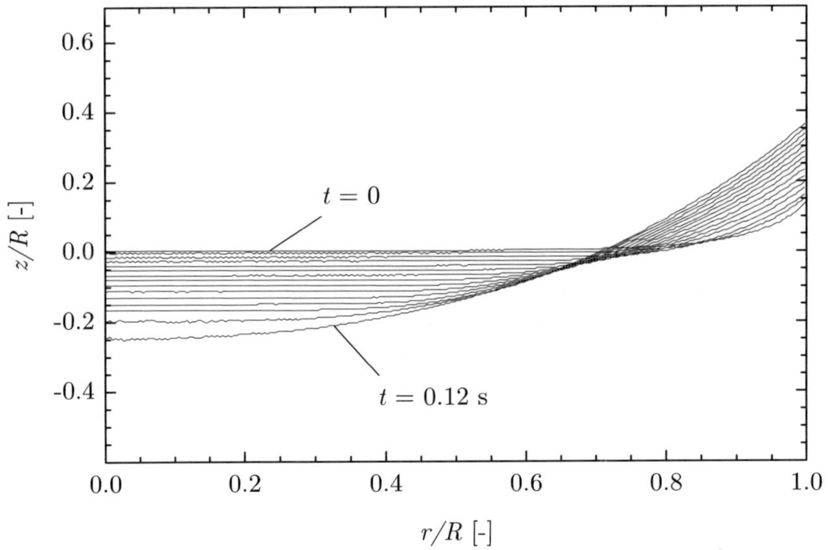

Fig. 6.1. Initial development of the free liquid interface upon step reduction from k_{zi} to k_z. The capillary rise along the cylinder wall is basically independent of the cylinder radius. The free surface contours are evaluated from test no. 32. The coordinates are scaled with R. Note, that the free surface in the middle region cannot be observed by reason of total reflection

$$\frac{\partial}{\partial r}(u_r r) + r\frac{\partial}{\partial z}(u_z) = 0 , \tag{6.9}$$

$$\frac{\partial u_r}{\partial t} + u_r\frac{\partial u_r}{\partial r} + u_z\frac{\partial u_r}{\partial z} = -\frac{\partial p}{\partial r} + \text{Mo}\left\{\frac{\partial^2 u_r}{\partial r^2} + \frac{1}{r}\frac{\partial u_r}{\partial r} + \frac{\partial^2 u_r}{\partial z^2} - \frac{u_r}{r^2}\right\} , \tag{6.10}$$

$$\frac{\partial u_z}{\partial t} + u_r\frac{\partial u_z}{\partial r} + u_z\frac{\partial u_z}{\partial z} = -\frac{\partial p}{\partial z} + \text{Mo}\left\{\frac{\partial^2 u_z}{\partial r^2} + \frac{1}{r}\frac{\partial u_z}{\partial r} + \frac{\partial^2 u_z}{\partial z^2}\right\} + k_z^* . \tag{6.11}$$

According to this scaling two dimensionless groups determine the flow, the acceleration ratio $k_z^* = k_z/k_{zi}$ and the MORTON number

$$\text{Mo} = \left(\frac{k_{zi}\mu^4}{\rho\sigma^3}\right)^{1/4} = \frac{t_{puLc}}{t_{vuLc}} \tag{6.12}$$

or

$$\text{Mo} = \left(\frac{\nu^4\rho^3}{\sigma^3}k_{zi}\right)^{1/4} , \tag{6.13}$$

if the kinematic viscosity is considered. Since the static contact angles γ_{s0} and γ_s describe the initial and end boundary condition at the cylinder wall

$$\left.\frac{\partial h}{\partial r}\right|_{r=\sqrt{\text{Bo}},\ t=0} = \cot\gamma_{s0} \tag{6.14}$$

$$\left.\frac{\partial h}{\partial r}\right|_{r=\sqrt{\mathrm{Bo}},\ t=\infty} = \cot \gamma_s \tag{6.15}$$

these values have to be considered as well (please note, that the dimensionless length $\sqrt{\mathrm{Bo}}$ results from the scaling R/L_c).

The MORTON number (in a notation without the exponent 1/4) is generally applied in the multiphase flow literature and gives a relation for the motion of drops and bubbles. It describes the ratio of the acceleration k_{zi} to an acceleration $\sigma^3/(\rho^3\nu^4)$ generated by the mobility of the liquid molecules [1]. This MORTON number decreases with a higher mobility of the molecules. Here the MORTON number formally corresponds to an OHNESORGE number with the capillary length used as appropriate length scale. It can be considered to be a measure for the velocity of the initial capillary rise at the cylinder wall (compare Sect. 9.1.1). Using the capillary length L_c for the scaling of the initial time period, the BOND number (5.3) (substituting R with L_c) becomes $\mathrm{Bo} = 1$, thus only the acceleration ratio k_z/k_{zi} remains, which is $k_z/k_{zi} \sim 10^{-6}$ for the drop tower experiments. Nevertheless, choosing the capillary length to be the relevant length scale for the initial time period is only reasonable if the capillary length $L_c \ll R$, thus $\left(\sigma/\rho k_{zi} R^2\right)^{1/2} = \mathrm{Bo}^{-1/2} \ll 1$. Consequently only for higher initial BOND numbers $\mathrm{Bo} \gg 1$ the capillary length is the appropriate length scale with no influence of the cylinder radius on the flow.

6.3 Scaling of the Surface Oscillation (Global Time Domain)

While for the initial time period obviously the cylinder radius R cannot be the appropriate length scale to describe the flow, R is chosen in the second time period as the characteristic length scale, thus $L = R$. The time period begins when the disturbance, induced by the initial capillary rise along the cylinder wall, reaches the cylinder axis. Then the whole system is affected by the disturbance (compare Fig. 6.2). Therefore the time scale t_{pu} (2.18) is chosen to scale the system. Thus the characteristic velocity becomes

$$u_{pu} = \frac{R}{t_{pu}} = \left(\frac{\sigma}{\rho R}\right)^{1/2} . \tag{6.16}$$

Taking t_{pu}, u_{pu}, k_{zi} and the characteristic pressure $p_c = \sigma/R$ for the scaling of the mass balance (6.1) and the momentum balance (6.2) one obtains in analogy to the first time period the following equations in cylinder coordinates [2]

$$\frac{\partial}{\partial r}(u_r r) + r\frac{\partial}{\partial z}(u_z) = 0 , \tag{6.17}$$

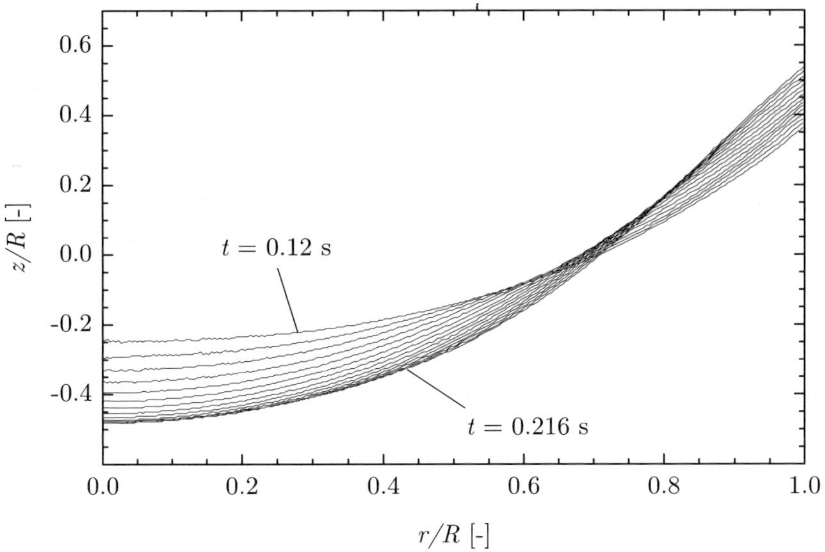

Fig. 6.2. Later development of the free surface. The disturbance affects the whole system and the free surface reaches its maximum deflection with a following oscillation around the equilibrium position. The free surface contours are evaluated from test no. 32. The coordinates are scaled with R

$$\frac{\partial u_r}{\partial t} + u_r \frac{\partial u_r}{\partial r} + u_z \frac{\partial u_r}{\partial z} = -\frac{\partial p}{\partial r} + \mathrm{Oh} \left\{ \frac{\partial^2 u_r}{\partial r^2} + \frac{1}{r}\frac{\partial u_r}{\partial r} + \frac{\partial^2 u_r}{\partial z^2} - \frac{u_r}{r^2} \right\}, \quad (6.18)$$

$$\frac{\partial u_z}{\partial t} + u_r \frac{\partial u_z}{\partial r} + u_z \frac{\partial u_z}{\partial z} = -\frac{\partial p}{\partial z} + \mathrm{Oh} \left\{ \frac{\partial^2 u_z}{\partial r^2} + \frac{1}{r}\frac{\partial u_z}{\partial r} + \frac{\partial^2 u_z}{\partial z^2} \right\} + \mathrm{Bo}\, k_z^* . \quad (6.19)$$

Here the determining parameters are the OHNESORGE number

$$\mathrm{Oh} = \left(\frac{\nu^2 \rho}{\sigma R} \right)^{\frac{1}{2}} = \frac{t_{pu}}{t_{vu}} , \quad (6.20)$$

and the BOND number given by (5.3) or by $\mathrm{Bo} = \rho k_z R^2/\sigma$, if k_{zi} is neglected. Since the residual acceleration k_z is very small, the influence of the BOND number can be neglected in the momentum transport equation. However, since the initial and final static surface configurations are determined by the BOND number and by the boundary conditions at the cylinder wall

$$\left. \frac{\partial h}{\partial r} \right|_{r=1,t=0} = \cot \gamma_{s0}, \quad \left. \frac{\partial h}{\partial r} \right|_{r=1,t=\infty} = \cot \gamma_s , \quad (6.21)$$

an influence of the BOND number due to the initial free surface curvature can be expected. The OHNESORGE gives a relation for the damping behavior of the flow. Small OHNESORGE numbers correspond to a low damping of the flow, while high OHNESORGE numbers lead to creeping flow.

References

1. H. Brauer: *Grundlagen der Ein- und Mehrphasenströmungen*, (Verlag Sauerländer, Aarau 1971) p. 303
2. J. Gerstmann, M.E. Dreyer, H.J. Rath: AIP Conference 504, 847 (2000)

7

Experiments on Surface Oscillations

Two series of experiments were carried out to investigate the reorientation behavior of a liquid interface in a cylindrical container exposed to a sudden change in gravity from $k_{zi} = g_0$ to $k_z = 10^{-5} g_0$. The first series is aimed at the investigation of the entire reorientation process with a focus on the frequency and the damping behavior of the surface oscillation. Those experiments require quite large OHNESORGE numbers with rather small cylinder radii (10 mm $< R <$ 20 mm) for the selected test liquids. Results for these experiments are presented in Sect. 9 for the surface center point (amplitudes, frequency and damping) as well as for the three phase contact line (rise velocity, the amplitudes and the fixing time).

The second series of experiments was focussed on the investigation of the initial period of the reorientation process, especially on the behavior of the contact line during the initial rise. To enable the observation of those characteristics over a wide time span, the cylinder radius was increased (40 mm $< R <$ 70 mm) for these experiments yielding much smaller OHNESORGE numbers compared to the first series of experiments. The MORTON number was comparable to the values in the first series. Results for these experiments are presented only for the contact line (compare Sect. 9.1.1), where the initial rise velocity and the first transit through equilibrium are considered.

7.1 Experimental Setup

7.1.1 Apparatus

The experimental setup was integrated in a drop capsule, which was released in the drop tower Bremen to yield microgravity conditions for an experiment time up to $t = 4.7$ s. The main components of the setup are a circular cylinder, a high-speed digital recording system and a background illumination device. The test setup is schematically shown in Fig. 7.1. To minimize optical disturbances resulting from general light refraction and the different refractive

Michael E. Dreyer: *Free Surface Flows under Compensated Gravity Conditions*
STMP **221**, 63–90 (2007)
DOI 10.1007/3-540-44628-1_7

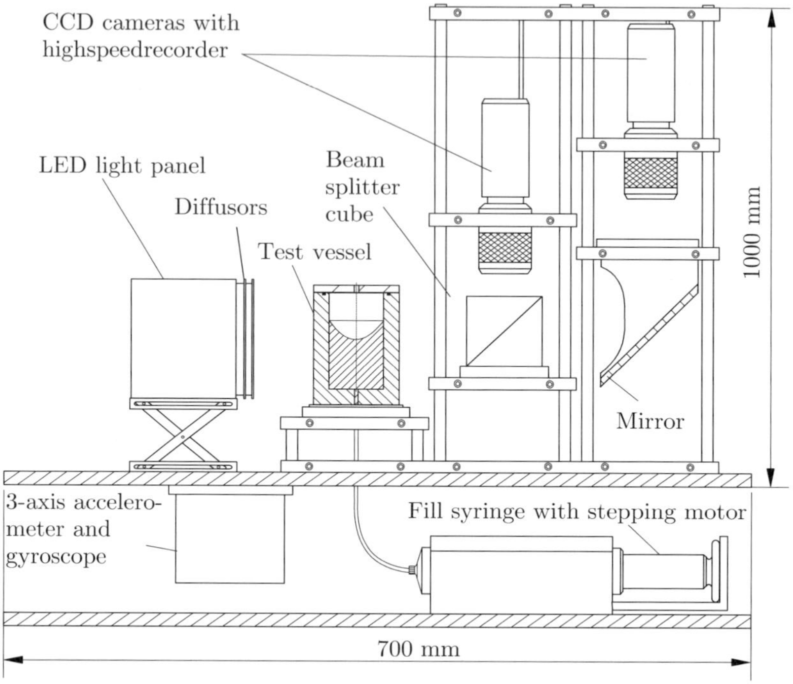

Fig. 7.1. Schematical sketch of the experimental setup, which was integrated in the drop capsule

indices between vessel material and the test fluid, the cylinder is manufactured from a solid PMMA cube. The outer dimensions of the cuvette are 80 mm $\times$ 80 mm $\times$ 50 mm (height, breadth and depth). The surface roughness is smaller than 1.6 µm, which was measured using a white light interferometer.

The alignment of the cylinder axis with the gravity vector is adjusted by using a leveling system on which the vessel is mounted to keep the flow axis-symmetric and therefore two-dimensional. Additionally the experiment platform is leveled to the 4 stringers of the drop capsule. In order to prevent contamination of the test fluid and the cylinder wall, the vessel is sealed with a cover plate and a breather hose after preparation of the experiment. Light diffusers between the LED panel and the vessel enable a homogenous light distribution from behind. To be able to observe a contour of the free liquid interface of very high contrast (which is necessary for a reliable and high precision edge detection of the liquid interface), a screen of black cardboard with a triangle cutout was mounted between the light source and the vessel. The distance between the light source and the vessel was varied for an optimal light distribution and contrast of the liquid interface. Moreover black cardboard mounted at the lateral faces of the vessel prevent disturbances from lateral light refraction.

The filling of the vessel with test liquid is obtained by interconnecting the inlet at the bottom of the vessel with a large glass syringe (acting as a reservoir) located on a lower platform. A stepping motor ensures a uniform and slow filling of the vessel just prior to the drop of the capsule to guarantee a uniform initial condition at the contact line of a non-wetted cylinder wall above the contact line. The fill height of the liquid was always $h_0 > 1.7R$ to prevent interactions of the flow with the bottom boundary of the vessel (compare Table 7.2). A high speed digital recording system (Kodak Motion Corder Analyzer, Series SR) with two CCD cameras (with Nikon MICRO-NICCOR lenses: $f = 105$ mm/1 : 2.8 and $f = 55$ mm/1 : 2.8) with a 512×480 pixel array was used to observe the reorientation process with a recording frequency f_r of 250 or 500 frames per second (compare Table 7.2). The application of a beam splitter cube enabled the observation of two different areas of the interface contour: A detail view on the moving contact line and a general view on the meniscus with a pixel resolution of $22 - 85\,\mu$m, depending on the chosen vessel diameter. The lenses of the camera system were focussed on the cylinder axis by accurately aligning a high precision glass measure into the cylinder. A series of video images, recorded with the high-speed digital video system, show the development of the sharp fluid interface after step reduction in gravity in Fig. 7.2. The constant time step between consecutive images is $\Delta t = 0.02$ s. The test liquid appears bright, the gas above the meniscus appears dark due to total reflection.

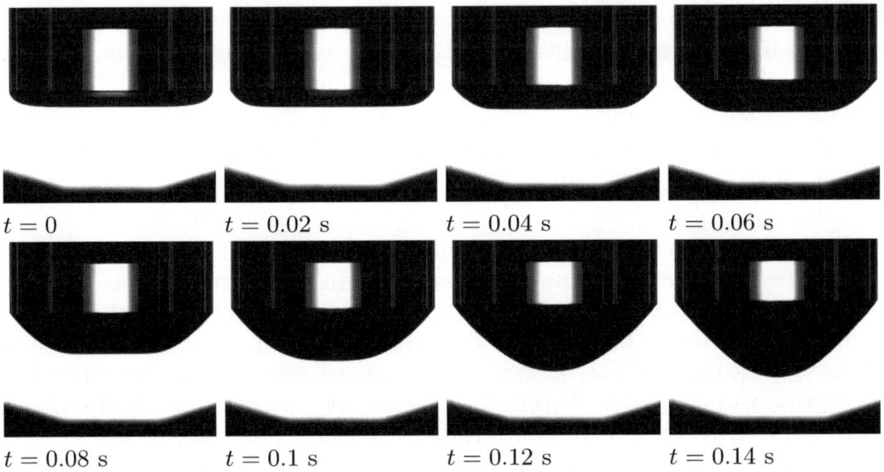

Fig. 7.2. Series of video images for test no. 46 showing the reorientation to the first minimum at the center point. The liquid appears bright, the gas above the free surface appears dark due to total reflection at the inner cylinder walls. The triangles at the bottom of the images are caused by a screen mounted between the light panel and the vessel, to increase the contrast of the free surface

$t = 0; j = 43$ $t = 0.008$ s; $j = 45$ $t = 0.016$ s; $j = 47$

Fig. 7.3. Series of video images for test no. 36. The pictures show the detail view on the contact line and the contact angle. The liquid appears bright, the gas appears dark due to total reflection on the inner cylinder walls

A measurement of the residual accelerations and the angular velocities acting on the drop capsule was carried out with a 3-axis accelerometer and a cyberoptical gyroscope, which was mounted beneath the experiment platform. Temperature sensors (PT-100) were placed on the bottom of the experiment platform, at the slide with the glass syringe and at the vessel to monitor and record the temperature distribution during the experiment. The sensors were calibrated in a temperature band of 15°C–35°C.

In order to obtain reliable and reproducible results an accurate cleaning of the test equipment was performed by rinsing the vessel and the glass syringe several times with toluene and ethyl alcohol with a following drying at ambient air. To reduce the water content solved in the border layers of the PMMA, the test cylinder was exposed to vacuum for 3 hours before the final cleaning procedure was carried out.

7.1.2 Test Liquids

A number of different silicone fluids (abbreviated with SF) with a nominal viscosity between 1.0 cSt and 10.0 cSt were used in this series. Due to different refractive indices between silicone fluid and the vessel material, the vicinity of the moving contact line could not be observed accurately. To avoid disturbances from refraction two liquids were used, for which the refractive index matches with the value of PMMA ($n = 1.49$). The first liquid is a solution of 1234-tetrahydronaphthalene and decahydronaphthalene, hereinafter referred to as detra and the second liquid is butylbenzene. Surface tension, viscosity, density and refractive index of these test liquids were measured with standard laboratory procedures and are given in Table 3.1.

7.1.3 Contact Angle

In order to vary the contact angle from $\gamma_s = 0$ for silicone fluids with PMMA, a thin film of a surface modifier (FC-732 by 3M Co.) with a low surface energy of 12×10^{-3} N/m (according to the manufacturer) was applied to the cylinder wall just prior to each drop after the cleaning procedure by filling the cylinder with FC-732 as a liquid and spilling it rapidly to achieve a homogenous film. The thin film of a thickness on the order of 1 µm was dried at ambient air. This surface modification and the assembly of the vessel was carried out in a glove box with air atmosphere to prevent impurities from dust. For experiments with detra and butylbenzene no surface modifier was applied. Thus two groups of experiments with low static contact angle ($\gamma_s < 10°$) and high static contact angle ($\gamma_s > 10°$) can be distinguished. For the silicone fluids (with surface modifier) a static contact angle $48° < \gamma_s < 60°$ was obtained while for experiments with detra and butylbenzene the contact angle varied between $0 < \gamma_s < 5°$.

Mean values for γ_s and γ_{s0} for the test liquids are given in Table 7.1. It can be seen for silicone fluids that the (advancing) contact angle of the initial free liquid interface is some degrees higher than the measured final contact angle caused by the initial filling procedure. A more detailed description of the evaluation of the static contact angle is given in Sect. 7.2.3.

Table 7.1. Initial and final contact angles of test liquids for $T = 25°$C

Test liquids	γ_{s0} (°) ±5°	γ_s (°) ±2°
SF 0.65	0	0
SF 1.0	55	48
SF 3.0	62	54
SF 10.0	65	60
Detra	6	5
Butylbenzene	3	0
FC-77	0	0

7.1.4 Test Parameters

Two series of test cases were carried out. 10 experiments with a small cylinder radius (cylinder radius R varied between 10 mm and 20 mm) were performed in the first series yielding quite high OHNESORGE numbers. The aim of these experiments was to investigate the entire reorientation of the system. In the second series, 8 experiments with a larger cylinder radius between 40 mm and 70 mm were carried out with a focus on the investigation of the first time period.

Experiments in the Global Time Regime

The following test cases were carried out by varying the OHNESORGE number, the initial BOND number, the MORTON number and the static contact angle γ_s of the test liquid with the cylinder wall. The different test cases are tabulated in Table 7.2 and Table 7.3. The OHNESORGE number for these test cases varies between $1.6 \times 10^{-3} < \text{Oh} < 20.2 \times 10^{-3}$ while the BOND number was varied in a range between $33.2 < \text{Bo} < 188.7$. The MORTON number, which depends basically on the properties of the test liquids, ranges between $\text{Mo} = 5.1 \times 10^{-3}$ for butylbenzene and $\text{Mo} = 52.7 \times 10^{-3}$ for SF 10.0. The aim of this series of experiments was to investigate the entire reorientation process over time to be able to evaluate the main characteristics such as amplitudes, frequency, and damping behavior of the surface oscillation.

Table 7.2. Experiment parameter for the global time regime

No	Test liquid	T (°)	D (mm)	h_0 (mm)	Coating	f_r (fps)
27	SF 1.0	24.6	25	49	FC − 732	500
30	SF 10	25.4	35	41	FC − 732	500
31	SF 3.0	24.9	35	44	FC − 732	500
32	Butylbenzene	24.4	35	38	−	250
36	Butylbenzene	24.6	20	39	−	250
37	SF 3.0	25.7	20	40	FC − 732	500
40	SF 1.0	24.7	20	42	FC − 732	500
41	SF 10	23.3	20	42	FC − 732	500
46	Detra	23.8	30	37	−	250
52	SF 3.0	22.9	40	34	FC − 732	500

Table 7.3. Determining parameter of experiments in the global time regime

No	γ_s (°)	γ_{s0} (°) ±3°	Oh (-)	Bo (-)	Mo (-)
27	48 ± 2	51	2.1×10^{-3}	7.5×10^{1}	6.1×10^{-3}
30	59 ± 2	64	1.5×10^{-2}	1.4×10^{2}	5.1×10^{-2}
31	53 ± 2	59	4.9×10^{-3}	1.5×10^{2}	1.7×10^{-2}
32	$0 + 2.5$	3	1.6×10^{-3}	1.0×10^{2}	5.1×10^{-3}
36	$1 + 2.5$	3	2.1×10^{-3}	3.3×10^{1}	5.1×10^{-3}
37	54 ± 2	59	6.5×10^{-3}	4.8×10^{1}	1.7×10^{-2}
40	47 ± 2	52	2.3×10^{-3}	4.8×10^{1}	6.1×10^{-3}
41	60 ± 2	65	2.0×10^{-2}	4.7×10^{1}	5.2×10^{-2}
46	5 ± 2.5	6	3.8×10^{-3}	7.4×10^{1}	1.1×10^{-2}
52	54 ± 2	59	4.8×10^{-3}	1.9×10^{2}	1.8×10^{-2}

Table 7.4. Experiment parameter for the initial time domain. No coating was applied on the cylinder walls. The recording frequency was 250 fps in all cases

No	Test liquid	T (°C)	D (mm)	h_0 (mm)
7	SF 0.65	23.6	80	80
8	SF 0.65	23.1	80	80
9	FC-77	23.6	80	80
10	FC-77	22.7	80	80
11	SF 0.65	22.6	140	140
12	SF 0.65	22.3	140	140
13	FC-77	22.2	140	140
14	FC-77	22.8	140	140

Experiments in the Initial Time Domain

The aim of this experiment series was to investigate solely the initial rise of the contact line at the cylinder wall. All experiments of this series are listed in Table 7.4 with the determining parameters in Table 7.5. For these test cases the OHNESORGE number was varied between $0.45 \times 10^{-3} < \mathrm{Oh} < 1.62 \times 10^{-3}$ where the BOND number ranges between $766 < \mathrm{Bo} < 5913$. The MORTON number is $\mathrm{Mo} = 3.1 \times 10^{-3}$ for SF 0.65 and $\mathrm{Mo} = 10.7 \times 10^{-3}$ for FC-77.

Since the oscillation cycle is not completed during the experiment duration of 4.7 s, frequency and damping cannot be evaluated here. The experiments described in this section are performed with a setup which is in general comparable to the setup used for the small cylinders. Two rectangular containers with a cylindrical bore are used. One has a base area of 150 mm × 160 mm, 310 mm height, a bore diameter of 80 mm, and the filling height of the test liquid is 80 mm. The second one has a base area of 160 mm×160 mm, 310 mm height, a bore diameter of 140 mm and the filling height of the test liquid is 140 mm.

Table 7.5. Determining parameters for the experiments in the initial time regime

No	γ_s (°)	γ_{s0} (°)	Oh (–)	Bo (–)	Mo (–)
7	0	0	5.9×10^{-4}	7.6×10^{3}	3.1×10^{-3}
8	0	0	5.9×10^{-4}	7.6×10^{3}	3.1×10^{-3}
9	0	0	1.6×10^{-3}	1.9×10^{4}	1.1×10^{-2}
10	0	0	1.6×10^{-3}	1.9×10^{3}	1.1×10^{-2}
11	0	0	4.5×10^{-4}	2.3×10^{3}	3.1×10^{-3}
12	0	0	4.5×10^{-4}	2.3×10^{3}	3.1×10^{-3}
13	0	0	1.2×10^{-3}	5.9×10^{3}	1.1×10^{-2}
14	0	0	1.2×10^{-3}	5.9×10^{3}	1.1×10^{-2}

Due to the large dimensions of the test containers, the LED panel is not mounted behind the containers. Instead, the LED panels are mounted above the containers and reflector plates are used to enable a homogenous lighting from behind. The same digital recording system as described above is used here. A recording frequency of $f_r = 250$ Hz is used and the pixel array is 512×480 pixel. Due to the large area of observation a pixel resolution of $0.1 - 0.2$ mm/pixels can be achieved. As test liquids SF 0.65 and FC-77 were used. No coating was applied for the experiments with the large cylinders, thus the static contact angle of the two fluids with the PMMA is $\gamma_s = 0$.

Figure 7.4 shows the liquid surface at an inclined view from above. Immediately after the release of the drop capsule the contact line starts to move upwards inducing an annular surface wave (visible for $t = 0.24$ s) which moves towards the center of the vessel. While the contact line proceeds to rise, the annular wave reaches the center and is reflected ($t = 0.44$ s). At $t = 0.88$ s the waves merge into a single wave crest of significant height. After this wave crest reaches its maximum position, it rapidly moves downwards elongating

Fig. 7.4. Initial behavior of the liquid surface after entering microgravity. Sequence of video frames for $R = 50$ mm and FC-77 test liquid. Pictures were taken at an inclined view to observe the liquid surface from above

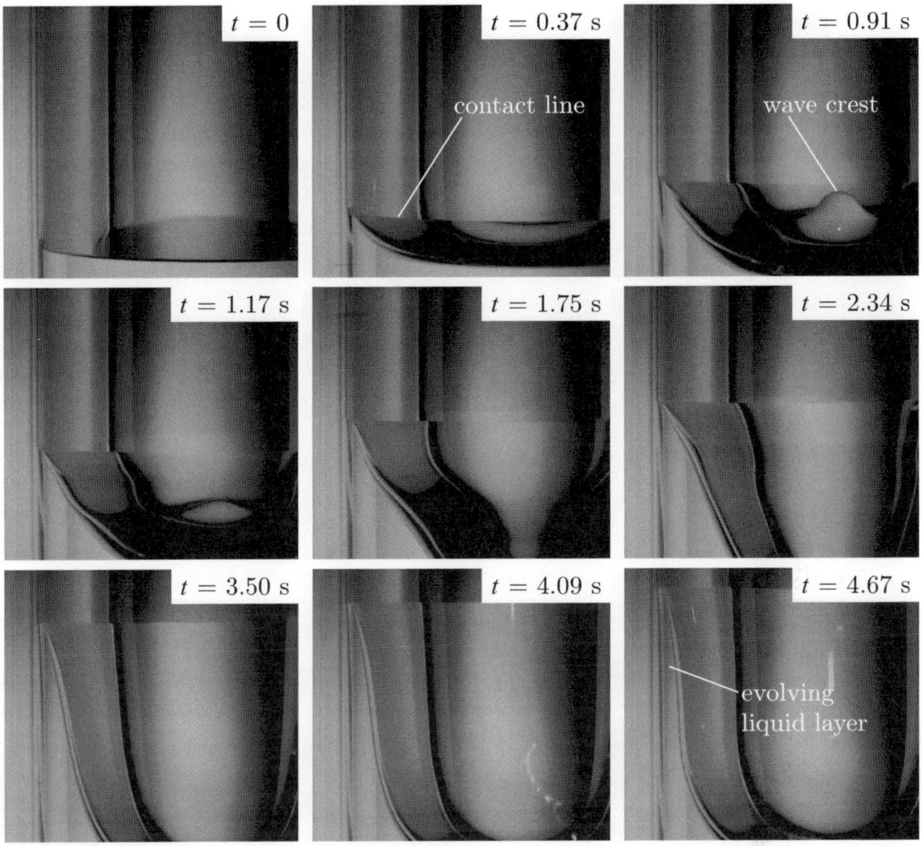

Fig. 7.5. Sequence of video frames taken with a high speed digital video system for $R = 70$ mm and SF 0.65 test liquid. Pictures were taken from the side to enable the evaluation of the contact line position by digital image processing

the entire liquid surface ($t = 1.88$ s). The rising contact line stagnates for a short moment at the same time when the center point of the liquid surface reaches its minimum position.

Figure 7.5 shows still frames of a complete experiment sequence taken from the side for the quantitative evaluation of the contact line position which is marked for $t = 0.37$ s. In this view the initial annular surface waves are not visible. The wave crest becomes visible at $t = 0.91$ s although observed through the curved liquid surface. The frames for $t = 3.5$ s, 4.09 s and 4.67 s show that the curvature of the surface changes its sign in the region near the contact line. At $t = 4.67$ s the development of a liquid layer becomes visible. Due to the experiment duration of 4.74 s the development is not completed here.

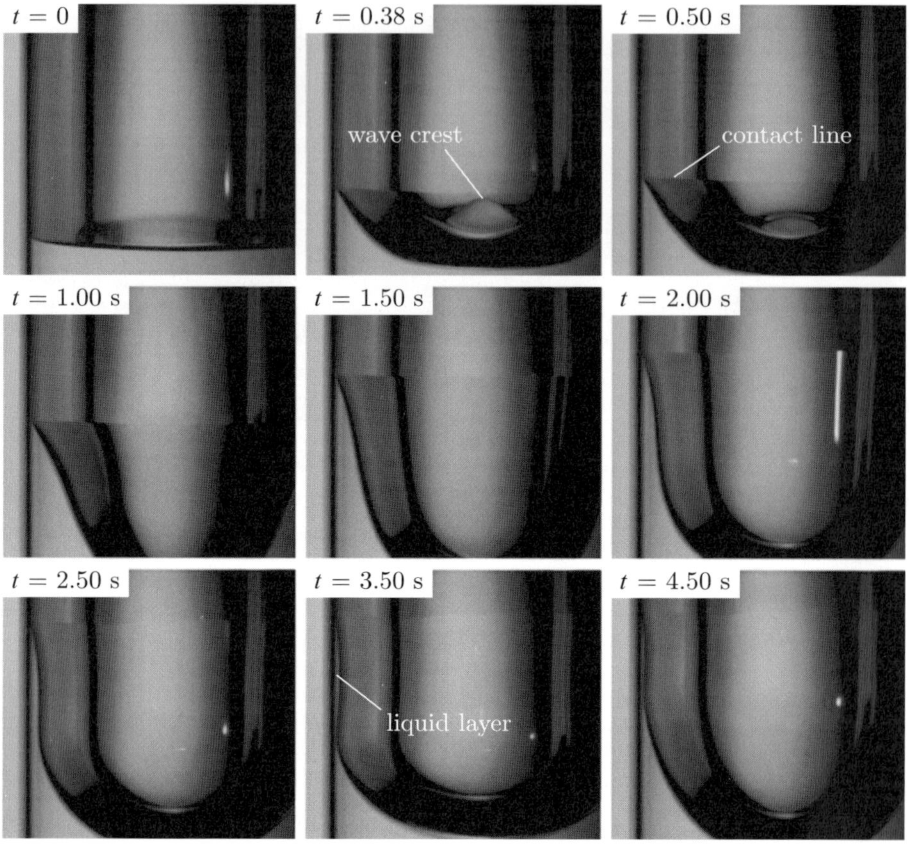

Fig. 7.6. Sequence of video frames taken with a high speed digital video system for $R = 40$ mm and SF 0.65 test liquid. The development of the liquid layer is visible for $t = 2.0$ s to $t = 4.4$ s

Figure 7.6 shows still frames for $R = 40$ mm and SF 0.65. Due to the smaller vessel radius (compared to Fig. 7.5) the formation of the liquid layer already starts at $t = 2.0$ s. The following pictures show the further development of the layer. At $t = 3.5$ s the liquid surface almost touches the wall at a second position approximately 30 mm below the contact line enclosing the liquid in between. At $t = 4.5$ s the liquid layer has become smaller since liquid leaves downwards back into the bulk.

7.2 Data Evaluation

The following section describes the evaluation of the experimental data and the corresponding uncertainties of the measuring techniques. After the definition of the experimental time, the digital image processing of the experiment videos for evaluation of the free liquid interface over time was carried out. The knowledge of the contour histories gives the foundation for the characterization of the free surface reorientation. Two specific points on the free liquid interface are used for the characterization of the surface reorientation, the center point $z_c = h\,(r = 0, t)$ and the three phase contact point $z_w = h\,(r = R, t)$.

7.2.1 Definition of the Experimental Time

Since there is no time signal displayed in the video images, the experimental time has to be defined after the experiment was carried out. The most accurate way is to observe the video images when the free liquid interface shows the first response after the drop capsule has been released. Having found the first movement of the meniscus in frame j, then the time $t = 0$ s is assigned to frame $j-1$ as it is shown in Fig. 7.3 for test no. 36, where the first movement of the meniscus can be observed in frame $j = 44$. The accuracy of this method for experiments with $f_r = 250$ Hz is about ± 1 frame (corresponding to $\Delta t = \pm 0.004$ s) and for experiments with $f_r = 500$ Hz about ± 2 frames ($\Delta t = \pm 0.004$ s).

7.2.2 Evaluation of the Liquid Interface

The recorded frames of the digital video system were stored as 8-bit intensity images in Windows Bitmap (bmp) format in a 512×480 or a 512×240 pixel array depending on the recording frequency. The maximum optical resolution was only obtainable with a recording frequency $f_r = 250$ Hz, while for the highest time resolution the images are limited to the smaller 512×240 pixel array. The evaluation of the liquid interface $h\,(r, t)$ is structured in the detection of the liquid interface, the calibration of the image pixels and the transformation from pixel coordinates to real coordinates.

Detection of the Liquid Interface

A detection of the liquid interface was carried out for each frame by applying an image processing routine programmed in Matlab (Version 5.3.0.10183 (R11)). For an 8-bit intensity (black and white) image a white region corresponds to a bit value of 255 (which corresponds in Matlab to 256 with the origin of 1), while a black region gives a bit value of 0 (Matlab: 1 with origin of 1). The strategy of the edge detection routine is to search for the maximum (from a dark region to a bright region) vertical gradient of each column per image, starting at the cylinder axis where the interface is definite and easy to find. For edge coordinates found in adjacent columns, a limited derivation from the latter edge coordinate is accepted to ensure that a coherent interface contour is found. The derivation limit varies between columns in the cylinder axis and columns at the cylinder wall. In this way the interface contour can be detected in less than a second per image.

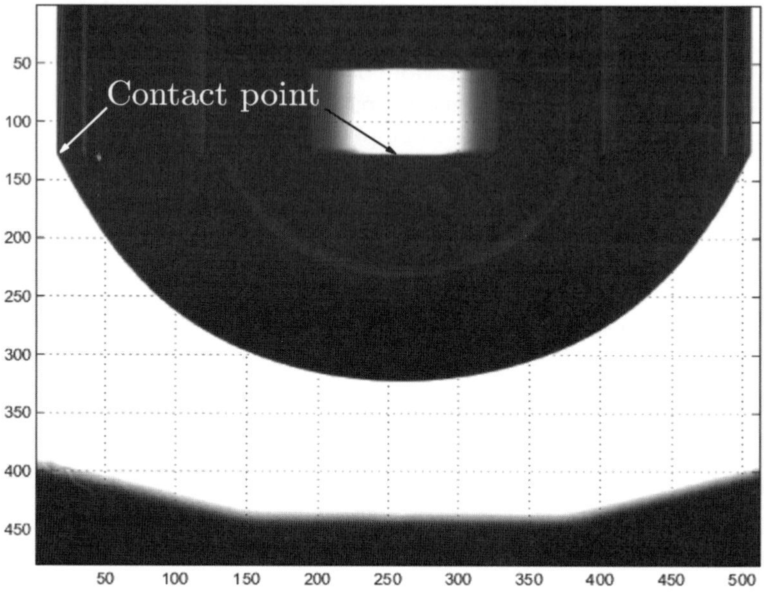

Fig. 7.7. Video image of test no. 46 (Detra, $R = 15$ mm, $j = 100$)

Detection of the Contact Point

The contact point is detected either near the cylinder wall, when total reflection for higher contact angles leads to a sharp edge (as is can be observed in Fig. 7.7), or near the cylinder axis using a stripe of reflection above the liquid interface if the contact angle is too small and total reflection is not present. Those reflections near the cylinder axis can also be observed in Fig. 7.7. However if the contact point exceeds the final equilibrium position z_{we} during the initial capillary rise, a very thin layer of liquid is left at the cylinder wall, which is very difficult to observe and to detect. For those experiments it is only possible to detect the contact point until the first maximum deflection z_{wp} (compare Fig. 7.13) occurs.

Calibration

To calibrate the two camera views, a transformation from pixel coordinates into real coordinates is necessary. Moreover the cylinder wall and the initial free surface have to be defined in the video images, which is described in the following sections.

For the calibration of the horizontal and vertical pixel size, a high precision glass measure (pitch 1 mm and line thickness of 30 μmm) was positioned inside the test cylinder and aligned in the cylinder axis of the vessel during the preparation of the experiments. It was then dipped into the test liquid and recorded with the recording system. For this procedure test liquid from the same batch was taken as it was used for the experiments. After the recording of the calibration images all adjustments of the camera system were kept diligently in order to ensure a proper pixel calibration. In order to prevent impurity of the test fluid, the procedure was carried out prior to the final cleaning procedure and preparation of the drop of the capsule.

Contour History Plots

After detection of the liquid interface and transformation to real coordinates, contour history plots $z^*(r^*, t)$ for the total view and the detail with high resolution are obtained. Exemplarily the contour histories for test no. 32 to the first minimum of oscillation are shown in Fig. 7.8 for the total view and in Fig. 7.9 for the detail view. With the contour history plots any point on the liquid interface can be plotted over time, which is done for characteristic points in Sect. 7.3. The uncertainties of the contour histories are described in the following section.

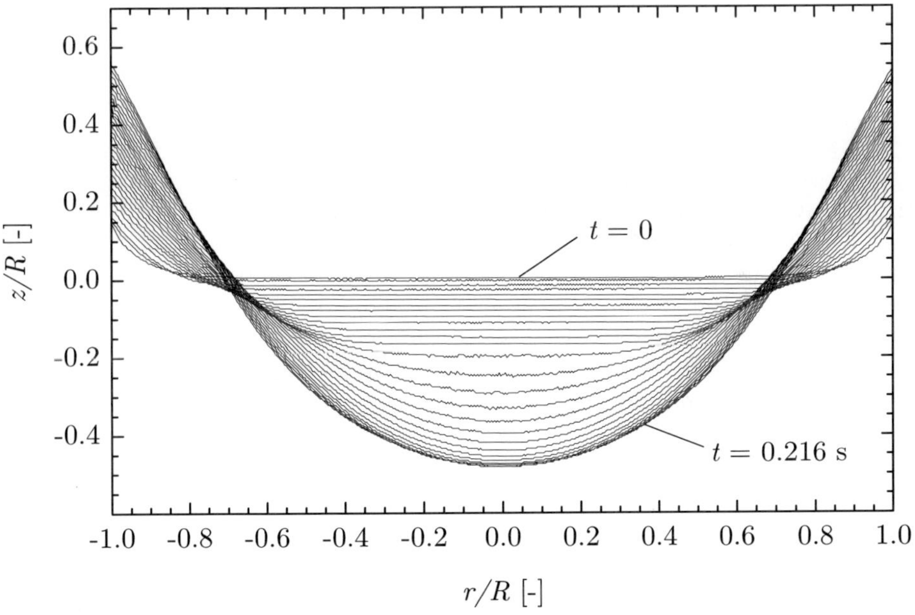

Fig. 7.8. Dimensionless contour history plots in the total view for test no. 32. The contour histories are plotted only to the first maximum z^*_{cp} of oscillation. Please note that the center part of the surface profiles does not represent the real surface

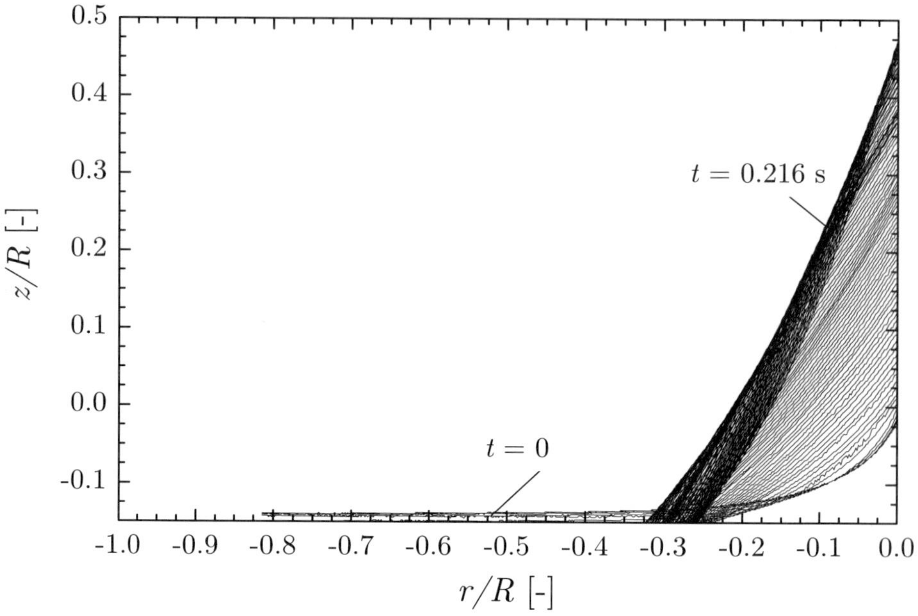

Fig. 7.9. Dimensionless contour history plots in the detail view for test no. 32. The contour histories are plotted only up to the first maximum z_{wp} of the oscillation

7.2.3 Error Analysis

Interface Contour

As described above, immediately after transition to reduced gravity, the meniscus starts to rise along the cylinder wall and induces a disturbance traveling towards the cylinder axis. This can be seen clearly with a camera from above (see Fig. 7.4), but not from the side. Due to total reflection at the first minimum (counting from the wall) of the free surface, the remainder of the profile towards the center cannot be detected. But a top view allows a qualitative observation of the free surface only. The disadvantage of the side view which disturbs in the initial time period only, has been accepted to enable a precise evaluation of the whole surface.

The accuracy of the detection of the liquid interface near the cylinder axis is on the order of ± 1 pixel for all experiments. However for experiments with small static and dynamic contact angles higher inaccuracies result at the vicinity of the wall from the detection of the liquid interface column by column. Thus near the cylinder wall the accuracy for those experiments (test no. 32, 36 and 46) is on the order of ± 8 pixels. On the other hand for experiments with higher static contact angles, which were carried out with silicone fluids, disturbances due to light refraction lead to uncertainties on the order of ± 4 pixels in the vicinity of the cylinder wall.

Contact Point

As described above the contact point is detected near the cylinder axis or near the cylinder wall using a reflection in the background if total reflection is absent. The uncertainties which appear during the detection of the contact point can be distinguished between the initial capillary rise to the first maximum and the final position of the contact point.

For the initial capillary rise a higher dynamic contact angle, caused by a high velocity of the contact point, leads to a contact point with high contrast due to total reflection, which can be detected with an accuracy of ± 1 pixel independent of the static contact angle of the test liquid. However for some experiments the capillary rise of the contact point along the cylinder wall is unsymmetrical, which makes it necessary to consider an error of ± 4 pixels for the initial capillary rise.

The error for the final position must be distinguished between experiments with high static contact angles and low static contact angles. For experiments with higher static contact angles ($\gamma_s > 10°$) total reflection is present at the contact line during the entire reorientation process and thus a detection is easy. Therefore an error of ± 4 pixels can also be considered for the final position. For experiments with low static contact angles ($\gamma_s < 10°$), the contact point is difficult to observe after the first maximum is reached and the velocity of the contact point is low. Due to the low contact angle total reflection is absent. This can be observed in particular for test no. 32, when the initial capillary rise at the cylinder wall exceeds the equilibrium position z_{we} (compare Fig. 7.13). For these experiments an error of ± 8 pixels must be considered for the final position of the contact point.

Calibration and Transformation

The calibration procedure of the video images causes some errors which can be summarized by:

1. Error due to inaccurate definition of the calibration points, which can be assumed to be on the order of ± 1 pixel.
2. Mechanical oscillations of the video cameras during the initial capillary rise, caused by the release of the drop capsule, induce a shifting of the liquid interface within the video image. These errors are eliminated by arranging a steady measure inside the camera view and using the measure as a reference in the digital image processing routines.
3. A non-central arrangement of the calibration measure inside the cylinder, which causes an error for the calibration of the pixel resolution.

These errors can be treated to be minor and cause much smaller inaccuracies than the other errors reported above.

Contact Angle

Hysteresis of the contact angle must be considered for the definition of the static contact angle. Due to the filling procedure of the test liquid from the bottom of the vessel, the meniscus advances over a non-wetted wall. The velocity of the meniscus is about 0.03 mm/s, which corresponds to a capillary number Ca $\approx 1.3 \times 10^{-5}$ (see (4.11)) for SF 10. Thus the initial static contact angle γ_{s0} is determined by an advancing contact point. The final static contact angle γ_s depends on the final moving direction of the contact point. For all experiments it can be observed that the OHNESORGE number determines whether there is an overshoot of the contact point during the first capillary rise along the cylinder wall over the equilibrium position z_{we} or not (compare Sect. 9.1.1). For experiments with small static contact angles an overshoot over z_{we} causes a receding contact point, while for the case with no initial overshoot an advancing contact point (a) to z_{we} can be observed (Fig. 9.1). On the other hand experiments with high static contact angles show, also depending on the OHNESORGE number, an overshoot of the contact point over z_{we} with further oscillations until the contact point is fixed. Depending on the final moving direction on the finally pre-wetted cylinder wall an advancing (a) or receding (r) contact point was observed.

Final Static Contact Angle

Assuming a spherical shape of the final liquid interface, the static contact angle γ_s was determined geometrically from the video images with the length α (compare Fig. 5.2 in (5.10)). The uncertainties for the evaluation of γ_s are mainly determined by inaccuracies in the digital image processing detecting the contact point and the center point.

For experiments with silicone fluid using the surface modifier FC-732 ($\gamma_s > 10°$), the accuracy of the digital image processing routine detecting the contact point is on the order of ± 4 pixels, while the contour of the liquid interface in the cylinder axis can be detected with an accuracy of ± 1 pixel. Considering the corresponding uncertainties Δz_w and Δz_c by varying α with $\pm (\Delta z_w + \Delta z_c) / R$ leads to an uncertainty of the static contact angle of $\Delta \gamma_s = \pm 2°$ using (5.10).

The uncertainties for experiments with detra and butylbenzene ($\gamma_s < 10°$) are higher for the detection of the contact point, since due to a lack of total reflection for low contact angles the contact point is difficult to detect. Thus the accuracy for the detection of the contact point is estimated to be ± 8 pixels and for the center point ± 1 pixel. Consideration of these pixel uncertainties leads to an error of maximal $\pm 2.5°$ for the final static contact angle.

Initial Static Contact Angle

The determination of the initial (advancing) contact angle γ_{s0} is more difficult since no analytical relation for γ_{s0} can be described. Therefore the initial free

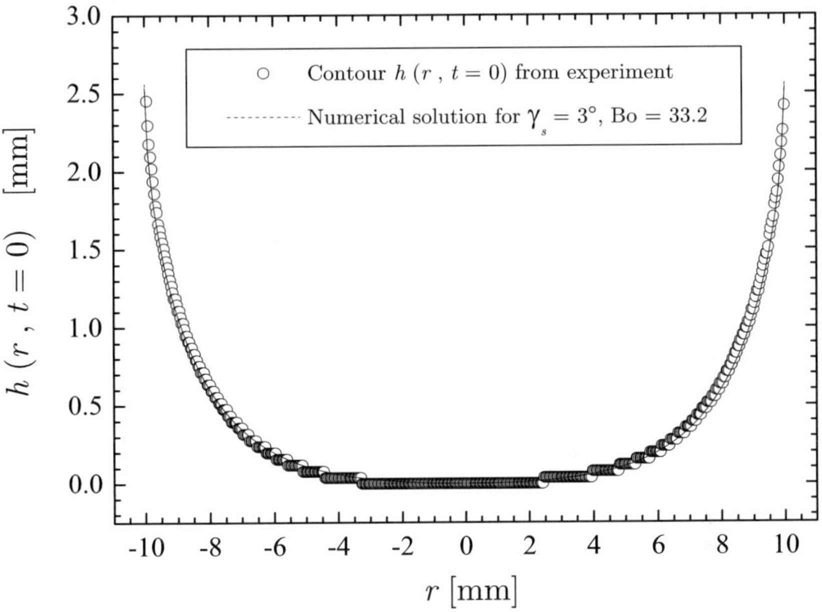

Fig. 7.10. Comparison of the initial contour of the liquid interface $h(r, t = 0)$ from experiment with the numerical solution (5.4) for test no. 36

surface evaluated from the experiments was compared with the numerically solution of (5.4) introduced in Sect. 5.1 by varying γ_{s0} and thus the boundary condition (5.5). The error for the initial static contact angle was estimated from the comparison of the free surface contours and is on the order of $\Delta\gamma_{s0} = \pm 3°$ (compare Fig. 7.10 and 7.11). The initial static contact angles of the test cases are given in Table 7.3.

7.3 Reorientation Characteristics

To describe the behavior of the surface oscillation, characteristic points on the free liquid interface may be observed. The first point is the surface center point $z_c(t) = h(r = 0, t)$, which lays on the free surface in the cylinder axis. The second point is the three phase contact point $z_w(t) = h(r = R, t)$ at the cylinder wall. The dynamic behavior of these characteristic points is analyzed in detail in the following section. The coordinate system for the center point (z_c, r) was introduced in Fig. 5.3. The coordinate system for the contact point (z_w, r) is located at position (z_{w0}, R) as shown in Fig. 5.1. The vertical difference between z_c and z_w is $z_{w0} \sim L_c\sqrt{2(1 - \sin\gamma_{s0})}$ and must be considered when comparing z_c and z_w.

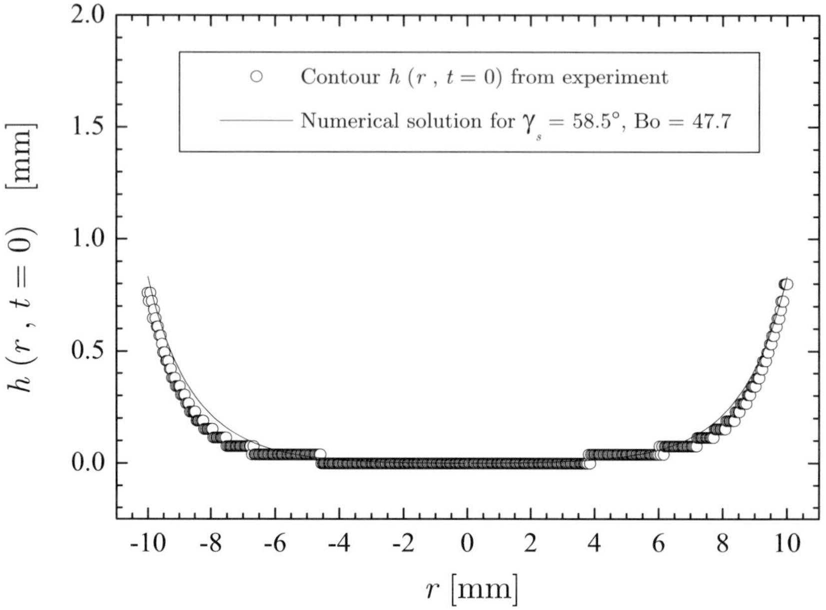

Fig. 7.11. Comparison of the initial contour of the liquid interface $h(r, t = 0)$ from experiment with the numerical solution (5.4) for test no. 37

7.3.1 Center Point

For all experiments the center point reveals an oscillatory motion to the final equilibrium position z_{ce} as it is schematically shown in Fig. 7.12 for test no. 32. The damping and the frequency of the center point oscillation depends both on the static contact angle and the OHNESORGE number. In the very initial first period the center point cannot be observed due to total reflection as described in Sect. 7.2.3. However, at $z_c = z_{cd}(t_{cd})$ a characteristic kink can be observed in the history. Following the history, at time t_{cr} the equilibrium position is reached the first time and at $z_c = z_{cp}(t_{cp})$ the history shows the first peak. The reorientation process is finished, when the center point reaches its final position at time t_R. The definition and evaluation of the characteristic points are explained in more detail in the following.

Disturbance Arrival

After transition to reduced gravity the meniscus at the cylinder wall immediately starts to rise along the wall. This induces a disturbance traveling towards the cylinder axis. When the disturbance affects the whole surface, the real center point becomes visible. This causes a weak kink at t_{cd} in the history of the center point (see Fig. 7.12).

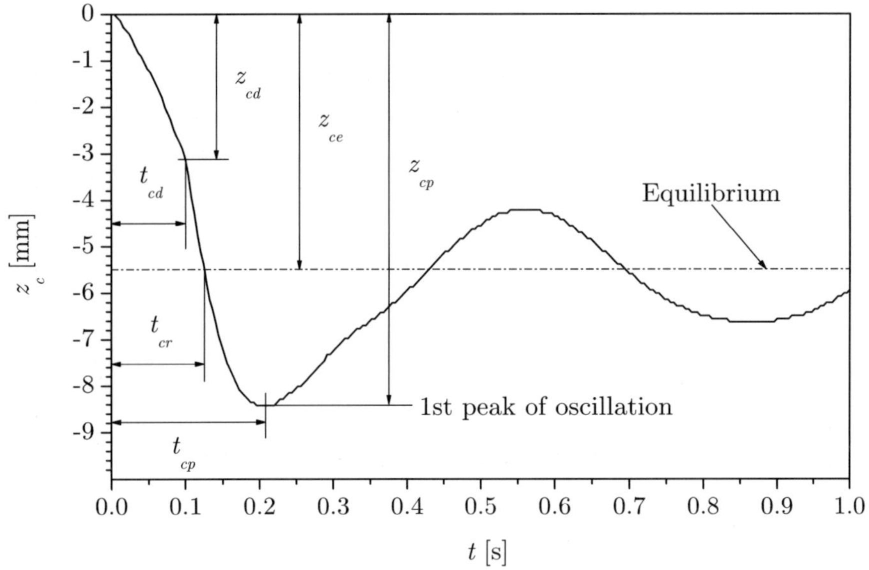

Fig. 7.12. Definition of reorientation characteristics in the history of the center point $z_c = h\,(r = 0, t)$. The line from (0,0) to (t_{cd}, z_{cd}) is the results of the image processing but it does not reflect the real behavior of the center point due to total reflection

First Transit through Equilibrium

The second characteristic point of time t_{cr} (time of center rise) of the free surface reorientation is defined, when the center point passes the equilibrium position of the interface given by Bo and γ_s ($\neq \gamma_{s0}$) the first time, if Bo is defined with the residual acceleration k_z ($\neq k_{zi}$) during the drop. Obviously t_{cr} depends on the cylinder radius since the final interface shape can be considered to be of spherical shape with the radius $R_K = R/\cos\gamma_s$ for negligible residual accelerations k_z during the drop. As described in Sect. 4.5, Siegert et al. [3] presented an empirical correlation for the time t_{cr} given in (4.24), which is based on drop tower experiments with experiments with a static contact angle $\gamma_s = 0$.

First Peak of Oscillation

Another characteristic point is defined, when the center point achieves its first maximum of oscillation, as shown in Fig. 7.12. The deflection of the center point is defined with z_{cp}, while the time is denoted with t_{cp} (index for center peak). Further peaks of the center point are denoted with z_{cp2}, z_{cp3}, for the second, third peak and so on. In all experiments the OHNESORGE number was chosen to be small enough to lead to an overshoot of the center point history.

Arrival of Equilibrium

For the definition of the reorientation time t_R Weislogel & Ross [4] introduced a criterion that the reorientation time (or settling time) is reached when a movement of the center point is no longer observable. This criterion is rather ambiguous since it depends on the optical resolution of the camera systems and thus on the experimental setup. Furthermore, an additional error on the order of $\Delta t_R \approx \pi/\omega_d$ must be considered.

Another method to describe a reorientation time is to take the point when the envelope curves of the damped center point oscillation enter a lower limit, which is often defined to be 2% of the final value [5]. However this method requires the functional correlation for the envelope curves to be known.

7.3.2 Contact Point

Another characteristic point on the free surface is the contact point $z_w = h(r = R, t)$. In analogy to the definition of characteristics for the reorientation of the center point, the following flow characteristics are defined, where the aim of the study is to describe the influence of the MORTON number for the initial time period and the OHNESORGE number as well as the static contact angle on the general behavior of the contact point.

General Behavior

The general behavior of the contact point for small and higher static contact angles is different. For experiments with higher static contact angles ($\gamma_s > 10°$) the contact point shows - in the majority of cases - an overshoot over the final equilibrium position with an early pinning of the contact point. If the OHNESORGE number is increased to rather high values a reduction of the overshoot occurs with immediate pinning of the contact line. For experiments with small static contact angle ($\gamma_s < 10°$) an overshoot of the equilibrium position occurs only for quite small OHNESORGE numbers. A thin liquid layer is then left at the cylinder wall. For higher OHNESORGE numbers without an initial overshoot over the equilibrium position the contact point moves unidirectionally towards the equilibrium position.

Contact Angle Dependence

For experiments with small static contact angles ($\gamma_s < 10°$) the behavior of the contact point strongly depends on the OHNESORGE number. For low OHNESORGE numbers the first amplitude of the contact point exceeds the equilibrium position z_{we} to the maximum value z_{wp} and a thin layer of liquid remains at the cylinder wall, which is difficult to detect due to the lack of total reflection (compare Fig. 7.13). After the maximum z_{wm} is reached, the contact point may be characterized by a free contact line condition.

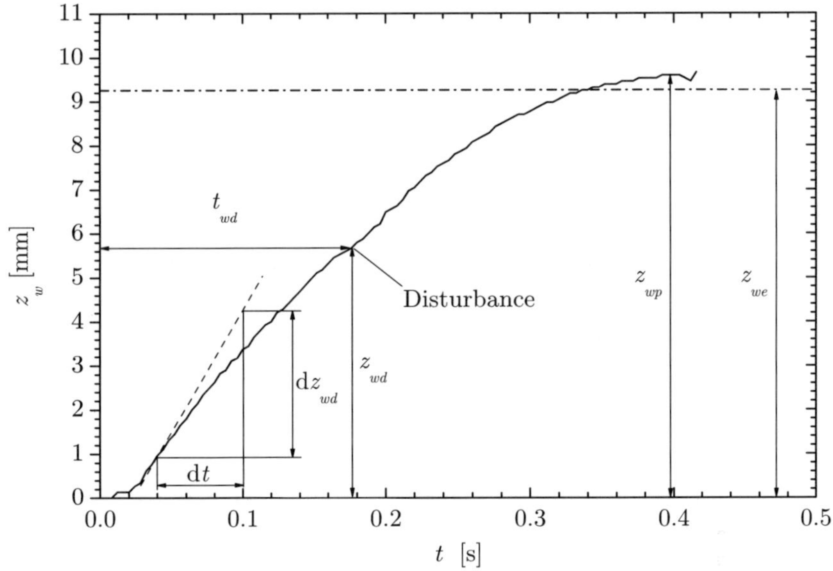

Fig. 7.13. Behavior of the contact point for low static contact angles ($\gamma_s < 10°$) with low OHNESORGE number. The contact point shows an overshoot over the equilibrium position z_{we}

For higher OHNESORGE numbers the first rise of the contact point ends below the equilibrium position z_{we}. A resting follows at that position and a new rise starts towards the equilibrium position z_{we} as shown in Fig. 9.1. In this case the fixed contact line condition may be assumed for the contact point. The behavior of the contact point for higher static contact angles ($\gamma_s > 10°$) was found to be different compared to experiments with low static contact angle. Depending on the OHNESORGE number an overshoot of the contact point to $z_{wp} > z_{we}$ was observed for most cases, while higher OHNESORGE numbers lead to pinning the first time z_{we} is reached (i.e. no overshoot). When the contact point recedes after the first maximum z_{wp}, no obvious remaining layer of liquid above z_{we} can be observed.

First Transit through Equilibrium

The time t_{wr} denotes the first arrival of the contact point of the final equilibrium position z_{we}. For experiments with low static contact angles the final position z_{we} was only overshot in test no. 32. Experiments with higher static contact angle show always an overshoot over z_{we}, which reduces with increasing OHNESORGE number.

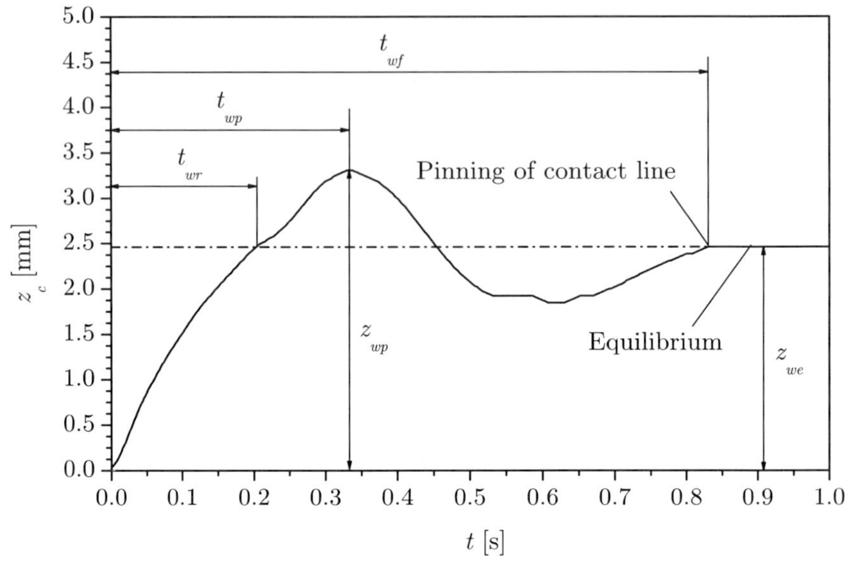

Fig. 7.14. Definition of characteristic times in the history of the contact point for higher static contact angles

First Maximum

The arrival of the contact point at the first maximum is distinguished for experiments with small static contact angle ($\gamma_s < 10°$) and high static contact angles ($\gamma_s > 10°$). As described above, for the latter experiments an overshoot of the contact point over the equilibrium position could be observed. Hence, the arrival time at the maximum position is denoted with t_{wp} (index p for peak), as marked in Fig. 7.14, while the deflection of the contact point at the maximum is denoted with z_{wp}. For experiments with small static contact angle the first detectable stagnation or resting point of the curve was used for the definition of t_{wp} and z_{wp}, as indicated in Fig. 7.13. An initial overshoot of the contact point over the equilibrium position is denoted for small static contact angles with z_{wp}.

Pinning of the Contact Point

The time t_{wf} denotes the time until the contact point is fixed (compare Fig. 7.14). This time could only be evaluated for experiments with a higher static contact angle. The uncertainties for t_{wf} are determined by an inaccurate definition of the experiment start time (compare Sect. 7.2.1 for Δt) and by an additional error for the definition of the point of time t_{wf} itself, which can be assumed to be on the order of ± 2 frames. This leads to an additional error of 0.004 s – 0.008 s, depending on the recording frequency f_r.

Rise Velocity

The rise velocity of the contact point was determined by two different methods. The first method is to define a mean initial rise velocity with the quotient of the coordinate z_{wp} and the time t_{wp} of the contact point at the first peak

$$\bar{u}_{wr} = \frac{z_{wp}}{t_{wp}} \; . \tag{7.1}$$

For experiments with small static contact angles ($\gamma_s < 10°$) the first peak corresponds to the first observable resting point as indicated in Fig. 7.13, while for experiments with higher static contact angle only the first peak was used for determination of $\bar{u}_{wr}$. The uncertainties for $\bar{u}_{wr}$ are governed by the accuracy of the experiment time determination (error $\Delta t = \pm 0.004$ s) and the detection of the contact point (error is about ± 4 pixels for the initial rise).

Secondly, a maximum rise velocity u_{wmax} was evaluated by differentiating the coordinate of the contact point $z_w\,(t)$

$$u_{wmax} = \max \frac{\mathrm{d}z_w}{\mathrm{d}t} \; , \tag{7.2}$$

where the contact point was smoothed before by adjacent averaging (11 data points for experiments with $f_r = 500$ Hz and 5 data points for experiments with $f_r = 250$ Hz).

7.3.3 Frequency and Damping

For the evaluation of the damping behavior of the oscillation of the center point the theory of a linear damped mechanical system is introduced and applied in the following section. Although the initial period of the free surface center point oscillation shows a nonlinear trend, the oscillation has a linear character after the contact point is fixed. Therefore the damping behavior of the surface oscillation may be characterized applying the theory of linear oscillation, which is introduced in the following section.

Theory of Linear Oscillation

Based on the linear, homogenous differential equation for a linear oscillator with a damping proportional to the velocity (analog to Newtonian damping) [6]

$$\ddot{z} + \frac{d}{m}\dot{z} + \frac{c}{m}z = 0 \; , \tag{7.3}$$

the introduction of the exponential decay coefficient $\delta = d/2m$ and the undamped natural frequency $\omega_0 = \sqrt{c/m}$ one obtains the following differential equation for the description of a damped linear oscillation[1]

[1] In this section the coordinate z describes the location of the center of mass of the oscillator, $\dot{z}$ the velocity and $\ddot{z}$ the acceleration.

$$\ddot{z} + 2\delta\dot{z} + \omega_0^2 z = 0 \; . \tag{7.4}$$

Inserting $z(t) = e^{\lambda t}$ in (7.4) leads to the characteristic equation $\lambda^2 + 2\delta\lambda + \omega_0^2 = 0$ with the solutions $\lambda_{1,2} = -\delta \pm \sqrt{\delta^2 - \omega_0^2}$, which can be transformed to $\lambda_{1,2} = -\delta \pm \omega_0\sqrt{1 - D^2}$, if the dimensionless damping ratio

$$D = \frac{\delta}{\omega_0} \tag{7.5}$$

is introduced [5]. For subcritical damping $D < 1$ the solutions for λ_1 and λ_2 are complex and can be simplified with the introduction of the natural frequency of the damped oscillation

$$\omega_d = \omega_0\sqrt{1 - D^2} \tag{7.6}$$

to $\lambda_1 = -\delta + \mathrm{i}\omega_d$ and $\lambda_2 = -\delta - \mathrm{i}\omega_d$ where the solution of (7.4) can be expressed by a linear combination of the two solutions

$$z\left(t\right) = Ae^{\lambda_1 t} + Be^{\lambda_2 t} = Ae^{(-\delta+\mathrm{i}\omega_d)t} + Be^{(-\delta-\mathrm{i}\omega_d)t} \; . \tag{7.7}$$

This equation can be transformed to

$$z\left(t\right) = e^{-\delta t}\left(C_1 \sin\omega_d t + C_2 \cos\omega_d t\right) \; , \tag{7.8}$$

with $C_1 = \mathrm{i}\left(A - B\right)$ and $C_2 = A + B$. By introducing the zero phase angle

$$\varphi_d = \arctan\frac{C_1}{C_2} \; , \tag{7.9}$$

and the constant

$$C = \sqrt{C_1^2 + C_2^2} \; , \tag{7.10}$$

Equation (7.8) can be written as

$$z\left(t\right) = Ce^{-\delta t}\cos\left(\omega_d t - \varphi_d\right) \; . \tag{7.11}$$

Substituting δ with the damping ratio D and the time t with the dimensionless time $\tau = \omega_d t$ leads to the dimensionless equation for the oscillation:

$$z\left(\tau\right)^* = Ce^{-\frac{D}{\sqrt{1-D^2}}\tau}\cos\left(\tau - \varphi_d\right) \; . \tag{7.12}$$

The initial conditions $z^*\left(t = 0\right) = z_{c0}/z_{ce} + 1 = z_{c0}^*$ (z_{c0}^* = initial dimensionless deflection) and $\dot{z}^*\left(t = 0\right) = \dot{z}_{c0}^*$ ($\dot{z}_{c0}^*$ = initial dimensionless velocity) lead to the definition of the constant

$$C = \sqrt{\left(z_{c0}^*\right)^2 + \left(\dot{z}_{c0}^* + \frac{D}{\sqrt{1 - D^2}} z_{c0}^*\right)^2} \; , \tag{7.13}$$

and the zero phase angle

$$\varphi_d = \arctan\left(\frac{\dot{z}^*_{c0}}{z^*_{c0}} + \frac{D}{\sqrt{1-D^2}}\right) . \tag{7.14}$$

Thus for the special case of $z^*_{c0} = 1$ and $\dot{z}^*_{c0} = 0$ Equation (7.12) reads with $C = 1/\sqrt{1-D^2}$ and $\varphi_d = \arctan\left(D/\sqrt{1-D^2}\right)$

$$z^*(\tau) = \frac{1}{\sqrt{1-D^2}} e^{-D\tau/\sqrt{1-D^2}} \cos\left[\tau - \arctan\left(\frac{D}{\sqrt{1-D^2}}\right)\right] . \tag{7.15}$$

For this case the upper envelope curve reads

$$z^*_u(\tau) = \frac{1}{\sqrt{1-D^2}} e^{-D\tau/\sqrt{1-D^2}} \tag{7.16}$$

and the lower envelope curve is

$$z^*_l(\tau) = -\frac{1}{\sqrt{1-D^2}} e^{-D\tau/\sqrt{1-D^2}} , \tag{7.17}$$

as plotted in Fig. 7.15.

Logarithmic Decrement

The dimensionless period of the damped oscillation is $\tau_d = \omega_d T_d = 2\pi$ and thus the ratio of two consecutive peak deflections gives

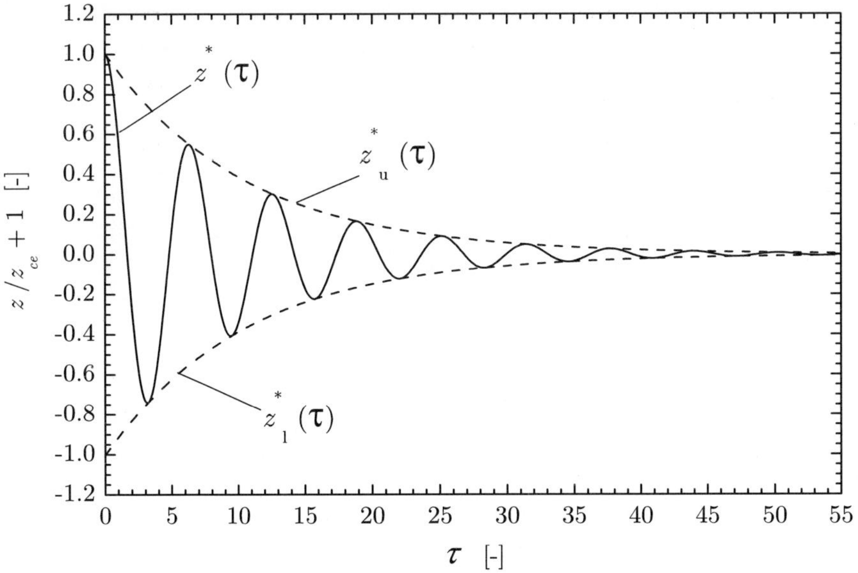

Fig. 7.15. Linear damped oscillation and its envelope curves in dimensionless notation

$$\frac{z^*_{cp1}(\tau)}{z^*_{cp2}(\tau+\tau_d)} = e^{-D\tau_d/\sqrt{1-D^2}} \;. \tag{7.18}$$

The logarithm of this ratio is defined as the logarithmic decrement

$$\Lambda = \ln \frac{z^*_{cp1}(\tau)}{z^*_{cp2}(\tau+\tau_d)} = \frac{D}{\sqrt{1-D^2}}\tau_d \;. \tag{7.19}$$

If Λ is known, the damping ratio D may then be determined by rearranging (7.19)

$$D = \frac{\Lambda}{\sqrt{4\pi^2-\Lambda^2}} \;. \tag{7.20}$$

Evaluation of the Damping Ratio

For the evaluation of the damping behavior of the system, the center point z_c is scaled with z_{ce} and transformed to 1. Thus the dimensionless coordinate z^*_c reads

$$z^*_c(\tau) = \frac{z_c}{z_{ce}} + 1 \;. \tag{7.21}$$

Figure 7.16 shows this transformation for test no. 32. The half cycles i of the oscillation are numbered serially to the last detectable cycle $i = 11$. Maximum peaks are denoted with even values for i while minimum peaks are denoted with uneven values. For a number of n consecutive maximum or minimum peaks the logarithmic decrement Λ may then be determined as described above if a linear damped oscillation is assumed

$$\Lambda = \frac{1}{n} \ln \frac{z^*_{cp1}}{z^*_{cpn+1}} \;. \tag{7.22}$$

The negative minima peaks z^*_{cpi} $(i = 1, 3, 5, ...)$ are transformed to positive values by multiplication with -1. The meaning of (7.22) is that the slope of the logarithmic maximum or minimum values z^*_{cpi} corresponds to the logarithmic decrement Λ. Thus a linear regression through the maximum and the minimum values leads to the corresponding logarithmic decrements Λ_{max} and Λ_{min}. For the linear part of the oscillation a mean value for the logarithmic decrement is then defined by

$$\Lambda = \frac{(\Lambda_{max} + \Lambda_{min})}{2} \;. \tag{7.23}$$

For all experiments with higher static contact angles ($\gamma_s > 10°$) the mean logarithmic decrement Λ was evaluated for all peaks after the contact point is pinned, which occurs at the time t_{wf}. Whereas for experiments with low static contact angle ($\gamma_s < 10°$), where no early pinning of the contact point occurs, only peak $i = 1$ was skipped (test no. 32, 36 and 46), since the frequency is almost constant over time (compare Sect. 9.3.1). With the mean value for the logarithmic decrement Λ, the damping ratio D can be determined with (7.20).

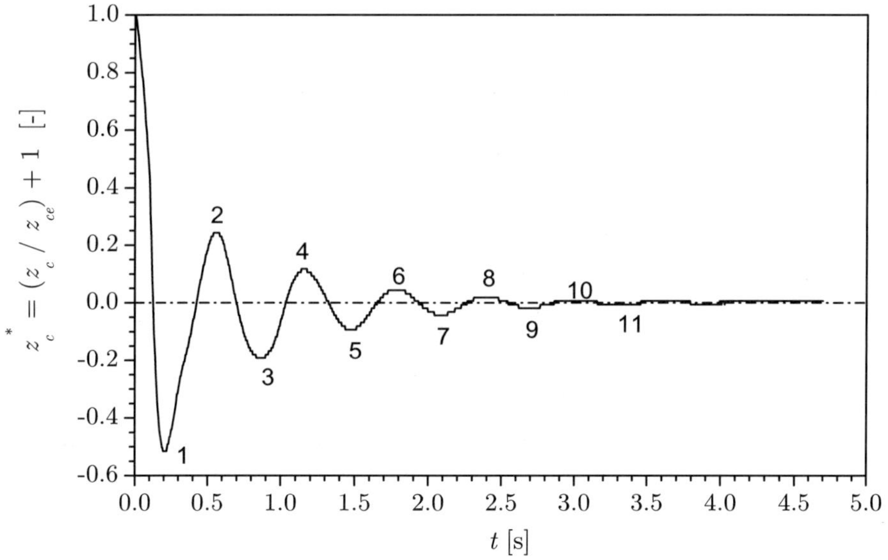

Fig. 7.16. Definition of peaks for determination of the logarithmic decrement

Frequency

For the evaluation of the frequency of oscillation for each half cycle of the liquid interface a value for the $T_i/2$ (where T_i denotes the period of oscillation) was determined from a peak i to the subsequent peak $i+1$ (which can be from a maximum to a minimum peak ($i = 1, 3, 5, ...$) as well as from a minimum to a maximum peak ($i = 2, 4, 6, ...$) compare Fig. 7.16 and Fig. 7.17). For each value of T_i the frequency is determined by $\omega_{di} = 2\pi/T_i$, and gives the value for the frequency of the damped oscillation (index d). It can be seen for all experiments with silicone fluid, that at the beginning of the experiments, when the contact line moves along the cylinder wall, a lower frequency occurs. This behavior can be observed in Fig. 7.17. After pinning of the contact line at t_{wf} an approximately constant frequency is established, which is determined by

$$\omega_d = \frac{1}{j - i_f} \sum_{i=i_f}^{j} \omega_{di} \ , \tag{7.24}$$

where i_f is the first peak after the contact line is fixed and j the total number of detectable peaks. The standard deviation

$$\Delta\omega_d = \left[\frac{1}{j - i_f - 1} \sum_{i=i_f}^{j} \left(\omega_{di} - \omega_d\right)^2 \right]^{1/2} \tag{7.25}$$

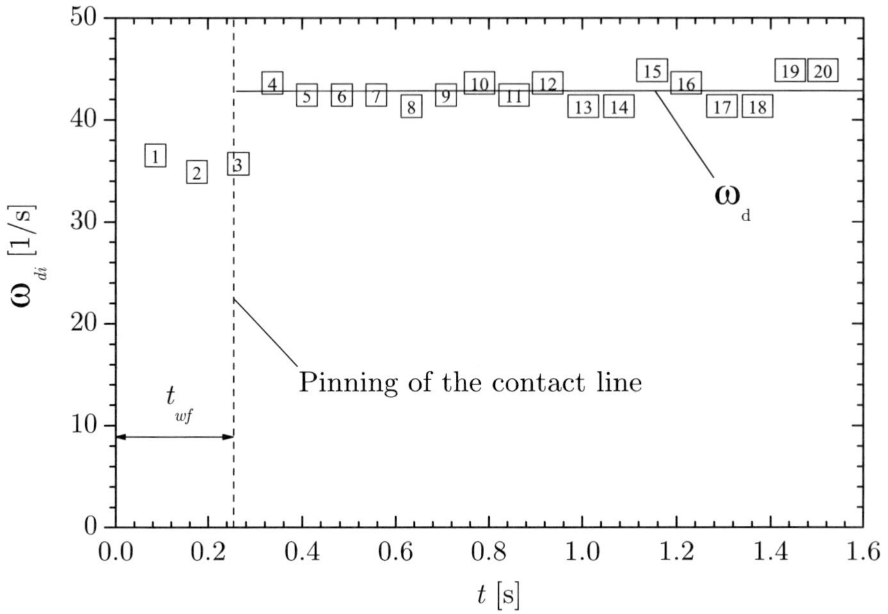

Fig. 7.17. Frequency ω_{di} over time with influence of contact line motion. After pinning of the contact point at $t_{wf} = 0.268$ s, a rise of the frequency to an averaged value of $\omega_d = 42.84\ \mathrm{s}^{-1}$ can be observed

describes the uncertainties of the frequency evaluation. Furthermore the evaluation of the damping ratio D allows the determination of the natural frequency of the undamped oscillation ω_0 with (7.6).

References

1. H. Brauer: *Grundlagen der Einphasen- und Mehrphasenströmungen*, (Verlag Sauerländer, Aarau 1971) p. 303
2. J. Gerstmann, M.E. Dreyer, H.J. Rath: AIP Conference 504, 847 (2000)
3. C.E. Siegert, D.A. Petrash, E.W. Otto: NASA TN D-2458
4. M.M. Weislogel, H.D. Ross: Microgravity Sci. Technol. **3**, 24 (1990)
5. K. Ogata: *System Dynamics*, 2 edn (Prentice Hall, Englewood Cliffs, New Jersey 1992)
6. K. Magnus: *Schwingungen*, (Teubner, Stuttgart 1976)

8

Numerical Calculations

The numerical simulations were performed with FIDAP version 8.61 and FLOW-3D version 7.7. FIDAP is based on the finite element method, FLOW-3D is based on the finite difference method and uses the volume of fluid method (VOF). Both codes solve the mass and momentum conservation equations as introduced in (6.1) and (6.2).

8.1 Numerical Model used in FIDAP

FIDAP was used with the deformable mesh option. The upper mesh boundary represents the free surface. Numerical simulations could be performed in the range of static contact angles from 10° up to 80°. Lower static contact angles yield strongly deformed meshes near the contact line and no convergence could be reached. Problems with moving contact lines require a slip condition at the contact point, otherwise the contact line is not able to move. FIDAP enables the possibility to incorporate a user defined slip model and a model for the dynamic contact angle. The computational region is shown in Fig. 8.1.

From previous numerical investigations [1, 2, 3] it is known that the elements must become progressively smaller the nearer they are located to the contact line. This condition is accomplished using a mesh refinement scheme. The smallest element is found directly at the point where the interface between liquid and gas meets the wall (Fig. 8.2). The mesh size of the computational domain is 60 nodes in the axial direction. The radial resolution varies between 30 and 120 radial nodes depending on the radius of the model. The basis element is a nine-node quadrilateral element.

The intention to describe the fluid motion mathematically encounters the difficulty to model the moving contact-line problem. In the numerical simulations the no-slip boundary condition must be removed at the contact line and slip must be allowed. Additionally a condition for the dynamic contact angle at the moving contact line must be defined. The node which represents the

Michael E. Dreyer: *Free Surface Flows under Compensated Gravity Conditions*
STMP **221**, 91–100 (2007)
DOI 10.1007/3-540-44628-1_8

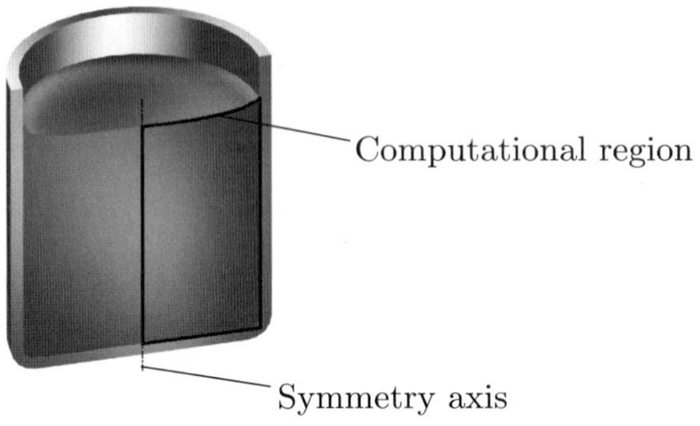

Fig. 8.1. Computational region of the free surface reorientation problem

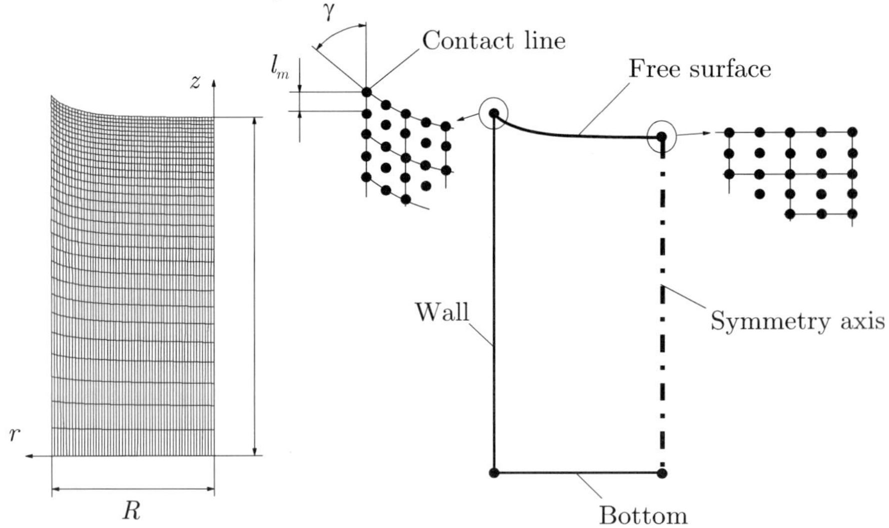

Fig. 8.2. Mesh for the numerical computations with FIDAP

contact line is strictly confined to motion along the z-direction and is allowed to move without restrictions.

Implementation of the Dynamic Contact Angle

No global dynamic contact angle model exists in the literature which is applicable to the settling phenomenon and includes all influencing factors. Unfortunately, the contact angle hysteresis must be taken into account. Contact angle hysteresis causes a pinning of the contact line as observed in the experiments. In the numerical calculation with FIDAP we used special models to simulate the behavior in the initial and global time domain.

- The first rise of the contact line is calculated with the static contact angle (Initial time domain).
- The behavior in the global time domain is simulated with a modified condition from Satterlee and Reynolds [4], where the dynamic contact angle depends on the z-coordinate of the contact line. The modified model from Satterlee and Reynolds enables the pinning or the free movement of the contact line.

The implementation of the contact angle model in FIDAP is made by linking a user defined subroutine, written in FORTRAN, to the solver of FIDAP. The structure of the subroutine is given by FIDAP.

No implementation is required for a constant contact angle. The contact angle is fixed to a constant value and does not vary during the simulation. This model is used to investigate the reorientation process in the initial time domain. The assumption of a constant contact angle yields a deviation from the real behavior. In reality the dynamic contact angle increases with increasing contact line velocity. Thus the curvature near the contact line and the pressure jump across the liquid interface decreases. The decrease of the pressure in the contact line region reduces the contact line velocity. Therefore in the numerical simulation the contact line velocities are higher than in the experiment. The boundary condition $\gamma_d = \gamma_s$ yields the upper limit values of the considered values in the initial time domain. This model was also used in the global time domain.

A modified model is based on the linear dynamic contact angle relation from Satterlee & Reynolds [4] (see also page 31) for static contact angles $\gamma_s = 90°$. The contact angle condition imposed at the contact line depends on the z_w coordinate of the free surface at the wall and is proportional to a free parameter Γ:

$$\left.\frac{\partial z}{\partial r}\right|_{r=R} = \Gamma\,(z_w - z_{we})\,, \tag{8.1}$$

where z_w represents the wall coordinate of the contact line. The origin of the coordinate system is defined at the initial static position of the contact point. In the case of static contact angles $\gamma_s = 90°$ z_{we} is equal to the initial wall coordinate, which is equal to zero. We modified the condition for static contact angles other than 90° to

$$\left.\frac{\partial z}{\partial r}\right|_{r=R} = \cot\gamma_d = -\Gamma(z_w - z_{we}) + \cot\gamma_s, \qquad \gamma_s \leq 90°\,, \tag{8.2}$$

where z_{we} represents the final coordinate of the contact line. Figure 8.3 shows the dynamic contact angle γ_d as a function of the wall coordinate z_w with Γ as parameter. For $\Gamma = 0$, the contact angle is constant and equal to the static contact angle γ_s, the contact line is free to move (free contact line condition).

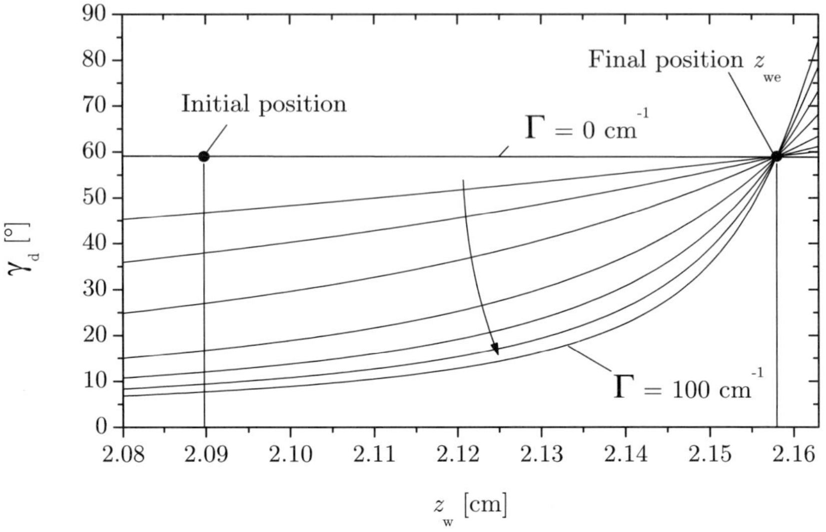

Fig. 8.3. Dependence of the dynamic contact angle γ_d on the z-coordinate of the contact line as function of the free parameter Γ for a liquid with a static contact $\gamma_s = 59°$

For $\Gamma \to \infty$, the contact line is fixed and the contact angle free (fixed contact line condition). The limits of the boundary condition $\Gamma = 0$ and $\Gamma \to \infty$ are also limits for the reorientation behavior in the global time regime [5, 1]. For the free contact line condition the frequency of the center point oscillation is minimal. With the fixed contact line condition the frequency is maximal. The numerical simulations with FIDAP are calculated with $\Gamma = 0$ for the free contact line condition and $\Gamma = 100$ cm^{-1} for the fixed contact line condition.

All calculations begin with an interface shape determined for the normal gravitational field instantaneously subjected to a zero gravity environment at a time $t = 0$. The initial free surface was determined by solving the Gauss-Laplace equation (5.1). The describing nonlinear differential equation (5.4) was solved with a Runge-Kutta algorithm. The calculated free surface configurations are used to design the initial mesh for the simulations. The initial fill height is set to $h_0 = 2R$. Thus the bottom of the container has no influence on the reorientation. The transition from the normal gravity environment to the micro gravity level is simulated with a step function. The final Bond number is equal $\mathrm{Bo} = 0$.

8.2 Numerical Model used in **FLOW-3D**

FLOW-3D provides the opportunity to simulate reorientation cases for complete wetting liquids and was employed for reorientation cases with static

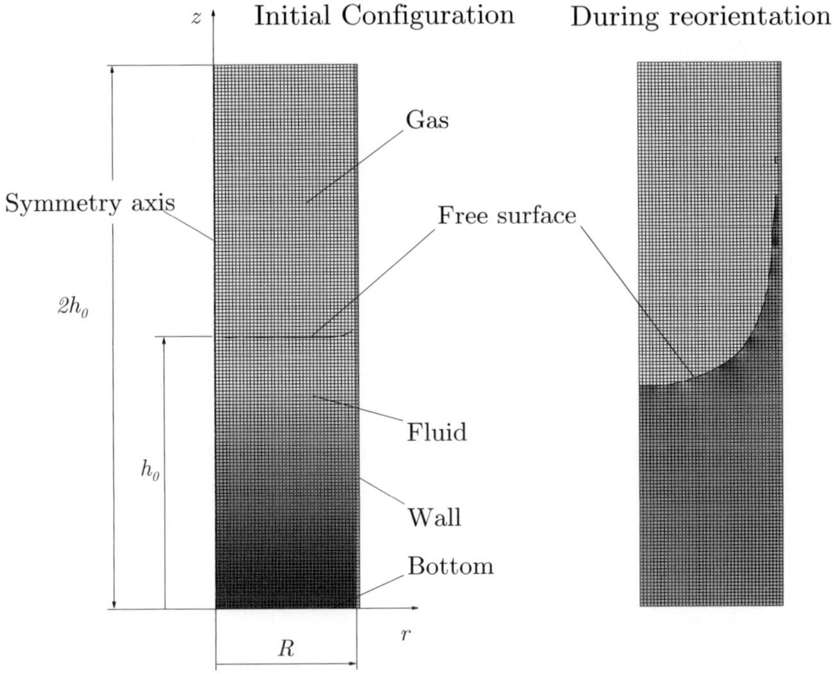

Fig. 8.4. 40 x 160 Mesh used in FLOW-3D calculations

contact angles $0 < \gamma_s < 5°$. To determine the initial surface configuration with FLOW-3D, the calculation is initiated with a flat surface and the initial acceleration. After an appropriate time a nearly steady state is reached. Thereafter the acceleration value is changed with a step reduction in gravity to the final acceleration level.

A high number of cells are needed for suitable resolution of the contact line. If the number of cells is too small, the movement of the contact line with time is uneven. Figure 8.4 shows an example of the cylindrical mesh used in the FLOW-3D calculations. In all the calculations the cells have the same length in r- and z- direction. With quadratic cells the curvature is calculated as accurately as possible. The radius R in the example in Fig. 8.4 is $R = 2$ cm, the initial fill height is $h_0 = 2R$. The mesh consists of 160 cells in the z-direction and 40 cells in the r-direction. Each cell size is quadratic and of the size 0.05 cm x 0.05 cm. The mesh consists of one cell in the φ-direction. The azimuthal angle of the cell boundaries is 30°. This value is strongly recommended by FLOW-3D. A symmetry-plane boundary was chosen on the front and back side of the mesh.

FLOW-3D does not allow the implementation of user defined models for the dynamic contact angle. The dynamic contact angle is determined with an intrinsic method. The method is based on a force balance and considers the change of the static contact angle due to the motion of the contact line.

The initial fill height h_0 is always equal $2R$ to eliminate the influence of fluid depth. The transition from the normal gravity environment to the micro gravity level is simulated with a step function. The final BOND number is $\mathrm{Bo} = 0$.

8.3 Data Evaluation

Surface Height

FIDAP

The result files contain the location of the nodes for each time step which represent the free surface including the node at the contact line and the center point.

FLOW-3D

FLOW-3D offers two possibilities to determine the position of the free surface, *fill fraction* and *surface height*. Using the fill fraction method, FLOW-3D determines the degree of filling of each mesh cell and calculates a mean fill height of the cell. The second method *surface height* determines the fluid height above user defined mesh cells. The fluid height is denoted from the center at the bottom of the defined mesh cells to the evaluated free surface position. The *surface height* method is more precise than the *fill fraction* method. In this study the *surface height* method is used. However, it must be taken into consideration that with this method the exact contact line and center point location could not be determined. The determined surface heights are related always to the center of the mesh cell and not to the boundary.

Damping and Frequency

FIDAP

For the evaluation of the damping behavior and the frequency of the center point oscillation the theory of a linear damped mechanical system is applied which was introduced in Sect. 7.3.3. With the dimensionless center coordinate $z^*_{num}(t)$

$$z^*_{num}(t) = \frac{z_c}{z_{ce}} ,$$

the solution of the differential equation for the description of a damped linear oscillation reads [6]

$$z^*_{num}(t) = 1 - e^{-D\omega_0 t}\left[\cos\left(\sqrt{1-D^2}\omega_0 t\right) + \frac{D}{\sqrt{1-D^2}}\sin\left(\sqrt{1-D^2}\omega_0 t\right)\right] , \tag{8.3}$$

with D the damping ratio and w_0 the natural frequency. The upper envelope reads

$$z^*_{num,u}(t) = 1 + \frac{1}{\sqrt{1-D^2}}e^{-D\omega_0 t} \tag{8.4}$$

and the lower envelope is

$$z^*_{num,l}(t) = 1 + \frac{1}{\sqrt{1-D^2}}e^{-D\omega_0 t} . \tag{8.5}$$

For the determination of the frequency the theoretical oscillation curve and the oscillation curve of the numerical simulation are compared. The parameters of the (8.3) w_0 and D are varied until the difference between both oscillation curves are minimal. As result of this procedure the natural frequency and the logarithmic decrement of the numerical simulation are assignable. Frequency and damping were evaluated from the first peak towards the 2% deviation from the final center point coordinate (Fig. 8.5).

FLOW-3D

Two examples of reorientation computations with FLOW-3D are shown in Fig. 8.6 for SF 0.65 and Fig. 8.7 for FC-77. The liquid volume is located in the lower half of the computational domain, the passive gas phase in the

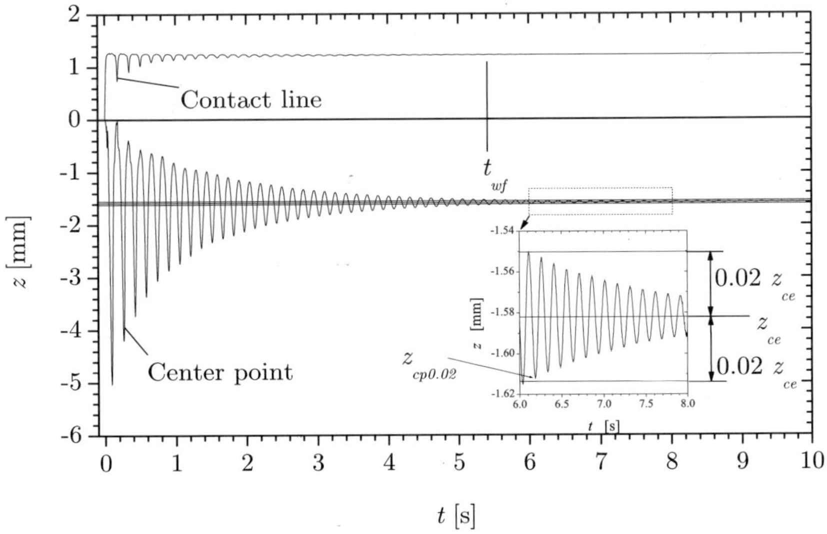

Fig. 8.5. Contact line and center point coordinate versus time from FIDAP simulations with the fixed contact line boundary condition

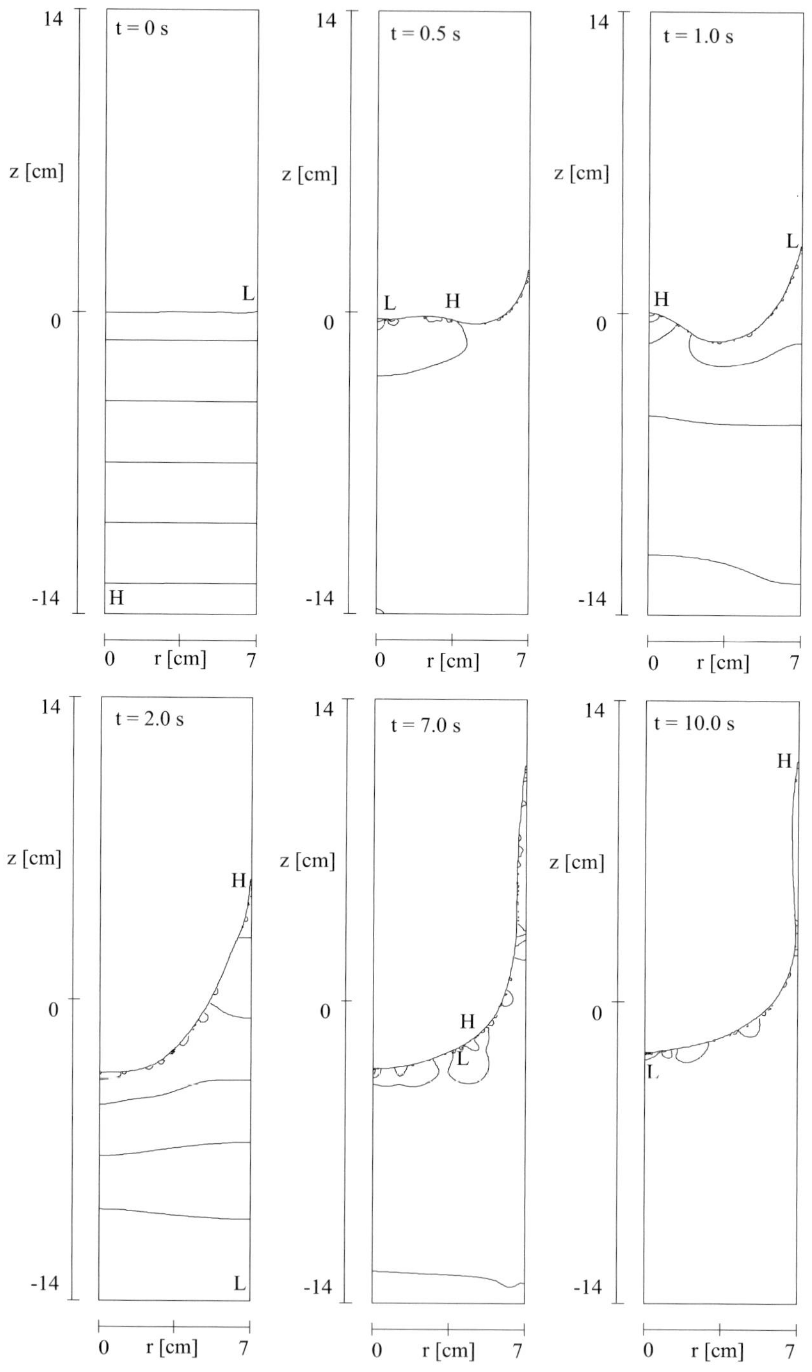

Fig. 8.6. Free surface and pressure contour plot of SF 0.65 for different times and with a g_0 surface as the initial condition

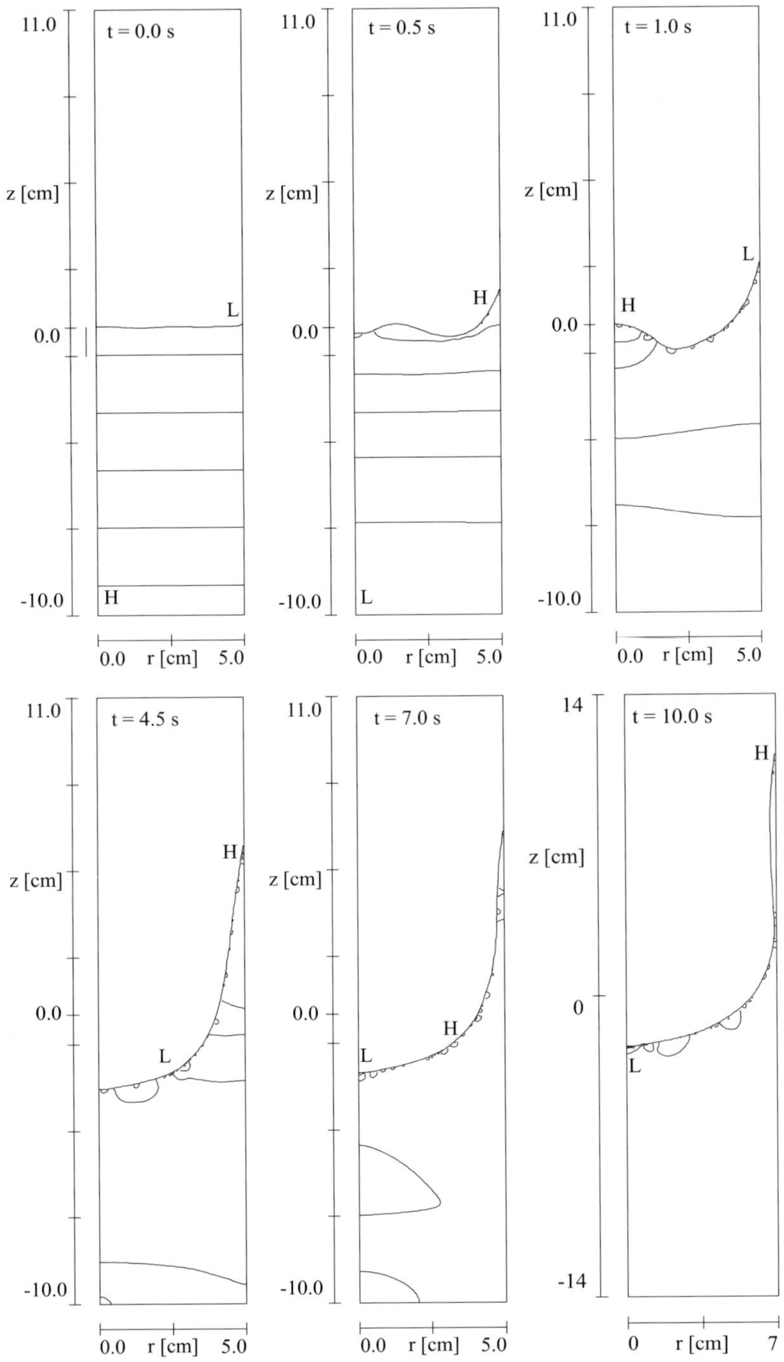

Fig. 8.7. Free surface and pressure contour plot of FC 77 for different times and with a g_0 surface as the initial condition

upper part. Both are separated by the free surface. Additional the pressure contour plot can be seen qualitatively with the regions of high (H) and low (L) pressure. The initial condition is a g_0 free surface.

References

1. J. Gerstmann, M.E. Dreyer, H.J. Rath: AIP Conference 504, 847 (2000)
2. J. Gerstmann, M.E. Dreyer, H.J. Rath: Z. Angw. Math. Mech. **80**, 717 (2000)
3. M.E. Dreyer, J. Gerstmann, M. Stange et al: Space Forum **3**, 87 (1998)
4. H.M. Satterlee, W.C. Reynolds: NASA TR LG-2, 1 (1964)
5. G. Wölk, M.E. Dreyer, M. M. Weislogel: J. Spacecr. Rockets **34**, 110 (1997)
6. K. Ogata: *System Dynamics*, 2 edn (Prentice Hall, Englewood Cliffs, New Jersey 1992)

9

Conclusions

The following sections summarize the main conclusions for the free surface reorientation. The focus of the conclusions is set on the initial behavior of the contact point (giving the initial behavior), and on the global behavior of the center point giving the general behavior in time (damping and frequency). The particular observations from the experimental and numerical studies are followed by a summary comparing both.

9.1 Contact Point Observations

9.1.1 Experimental Observations for the Contact Point

General Behavior

The general behavior of the contact point depends strongly on the static contact angle as well as on the OHNESORGE number. Different behavior occurs for high and low static contact angles. For low contact angles the contact point shows an overshoot over the final equilibrium position only for rather small OHNESORGE numbers on the order of $\mathrm{Oh} \lesssim 2\times10^{-3}$ (compare Fig. 9.1). In this case the contact point may be regarded to obey a free contact line condition with a following oscillation around the equilibrium position. This behavior is shown in detail in Fig. 9.2 for test no. 32. The corresponding history of the contact point is shown in Fig. 7.13.

For small contact angles and higher OHNESORGE numbers ($\mathrm{Oh} \gtrsim 2\times10^{-3}$) the contact point moves slowly to the final position with an advancing contact line (compare Fig. 9.3 and Fig. 9.1). As long as the final position is not reached, the liquid interface oscillates around a position corresponding to a higher static contact angle. This characteristic may be the reason for the higher dependence of the frequency on the OHNESORGE number for small static contact angles, as described in the previous sections.

Michael E. Dreyer: *Free Surface Flows under Compensated Gravity Conditions*
STMP **221**, 101–132 (2007)
DOI 10.1007/3-540-44628-1_9

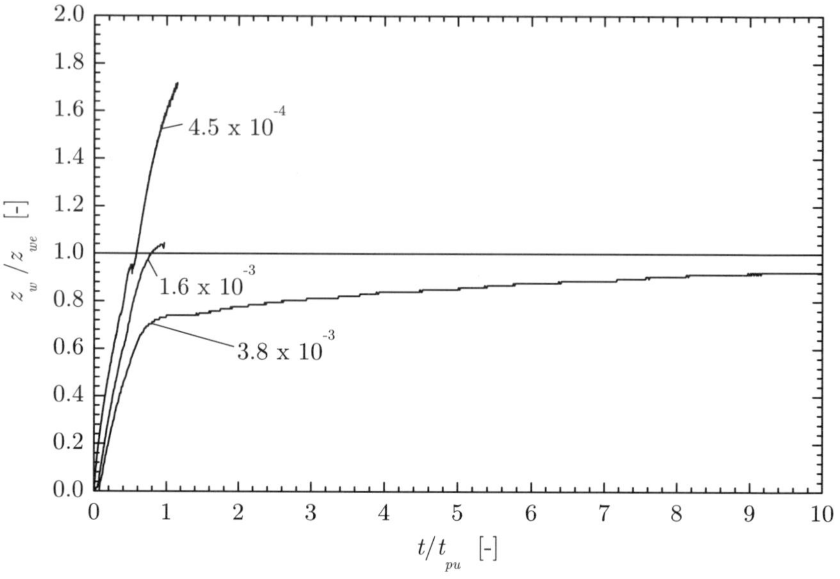

Fig. 9.1. Behavior of the contact point for experiments with low static contact angles ($\gamma_s < 10°$) for different OHNESORGE numbers. The dimensionless wall coordinate z_{wp}/z_{we} is plotted versus the dimensionless time t/t_{pu}. With decreasing OHNESORGE number an increasing overshoot over the equilibrium position during the initial rise occurs

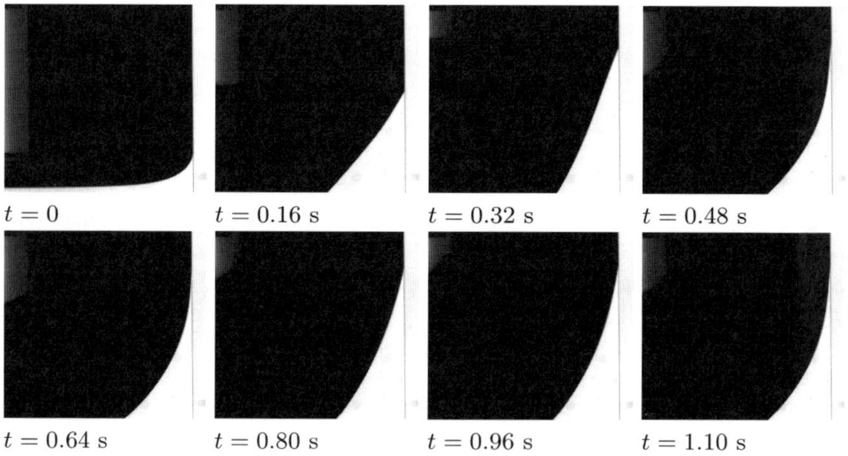

Fig. 9.2. Series of video images for test no. 32 shows the capillary rise of the contact point in detail (Oh = 0.0016, $R = 17.5$ mm). After the first maximum is reached at a position $z_{wp} > z_{we}$ a thin liquid layer is left at the wall ($t = 0.48$ s). This layer can also be observed after the second period at approximately $t = 1.1$ s

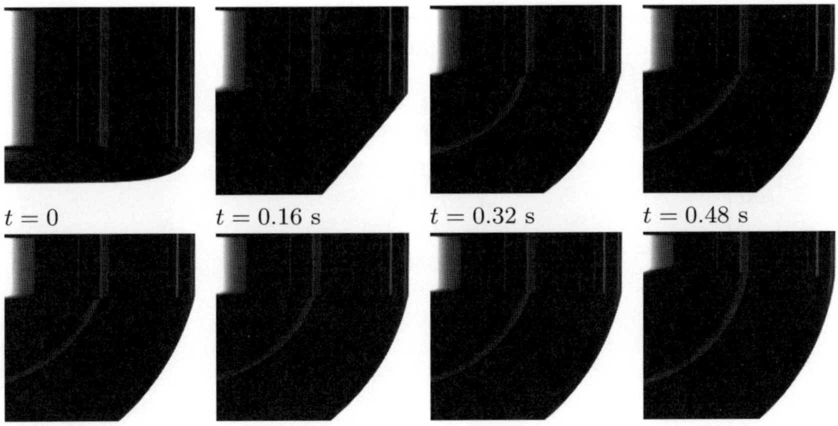

Fig. 9.3. Behavior of the contact point for test no. 46 ($R = 15$ mm). The equilibrium position is not exceeded during the initial capillary rise, which ends at approx. $t = 0.39$ s. This behavior is caused by a higher OHNESORGE number of ($\mathrm{Oh} = 0.0038$). After the first maximum is reached, the contact point creeps to its equilibrium configuration (compare the position of the contact point at $t = 4.7$ s)

For all experiments with higher static contact angles, an overshoot over the final equilibrium position could be observed in the range of OHNESORGE numbers investigated. However, a trend of decreasing z_{wp} with increasing Oh is found. Besides the difference between experiments with high and low static contact angles, in particular for small γ_s, the OHNESORGE number strongly affects the behavior of the contact point.

Rise Velocity

In the very initial period the contact point starts to move along the cylinder wall and reaches rapidly a maximum velocity u_{wmax} as shown in Fig. 9.4 for test no. 32 (compare also Fig. 7.13 for the history $z_w(t)$). After the maximum velocity is reached the velocity decreases and shows some characteristic oscillations (compare also Fig. 9.7 in the numerical results) until it reaches a minimum $\mathrm{d}z_w/\mathrm{d}t = 0$ with a maximum deflection of the contact point.

For all test cases the maximum velocity is scaled with u_{puLc} (6.7) and plotted versus the MORTON number in Fig. 9.13 (see page 112). A clear trend of an increasing dimensionless rise velocity for lower MORTON numbers can be observed. However, since the static contact angle rises for experiments with higher MORTON number the dependence on the MORTON number is less stronger than it appears from the experimental data. The reason is that higher static contact angles γ_s lead to an increasing dimensionless rise velocity. Considering this feature for the experimental results a dependence of u_{wmax}/u_{pulc} on the MORTON number for constant static contact angles is on the order of

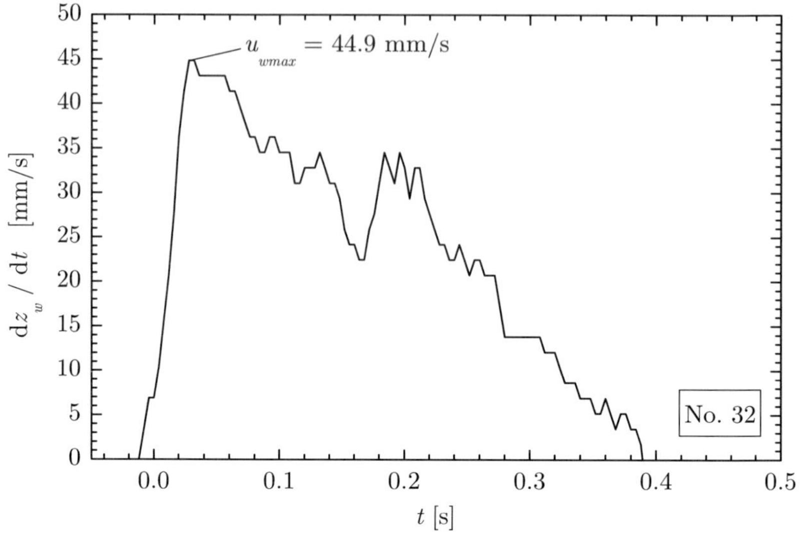

Fig. 9.4. Velocity of the contact point dz_w/dt versus time, which was determined by adjacent averaging (5 values) after calculation of the gradient. The maximum velocity is reached rapidly after transition to reduced gravity

$$\frac{u_{wmax}}{u_{puLc}} \sim \text{Mo}^{-1/4} . \tag{9.1}$$

As postulated above, the initial capillary rise at the cylinder wall does not depend on the cylinder radius if the initial BOND number is Bo $\gg$ 1. This behavior can be validated since experiments with much larger cylinder radii line up with the experiments with small cylinder radius. However the influence of a rather low BOND number can be observed for test no. 36 (Bo = 33), where the dimensionless velocity is remarkably lower than for test no. 32 with Bo = 102. This effect can also be seen for experiments with SF 3.0, where test no. 37 with the lowest BOND number shows a slightly lower dimensionless velocity.

Since u_{wmax} describes the velocity of the contact point only for the maximum value, an average velocity for the contact point during the initial capillary rise is defined by the mean rise velocity $\bar{u}_{wr}$ (7.1), which is shown in Fig. 9.14 scaled with u_{puLc} versus MORTON number. Basically the same characteristics as described for u_{wmax} can be observed. However the mean velocity $\bar{u}_{wr}$ is about 29% – 64% lower than the maximum initial rise velocity.

First Arrival at Equilibrium

In analogy to the description of the time t_{cr} in Sect. 9.3.1, the time t_{wr} for the first transit through equilibrium is scaled with t_{pu}. For experiments with small static contact angles the time t_{wr} can only be evaluated if the first maximum z_{wp} lays above the equilibrium position z_{we}. Two main groups of experiments can be distinguished: Experiments with small static contact angles and experiments with higher static contact angles. In both cases an increasing OHNESORGE number leads to higher dimensionless times. The dependence on the static contact angle is caused by higher rise velocities for small contact angles which would reduce t_{wr}/t_{pu}. However, the equilibrium deflection z_{we} increases for decreasing static contact angles yielding higher times t_{wr}.

First Maximum

The first maximum of the contact point may be investigated with regard to the time and the deflection. Since there are no obvious peaks in the histories of experiments with small static contact angles ($\gamma_s < 10°$), these experiments are skipped in this section (only test no. 32 shows an overshoot z_{wm} over the final position z_{we}). The time of the first peak t_{wp} scaled with the time t_{pu} is plotted versus OHNESORGE number in Fig. 9.16. In the range of static contact angles the ratio t_{wp}/t_{pu} is quite constant over the range of OHNESORGE numbers with a trend of decreasing t_{wp}/t_{pu} for increasing Oh. This might also be an influence of the static contact angle, which increases for test liquids with higher viscosities. The comparison of the times t_{wp} and t_{cp} in Fig. 9.31 shows, that the behavior is similar but t_{cp} lies slightly below the values of t_{wp}.

The maximum deflection of the contact point at the first peak for higher static contact angles is plotted in Fig. 9.15 scaled with the cylinder radius R. A higher OHNESORGE number leads to a reduction of the overshoot, where for lower OHNESORGE numbers a trend to a stagnating maximum deflection is observable.

Scaling of z_{wp} with the final equilibrium deflection z_{we} of the contact point shows the influence of the OHNESORGE number on the damping behavior of the system in Fig. 9.5. An increasing OHNESORGE number leads to a reduction of the overshoot to a value of $z_{wp}/z_{we} = 1$ for the test cases with a high static contact angle. This behavior can also be observed for experiments with low static contact angles, where only test no. 32 shows an overshoot over z_{we} with the maximum value z_{wp}. In test numbers 36 and 46 with higher OHNESORGE numbers Oh $\gtrsim 2 \times 10^{-3}$ no maximum z_{wp} (above z_{we}) appears. Thus the critical value for the OHNESORGE number (where $z_{wp}/z_{we} = 1$) is smaller for experiments with low static contact angles than for higher static contact angles.

Fixing Time of Contact Point

The time t_{wf} until the contact point is fixed could only be evaluated for experiments with higher static contact angle, where the results are shown in Fig. 9.6. As described above the number of oscillations around the equilibrium position reduces with increasing OHNESORGE number. Consequently a decreasing time

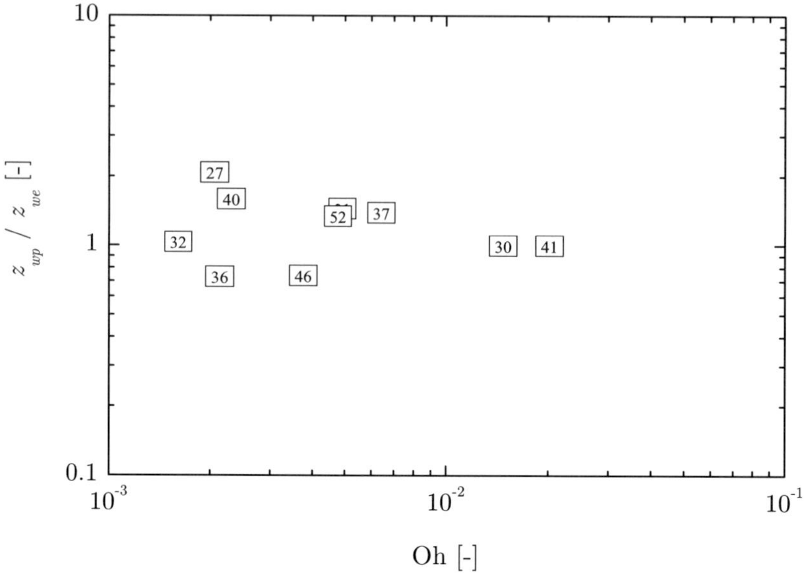

Fig. 9.5. Deflection of the contact point z_{wp} at the first maximum scaled with the equilibrium position z_{we} versus OHNESORGE number

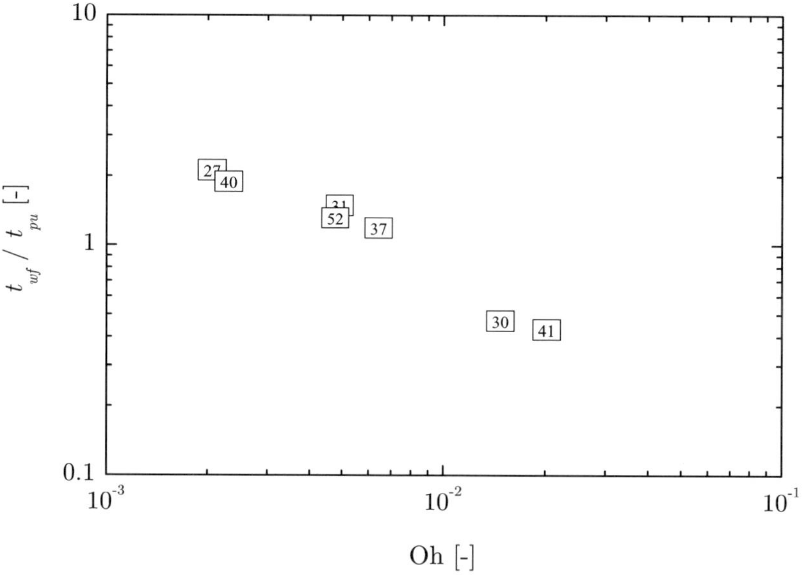

Fig. 9.6. Dimensionless time t_{wf} until the contact point is fixed versus OHNESORGE number

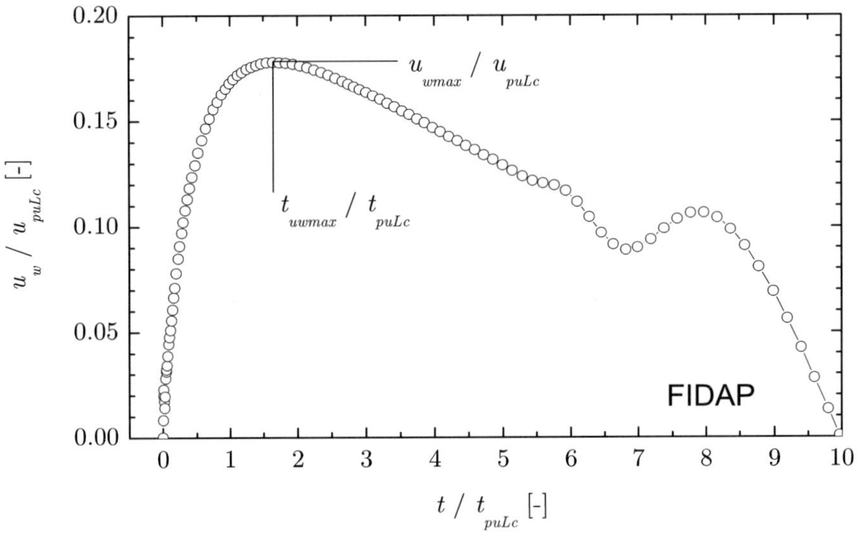

Fig. 9.7. Nondimensional contact line velocity versus dimensionless time. Mo $= 0.017, \gamma_s = 53.5°, u_{puLc} = 11.92$ cm/s, $t_{puLc} = 0.012$ s

t_{wf} is obtained for increasing OHNESORGE numbers. A clear dependence of the data points from the static contact angle cannot be observed. The dependence of t_{wf}/t_{pu} on the OHNESORGE numbers may be expressed by the trend

$$\frac{t_{wf}}{t_{pu}} \sim \mathrm{Oh}^{-3/4} . \tag{9.2}$$

9.1.2 Numerical Observation for the Contact Point

Rise Velocity

In the analysis of the contact line velocity two characteristic velocities can be determined, the maximum rise velocity u_{wmax} (Fig. 9.7) and the mean rise velocity $\overline{u}_{wr}$ (Fig. 9.8). Both velocities are nondimensionalized with the velocity u_{puLc} and the time t_{puLc}.

The experiments show that the advancing dynamic contact angle increases with increasing contact line velocity. The increase of the dynamic contact angle acts against the wetting process and reduces the rise velocity. The FIDAP calculation does not taken into account this effect due to the free contact line condition with a fixed static contact angle. Therefore the mean rise velocity calculated with FIDAP is higher than the mean rise velocities in the experiment or in the FLOW-3D calculations for the same initial static contact angle. This effect is taken into consideration for the comparison of the numerical results evaluated with FIDAP and FLOW-3D. The maximal rise velocity of the contact line is determined from the FIDAP calculations only. The FIDAP

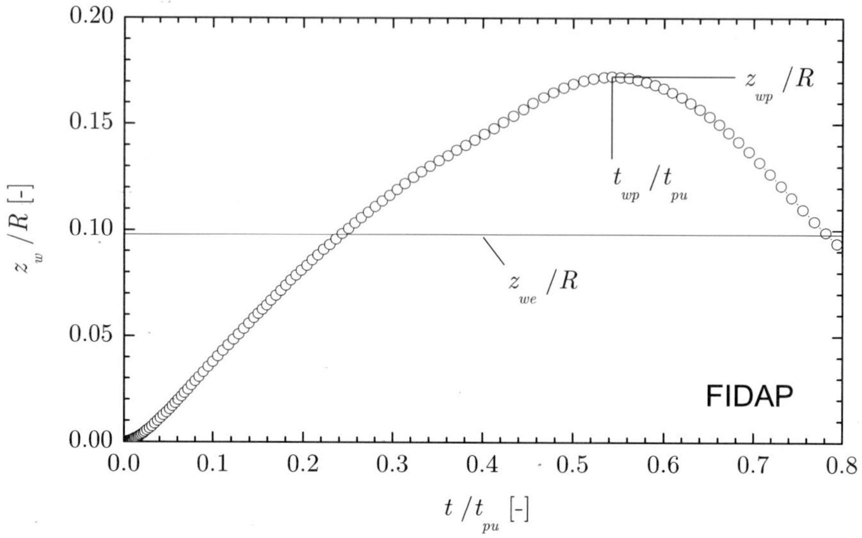

Fig. 9.8. Dimensionless contact line coordinate z_w/R versus dimensionless time t/t_{pu}. The mean rise velocity is defined as the ratio of the maximal wall coordinate z_{wp} and the associated time t_{wp}. z_{we}/R denotes the dimensionless final wall coordinate. $\mathrm{Mo} = 0.017, \gamma_s = 53.5°, u_{puLc} = 11.92$ cm/s, $t_{pu} = 0.22$ s

result file contains the velocity time history data of the node which represent the contact line. The FLOW-3D result file contains no velocity information of the free surface. The velocity must be calculated from the change of the free surface coordinate with time. This method leads to imprecise calculations of the maximal velocity. The time steps of the transient simulation are to large. Therefore the maximum velocity was not evaluated from the FLOW-3D calculations. The mean rise velocity, the quotient of the first wall peak and the associated wall peak time, was determined from FIDAP and FLOW-3D calculations (Fig. 9.8, Fig. 9.10). The FLOW-3D calculations show that with increasing OHNESORGE number the overshoot of the contact line with a significant wall peak decreases. For large OHNESORGE numbers the contact line creeps to the final position without overshoot. If no wall peak appears the mean rise velocity was determined from the quotient of a significant resting point and the associated resting point time.

In the FIDAP calculations the MORTON number and the static contact angle was varied. The free contact line condition was used as boundary condition for the dynamic contact angle. The FLOW-3D calculations were performed with static contact angles $\gamma_s = 0$ and varying MORTON numbers.

The mean wall rise velocity, calculated with FIDAP, increases with decreasing static contact angles (Fig. 9.10). For low MORTON numbers ($\mathrm{Mo} = 3.72 \times 10^{-3}$) the dimensionless mean rise velocity is nearly proportional to the static contact angle. The extrapolation to static contact angles $\gamma_s = 0$ yields

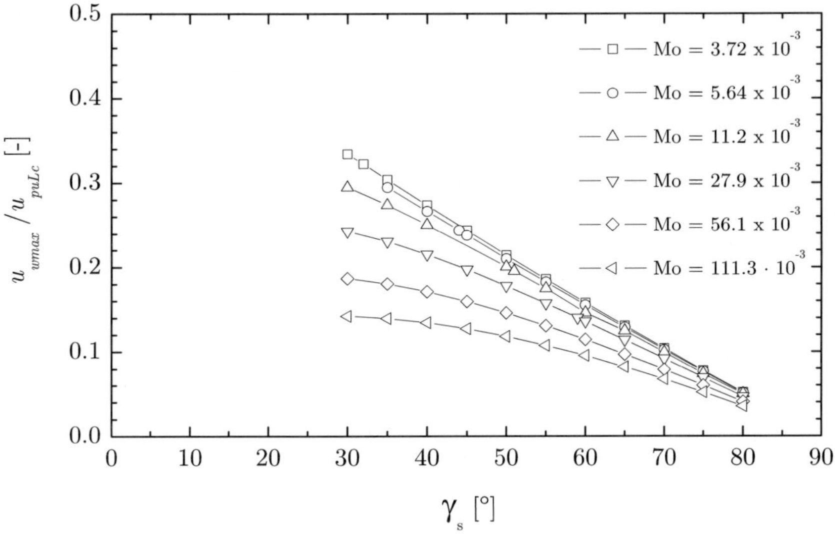

Fig. 9.9. Dimensionless maximum wall velocity versus the static contact angle for varying MORTON number (FIDAP)

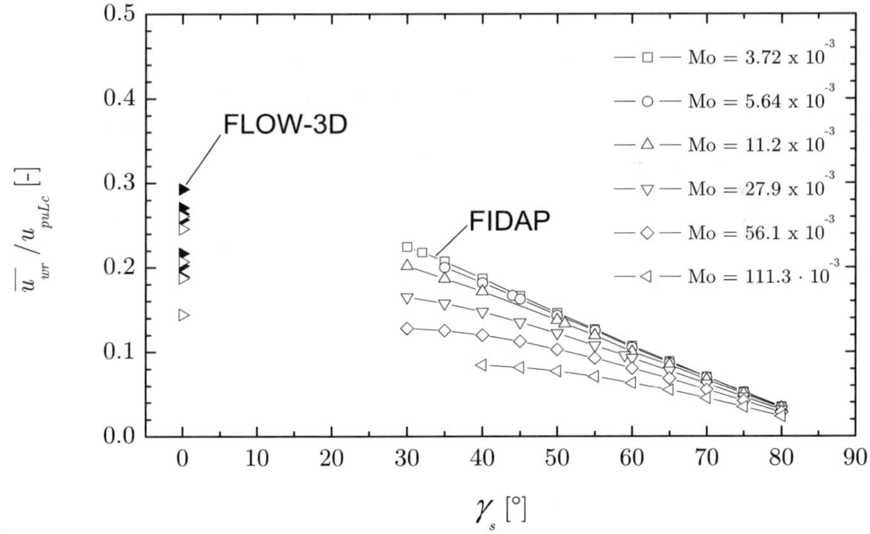

Fig. 9.10. Dimensionless mean rise velocity versus static contact angle and varying Mo number. The solid symbols of the FLOW-3D results mark the simulations with cell sizes less than 1 mm

a dimensionless mean rise velocity of approx. 0.35. The FLOW-3D results for contact angles $\gamma_s = 0$ show the same dependence on the static contact angle and the MORTON number. However, the FLOW-3D results depend on the mesh size. With increasing mesh density the mean rise velocity increases too. Due to a different internal handling of the dynamic contact angle in the FLOW-3D simulation the mean rise velocity is smaller compared to the FIDAP results.

The maximum rise velocity is evaluated only from the FIDAP calculations. The maximum wall velocity shows the same dependence on the static contact angle and the MORTON number as seen before by the mean rise velocity. The maximal wall velocity increases with decreasing static contact angle (Fig. 9.9) and decreasing MORTON number (Fig. 9.13). The dimensionless maximal wall velocity is about a factor of two larger than the mean rise velocity.

First Maximum

The first maximum of the contact line depends on the static contact angle and the OHNESORGE number. For high OHNESORGE numbers and low contact angles no precise overshoot exist. The FLOW-3D simulations have shown that for static contact angle $\gamma_s = 0$ no definable overshoot occur. Therefore the following data of the first peak contain no data from FLOW-3D simulations. The FIDAP simulations with high static contact angles and low OHNESORGE numbers show a distinct overshoot. In the numerical simulations the contact angle was varied between 30° and 80°. The OHNESORGE number was varied in the range of $1.46 \times 10^{-3} < \mathrm{Oh} < 44 \times 10^{-3}$. The dependence on the OHNESORGE is weak (Fig. 9.15). In general the dimensionless first peak z_{wp} decreases with increasing OHNESORGE number. The dimensionless wall peak coordinate for a static contact angle $\gamma_s = 30°$ is about 0.45 for the lowest considered OHNESORGE number of 1.46×10^{-3} and about 0.2 for $\mathrm{Oh} = 44 \times 10^{-3}$. The dependence on the static contact angle γ_s is more distinct (Fig. 9.11). The dimensionless wall rise z_{wp}/R decreases with increasing static contact angle. For static contact angles $\gamma_s = 80°$ the dimensionless wall peak coordinate is equal to 0.04. For static contact angles less than about 50° the dimensionless wall peak coordinate is proportional to the static contact angle. The extrapolation to a static contact angle $\gamma_s = 0$ for the lowest OHNESORGE number yields a dimensionless wall peak coordinate of about 1.

The associated wall peak time shows no dependence on the OHNESORGE number (Fig. 9.16). Again the static contact angle has a stronger influence on the reorientation behavior (Fig. 9.12). With decreasing static contact angle the wall peak time increases.

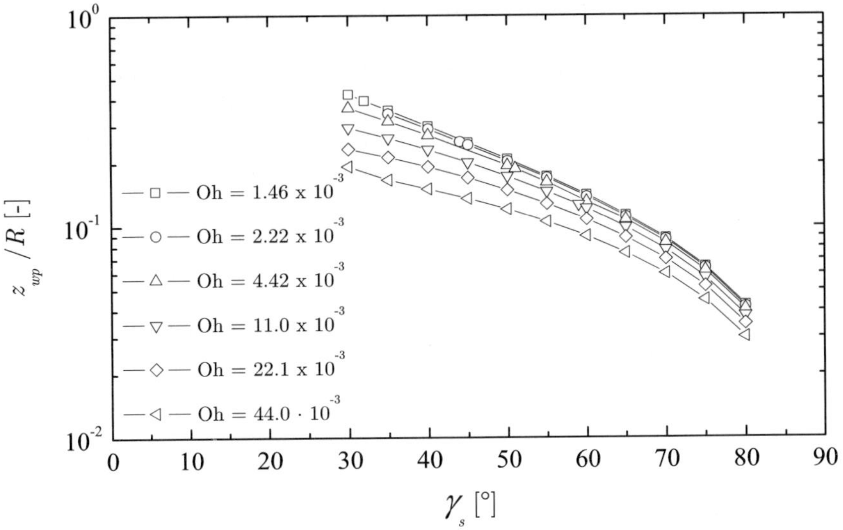

Fig. 9.11. Maximum contact line coordinate against the static contact angle for varying OHNESORGE number (FIDAP)

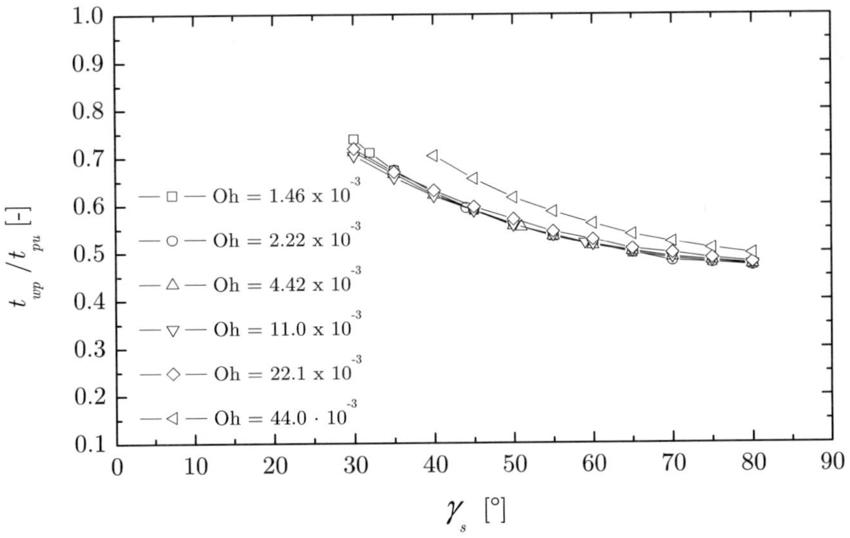

Fig. 9.12. Dimensionless time of the maximum wall coordinate versus the static contact angle for varying OHNESORGE numbers

9.2 Summary Contact Point Observations

A comparison of experimental results and numerical calculations for the dimensionless maximum rise velocity $u^*_{wmax} = u_{wmax}/u_{puLc}$ is presented in Fig. 9.13. The velocity u^*_{wmax}, which describes the maximum velocity of the contact point during the initial capillary rise along the cylinder wall, depends on the static contact angle γ_s as well as on the MORTON number.

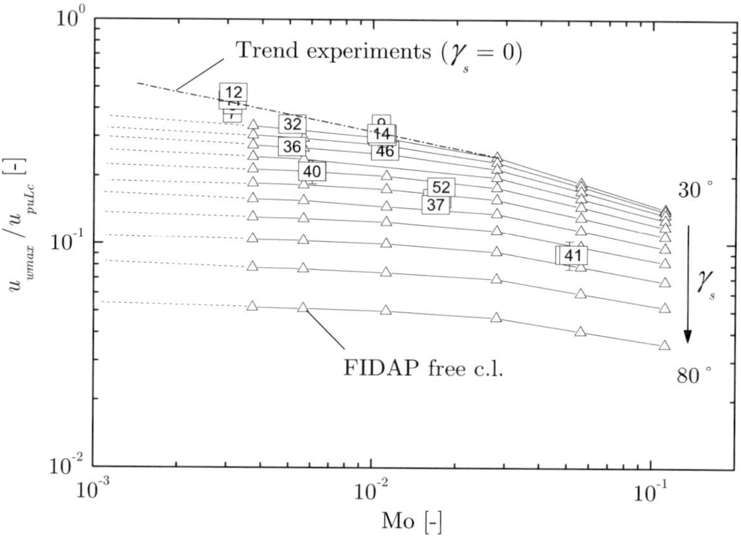

Fig. 9.13. Nondimensional maximum rise velocity versus MORTON number and varying γ_s. Indication of trend (*dashed line*), connected symbols for visualization only, trend from experiments for $\gamma_s = 0$ (*dash-dotted line*)

An increasing static contact angle leads to a decreasing velocity u^*_{wmax} for constant MORTON numbers. On the other hand u^*_{wmax} rises with decreasing MORTON number. As postulated above, the initial capillary rise at the cylinder wall does not depend on the cylinder radius, if the initial BOND number is $\mathrm{Bo} \gg 1$. The evidence is given by the fact that experiments with much larger cylinder diameter (compare Fig. 9.13 and 9.14) line up with the experiments with small cylinder diameter. A comparison of numerical results (△) with the experimental data (□) shows good agreement. Extrapolation of the trend given from the FIDAP calculations (with $30° < \gamma_s < 80°$) shows a trend $u^*_{wmax} \sim$ const for contact angles $\gamma_s > 80°$ and an increasing negative slope of the MORTON number dependence for decreasing γ_s. The extrapolation of the experimental data to lower MORTON numbers yields a relation

$$u^*_{wmax} \sim \mathrm{Mo}^{-1/4} \tag{9.3}$$

for the range $10^{-3} < \mathrm{Mo} < 3 \times 10^{-2}$.

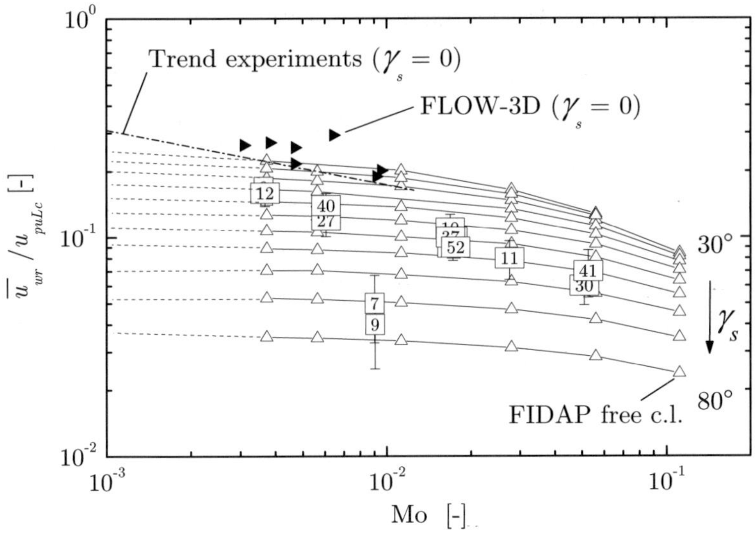

Fig. 9.14. Nondimensional mean rise velocity versus MORTON number and varying γ_s. Indication of trend (*dashed line*), connected symbols for visualization only, trend from experiments for $\gamma_s = 0$ (*dash-dotted line*)

The mean rise velocity $\bar{u}_{wr}$, which is given by (7.1) and describes the mean velocity during the initial capillary rise, is plotted in Fig. 9.14 scaled with u_{pulc} versus the MORTON number. The data points of $\bar{u}_{wr}/\ u_{pulc}$ are about 40 % – 75 % lower than the values for the maximum rise velocity. Further numerical calculations with FLOW-3D (▶) show a trend of overestimating the experimental results as indicated in Fig. 9.14.

The behavior of the contact point concerning the arrival at the first maximum (z_{wp}, t_{wp}) is presented in Fig. 9.15 and Fig. 9.16 by comparing experimental (□) and numerical results (△). Please note, that all data points are restricted to higher contact angles ($30° < \gamma_s < 80°$ for FIDAP and $\gamma_s > 10°$ for the experiments).

As indicated in Fig. 9.15, the deflection of the contact point at the first peak z_{wp}/R depends clearly on γ_s, where a trend to larger z_{wp}/R with decreasing γ_s is observable. This could be expected since the equilibrium position of the contact point also increases with decreasing γ_s. An increasing OHNESORGE number leads to a decrease of z_{wp}/R. This trend seems to be weaker for higher static contact angles. The agreement between experimental data and numerical results is remarkable.

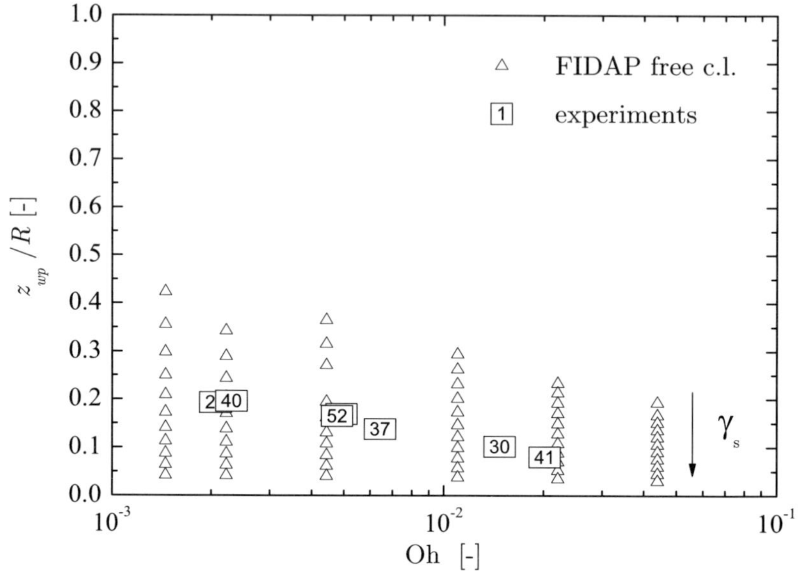

Fig. 9.15. Nondimensional maximum deflection of the contact point versus OHNESORGE number and varying γ_s. Experimental results for $\gamma_s > 10°$, numerical results for $30° < \gamma_s < 80°$, FIDAP (*free contact line*)

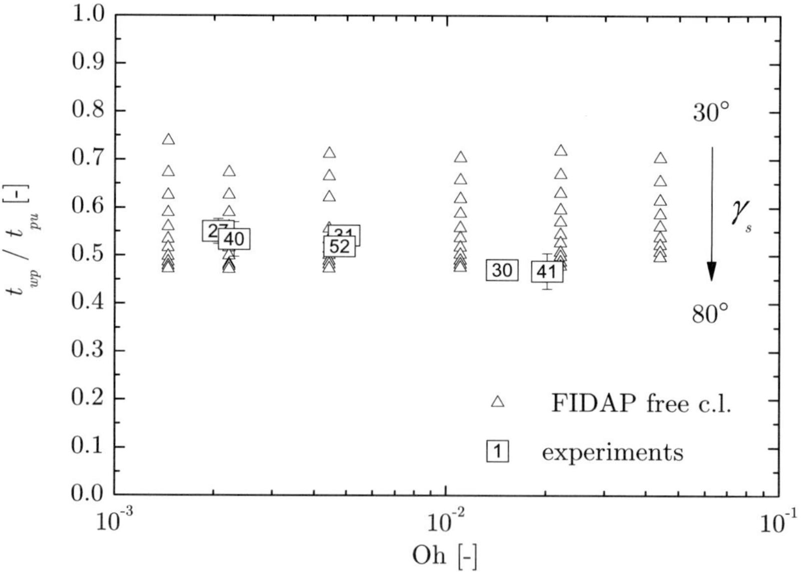

Fig. 9.16. Dimensionless time at the first maximum of the contact point versus OHNESORGE number and varying γ_s. Experimental results for $\gamma_s > 10°$, numerical results for $30° < \gamma_s < 80°$, FIDAP (*free contact line*)

The time t_{wp}/t_{pu} at the first maximum of the contact point is presented in Fig. 9.16 versus OHNESORGE number, where experimental (□) and numerical (△) results are compared. The time clearly depends on the static contact angle, where smaller contact angles lead to an increase of t_{wp}/t_{pu}. The influence of the OHNESORGE number can not be clearly identified. Obviously the dependence of the OHNESORGE number is rather weak. The FIDAP results (△ for free contact line) tend to lie slightly above the experimental results. The reason is most likely the free contact line condition which does not consider the variation of the dynamic contact angle.

9.3 Center Point Observations

9.3.1 Experimental Observations for the Center Point

Initial Period

As introduced in Sect. 6.2 the flow in the initial time period is independent of the cylinder radius. To identify a time from which the center point is affected by the reorientation, a characteristic time of the disturbance arrival has been chosen. The time t_{cd}, which depicts a characteristic kink in the history of the center point (as shown in Fig. 7.12) serves as a definition for the duration of the first period.

To scale the experimental results for the time t_{cd} defined in Fig. 7.12 it is assumed that the wave length of the disturbance is on the order of L_c (5.6) and therefore independent of the cylinder radius R. With the characteristic time scale t_{puLc} from (6.6) one obtains the characteristic velocity U_{puLc} for the disturbance from (6.7). Due to the radial movement of the disturbance, the arrival of the disturbance in the cylinder axis must be on the order of

$$t_{RLc} = \frac{R}{u_{puLc}} = R\left(\frac{\rho}{k_{zi}\sigma}\right)^{1/4} . \tag{9.4}$$

The characteristic time t_{cd} evaluated from the center point history, when a significant kink (change in slope) can be observed, is shown in Fig. 9.17 versus the characteristic time t_{RLc}. A linear dependence between t_{cd} and t_{RLc} with no obvious influence of the static contact angle can be detected. Experiments with silicone fluids and index matched liquids (thus different static contact angles) line up very well. However, a linear regression through the data points shows an offset, which could be caused by a delay between the intrinsic disturbance arrival and the occurrence in the observed contour history. Obviously the arrival of the disturbance leads to a crest on the free surface in the cylinder axis, which is difficult to observe (according to the above mentioned reasons). In fact, the contour history at $h(0,t)$ shows the minimum $z(r)$ of the travelling disturbance until it arrives in the cylinder axis. Upon arrival at the cylinder axis the center point is accelerated downwards, which might be the explanation for the observed kink in the history and the offset in Fig. 9.17.

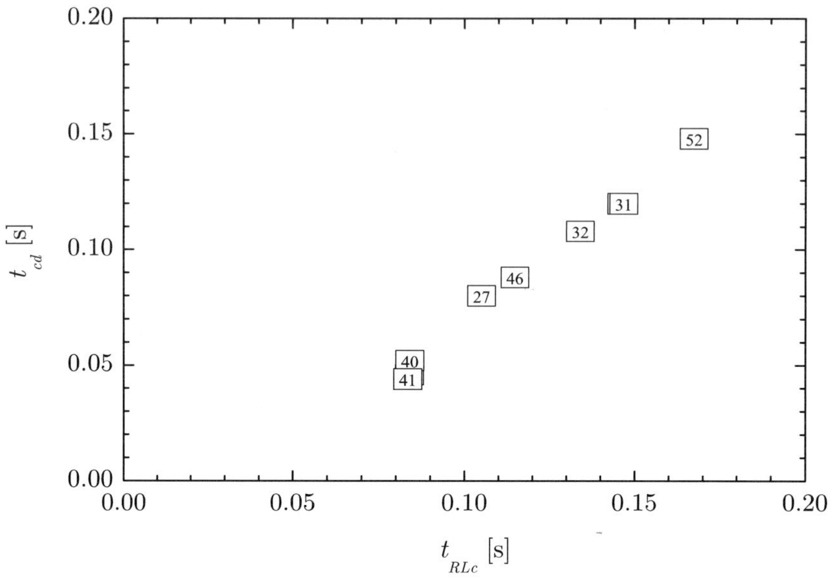

Fig. 9.17. Time t_{cd} until a characteristic kink is observable in the center point history versus t_{RLc}

Rise Time

The first transit of the free liquid interface through the equilibrium position for reduced gravity conditions (compare Fig. 7.12), denoted with the time t_{cr}, is plotted in Fig. 9.18 versus the time scale t_{pu}. As described above, an influence of both the cylinder radius and the static contact angle can be observed. A general trend shows, that the time t_{cr} rises with decreasing static contact angle γ_s, which is evident, since the coordinate of the equilibrium position z_{we} rises as well with decreasing static contact angle. The influence of the OHNESORGE number cannot be defined clearly from the presented data points shown in Fig. 9.19. If t_{cr} is scaled with t_{pu} and plotted versus the OHNESORGE number no dependence on the OHNESORGE number is revealed. The correlation from [1] (valid for experimental data with $\gamma_s = 0$) shows a good agreement with the data, if the trend to higher t_{cd} with decreasing γ_s is considered. However a slight overestimation of the correlation can be observed.

First Peak of Oscillation

The first peak is a characteristic point of the surface oscillation, since it describes the maximum deflection of the center point oscillation which is in the following reorientation damped to the equilibrium position. On the one hand, the time t_{cp} of the first peak can be assumed to correlate with the surface frequency. On the other hand, the deflection z_{cp} of the center point should

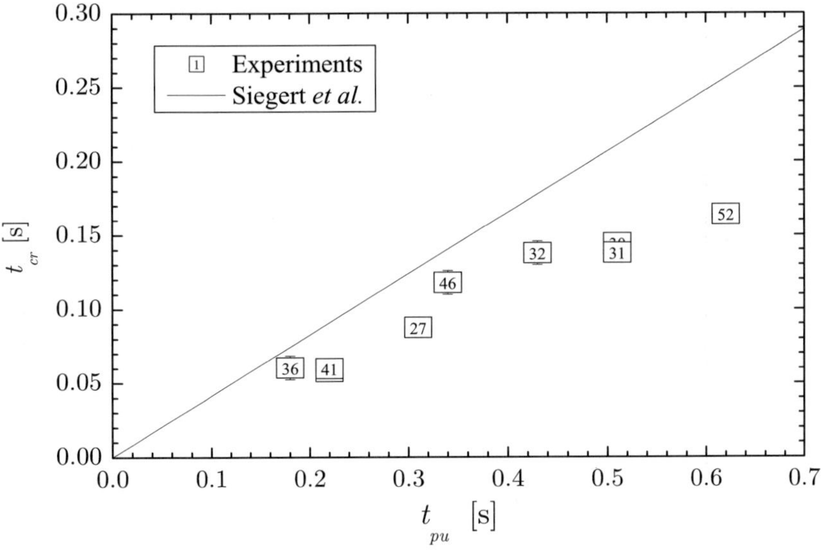

Fig. 9.18. Center rise time t_{cr} versus t_{pu} from experiments in comparison with the correlation (4.24) from [1]

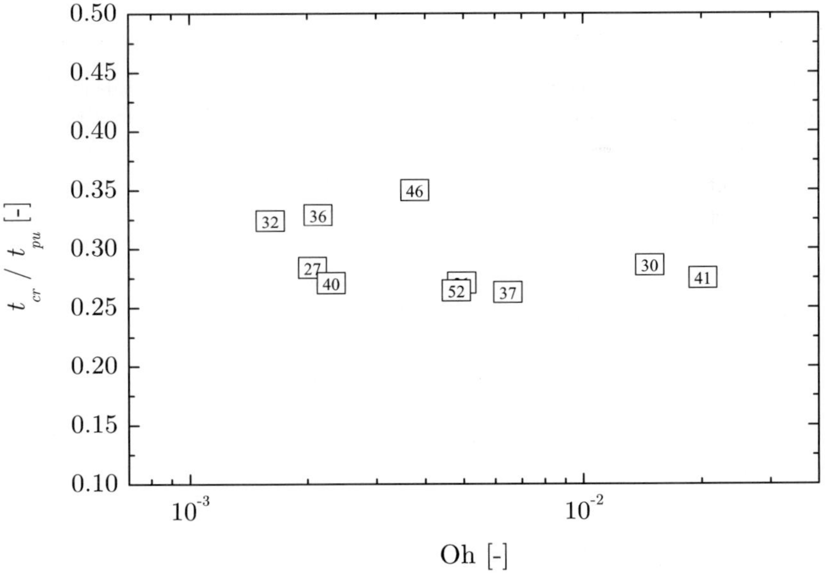

Fig. 9.19. Time t_{cr} scaled with t_{pu} versus OHNESORGE number

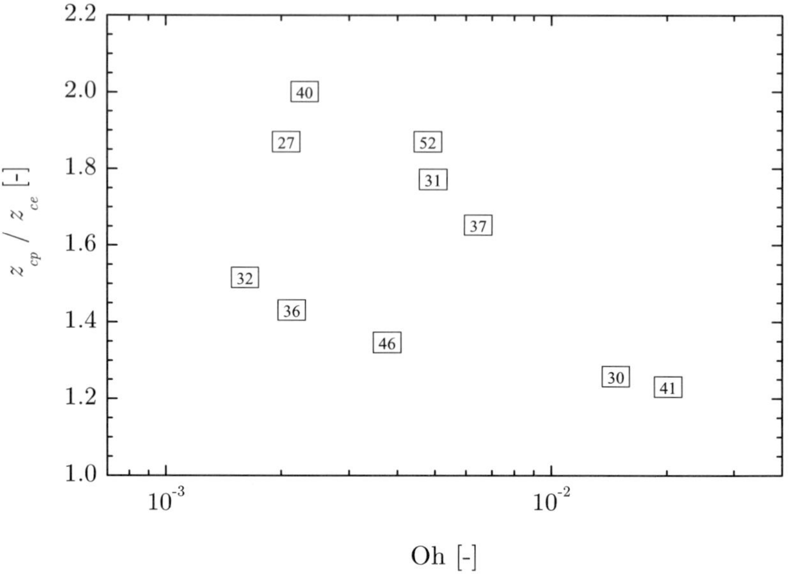

Fig. 9.20. Deflection of the center point at the first peak scaled with z_{ce} versus OHNESORGE number

give a relation to the damping behavior if the OHNESORGE number is varied. Figure 9.31 shows the time t_{cp} scaled with the time t_{pu} versus Oh. Analyzing the behavior of the flow one has to take into account that for silicone fluids a higher viscosity leads also to a higher static contact angle. Considering this, one can see that there is no obvious dependence of t_{cp}/t_{pu} on the OHNESORGE number in the investigated range of Oh numbers. A stronger influence on t_{cp}/t_{pu} is given by the static contact angle, where smaller contact angles lead to higher dimensionless times (compare test no. 32, 36 and 46).

The deflection of the center point z_{cp} at the first maximum scaled with the cylinder radius is plotted in Fig. 9.29 versus Oh. A higher OHNESORGE number causes a reduction of the maximum deflection, which describes the influence of the OHNESORGE number on the damping behavior of the system. The effect of the static contact angle on the center point maximum can be identified clearly from Fig. 9.30. By scaling z_{cp} with z_{ce} the influence of the static contact angle can also be observed in Fig. 9.20. Obviously small static contact angles (test no. 32, 36 and 46) lead to a lower overshoot over the final equilibrium position than high static contact angles. This might be caused by a higher damping behavior for experiments with small static contact angles.

Reorientation Time

As described in Sect. 7.3.1 the reorientation time t_R can be defined, when the envelope curves of the damped center point oscillation enter a lower limit which is set to 2% of the final value for z_c. By scaling and transformation of the center point coordinate with (7.21), the criterion for the reorientation time becomes

$$z_c^*(t_R) < 0.02 \; , \tag{9.5}$$

where the initial deflection is $z_{c0}^*(t=0) = 1$. With the relation for the upper envelope curve (7.16) the criterion becomes

$$0.02 > Ce^{-D\left(1-D^2\right)^{-1/2}\tau_R} \; , \tag{9.6}$$

where $\tau_R = \omega_d t_R = 2\pi$ is the dimensionless reorientation time. Rearranging of (9.6) and substitution of τ_R yields

$$t_R < -\frac{2\pi}{\Lambda\omega_d} \ln \frac{0.02}{C} \; , \tag{9.7}$$

or if Λ and C are substituted by (7.19) and (7.13)

$$t_R < -\frac{\left(1-D^2\right)^{1/2}}{D\omega_d} \ln \left[0.02\left(1-D^2\right)^{1/2}\right] \; . \tag{9.8}$$

Thus, if the damping ratio D and the natural frequency ω_d are known, the reorientation time t_R can be determined with (9.8). Thus the unknown functional relations for $D = f(\mathrm{Oh}, \gamma_s)$ and $\omega_d = f(\mathrm{Oh}, \gamma_s)$ could enable a general correlation for the reorientation time. To compare the experimental results for t_R (determined with (9.8)) the center point histories for randomly selected experiments are plotted in Fig. 9.21 for higher static contact angles and in Fig. 9.22 for small static contact angles. The time is scaled with t_R, thus according to the linear oscillation theory (assumed for the entire oscillation) and the evaluation of D and ω_d, the limit for the reorientation time is reached at $t/t_R = 1$. For experiments with higher static contact angles (Fig. 9.21) a trend to a slight overestimation of t_R can be observed, since the oscillations settle down earlier at $t/t_R \approx 0.7 - 0.8$. This behavior can be explained by the assumption of the linear oscillation theory for the entire reorientation. Since the initial period with the high damping behavior is neglected for the evaluation of Λ (and thus D) the theoretical approach with (9.8) must lead to slightly higher reorientation times.

On the other hand, the approach with the linear oscillation theory shows quite good agreement with the trend given from the center point oscillations for experiments with small static contact angles (Fig. 9.22). The good agreement is evident since for the evaluation of the damping behavior most of the amplitudes where used (compare Sect. 7.3.3). Thus it represents almost the entire oscillation.

Damping Behavior

The damping ratio D of the center point oscillation depends both on the static contact angle γ_s (compare Fig. 9.28) and the OHNESORGE number (compare

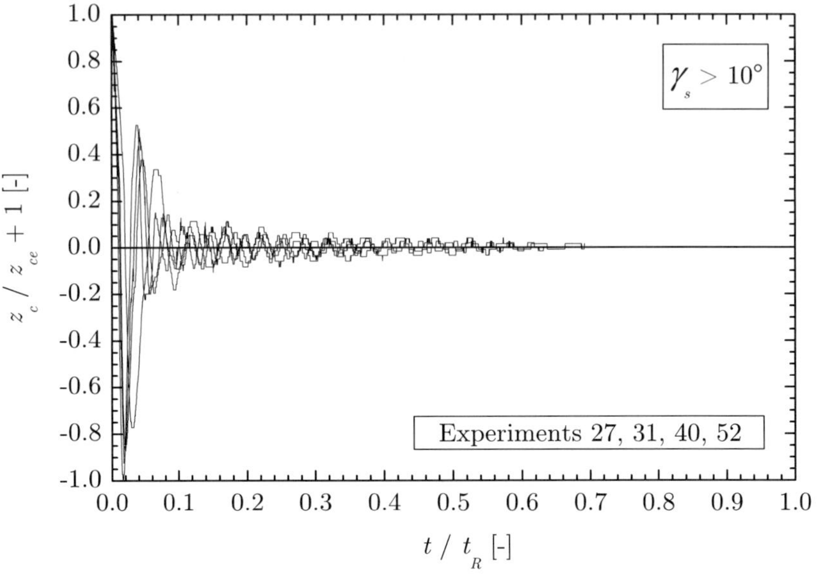

Fig. 9.21. Histories of the scaled center point versus dimensionless time t/t_R for experiments with high static contact angle. The reorientation time is determined by (9.8) with the values for D and ω_d evaluated from the experiments

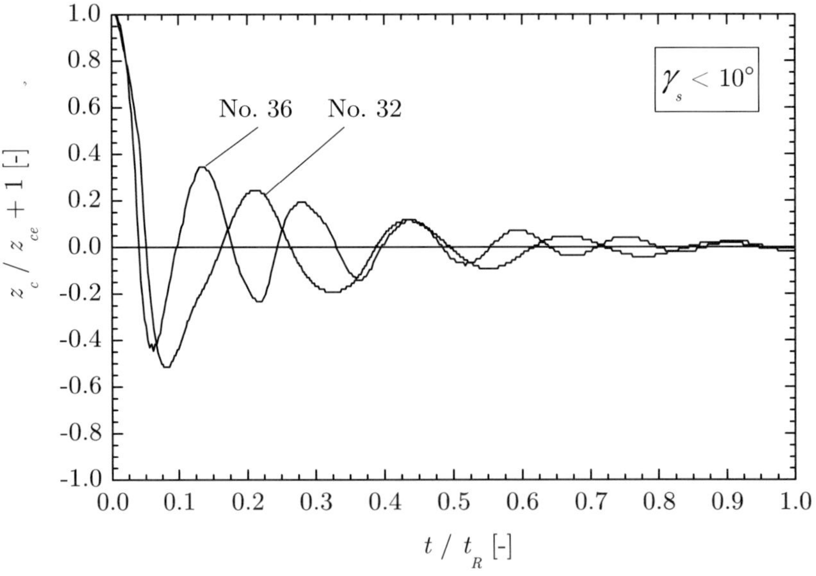

Fig. 9.22. Histories of the scaled center point versus dimensionless time t/t_R for experiments with low static contact angle. The reorientation time is determined by (9.8) with the values for D and ω_d evaluated from the experiments

Fig. 9.27). Plotting the damping ratio D versus the OHNESORGE number shows that for higher static contact angles the trend $D \sim \mathrm{Oh}^{0.5}$ is obvious. Some arguments for the exponent 1/2 are given in Sect. 9.4. Experiments with small static contact angles show significantly higher damping ratios. However, a definite trend for the damping ratio as a function of the OHNESORGE number cannot be observed. A more thorough discussion of the results for the damping behavior in conjunction with the behavior of the contact point and in comparison with numerical data is given in Sect. 9.4.

Frequency

The following section analyses the temporal behavior of the center point frequency as well as the dependence of the frequency on the static contact angle and the OHNESORGE number.

For all experiments with a high static contact angle ($\gamma_s < 10°$) the frequency varies in time as described in Sect. 7.3.3. In the beginning of the reorientation, when the contact point moves along the cylinder wall, a lower frequency ω_{di} can be observed. When the contact point reaches its final position and remains fixed, a nearly constant frequency of oscillation with the mean value ω_d is observable (compare Fig. 7.17). The difference between the first value ω_{d1} and the mean value ω_d varies between 14% and 22%. For experiments with a low static contact angle ($\gamma_s < 10°$) a different behavior for the frequency can be observed. The first frequency ω_{d1} shows a higher value for all experiments, which lies about 14% to 41% above the average value. All experiments with small static contact angle show in general a lower frequency, which leads to a lower number of periods usable for the evaluation of a mean frequency. Thus a lower accuracy is obtained for these experiments.

The dimensionless frequency $\Omega_0 = \omega_0 t_{pu}$ is plotted versus the static contact angle γ_s in Fig. 9.25. A general trend of decreasing frequency Ω with decreasing static contact angle can be observed. A comparison with numerical results and experimental results from [2] is given in Sect. 9.4.

The dimensionless frequency is plotted versus OHNESORGE number in Fig. 9.26. For similar OHNESORGE numbers a difference between experiments with high and low static contact angles can be observed. For higher static contact angles the dimensionless frequency Ω_0 is almost independent of the OHNESORGE number, with a trend of an increasing Ω_0 with rising γ_s as shown in Fig. 9.25. A different behavior can be observed for experiments with low static contact angles, where the data points show a stronger dependence on the OHNESORGE number. Although there are higher inaccuracies for these data points, a trend to an increasing frequency for increasing OHNESORGE number is observable, which was also reported by Weislogel & Ross [2]. A comparison with their data and a closer discussion of the results for the frequency is given in Sect. 9.4.

9.3.2 Numerical Observations for the Center Point

First Peak

The first peak of the center point characterizes the maximal deflection of the center point. Figure 9.23 shows a typical plot of the center point versus time calculated with FIDAP. The z-coordinate is nondimensionalized with R, the time with t_{pu}. Numerical simulations have been carried out with FIDAP and FLOW-3D. In the FIDAP simulations the OHNESORGE number was varied between $1.46 \times 10^{-3} < \mathrm{Oh} < 44.0 \times 10^{-3}$, and the contact angle in the range of $30° < \gamma_s < 80°$. The free contact line condition was used as boundary condition at the contact line. In the FLOW-3D calculations the static contact angle was set to $\gamma_s = 0$, and the OHNESORGE number was varied in the range of $7.256 \times 10^{-4} < \mathrm{Oh} < 2.033 \times 10^{-3}$.

The numerical results show that z_{cp} decreases with increasing OHNESORGE number (Fig. 9.29). The influence on the static contact angle is more distinct (Fig. 9.30). For static contact angles $\gamma_s = 0$ the dimensionless center peak coordinate is in the range of $z_{cp}/R = 0.6 \pm 0.1$. With increasing static contact angle the center peak value decreases. The z_{cp}/R value for a static contact angle equal 80° is less than 0.1. It can be expected that $z_{cp}/R = 0$ for a static contact angle of 90°.

The behavior of the center peak coordinate is coupled with the center peak time t_{cp}. The dependence of t_{cp}/t_{pu} on the OHNESORGE number is not clearly visible (Fig. 9.31). The dependence of the center peak time on the

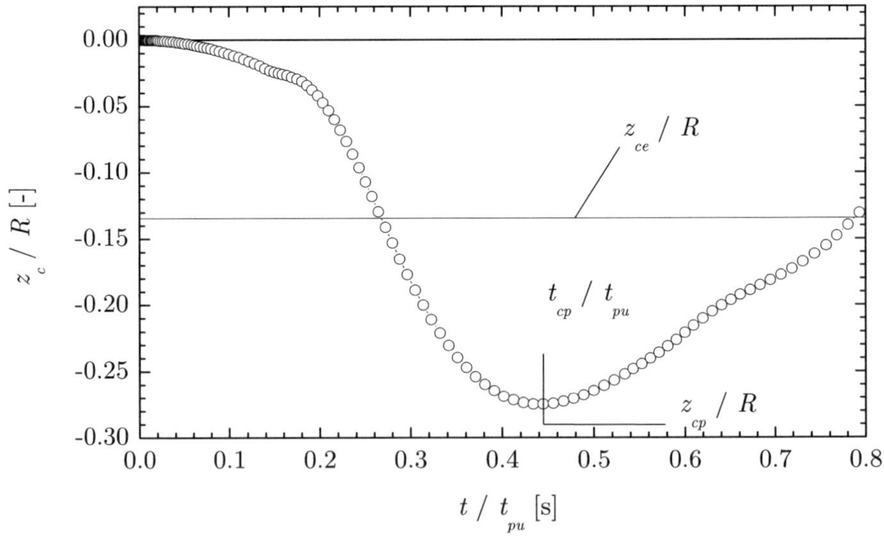

Fig. 9.23. Determination of the maximal center coordinate z_{cp} from FIDAP calculations with high contact angles. $\mathrm{Mo} = 0.017, \gamma_s = 53.5°, u_{puLc} = 11.92$ cm/s, $t_{puLc} = 0.012$ s

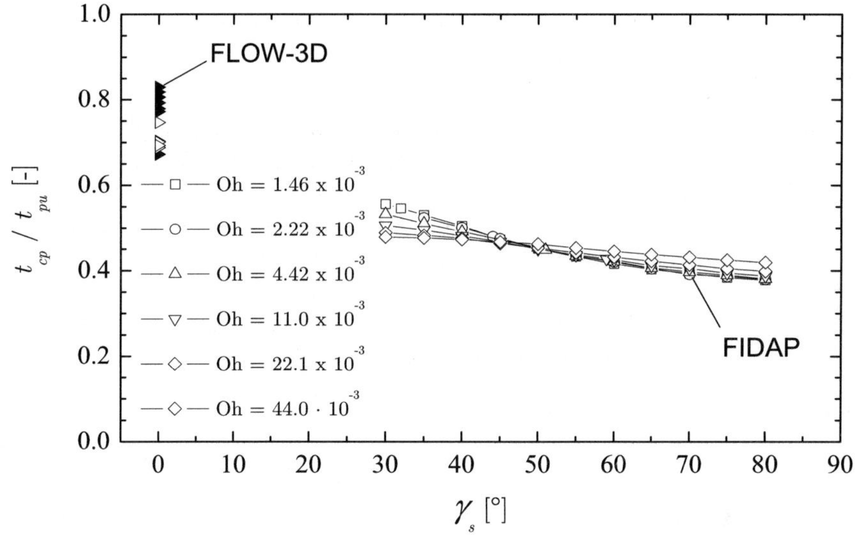

Fig. 9.24. Center peak time versus the static contact angle number for varying OHNESORGE numbers. Results of FLOW-3D and FIDAP simulations. The solid symbols of the FLOW-3D results mark the simulations with cell sizes less than 1 mm

static contact angle is more distinct (Fig. 9.24). For increasing static contact angles the dimensionless center peak time decreases. This effect increases with decreasing OHNESORGE number. For static contact angles $\gamma_s = 0$ the t_{cp}/t_{pu} value is ≈ 0.8. This value decreases to 0.4 for $\gamma_s = 80°$. The FLOW-3D results show an obvious dependence on the resolution of the mesh. Higher resolutions increase z_{cp} and t_{cp}, except for SF 0.65.

Damping

The data for the determination of the damping behavior in the global time period are based on the simulations with FIDAP. The damping behavior depends on the OHNESORGE number, the static contact angle and on the chosen boundary condition for the dynamic contact angle. The considered boundary conditions for the dynamic contact angle are the free and the fixed contact line condition. The free and the fixed contact line condition denote the border cases of the free surface oscillation. With the free contact line condition the contact angle is fixed, equal to the static contact angle, and the contact line is free to move. In the case of the fixed contact line condition, the contact line is fixed and the contact angle is allowed to vary. The OHNESORGE number varies between $1.46 \times 10^{-3} < \text{Oh} < 44 \times 10^{-3}$. The static contact angle γ_s varies in the range of $30° < \gamma_s < 80°$.

The numerical results reveal a significant influence of the boundary condition for the dynamic contact angle imposed on the contact line. The damping for the free contact line condition increases with increasing OHNESORGE number (Fig. 9.27) for a fixed value of static contact angle. The damping values for different static contact angles lie on parallel straight lines. The straight lines are moved to higher damping values for decreasing contact angles. With decreasing static contact angle the damping for a fixed value of the OHNESORGE number decreases (Fig. 9.28).

For the fixed contact line condition the damping depends on the OHNESORGE number only (see Fig. 9.27) and increases with increasing Oh number. The damping is less dependent from the static contact angle (Fig. 9.28). This seems to be obvious since the fixed contact line condition allows any contact angle independent of the initial situation.

Frequency

The data for the determination of the frequency in the global time period were calculated with FIDAP. The static contact angle was varied in the range $30° < \gamma_s < 80°$ and the OHNESORGE number in the range $1.456 \times 10^{-3} <$ Oh $< 44 \times 10^{-3}$.

At first the results with the free contact line condition are considered. The influence of the OHNESORGE number on the frequency is negligible for constant γ_s (Fig. 9.26). The frequency is nearly constant for a fixed static contact angle and a varying OHNESORGE number. The dependence on the static contact angle is weak (Fig. 9.25). The frequency decreases with decreasing static contact angle. The dimensionless natural frequency for a static contact angle of 80° is about 7. This value decreases to 5 for static contact angles equal to 30°. The extrapolation to a static contact angle $\gamma_s = 0$ yields a dimensionless frequency between 3 and 4.

The influence of the fixed contact line is contrary to the influence of the free contact line condition. For the fixed contact line condition the frequency does not depend on the static contact angle (Fig. 9.25). The frequency is constant for a fixed value of OHNESORGE number and a varying static contact angle. But the frequency depends on the OHNESORGE number and increases for increasing OHNESORGE number (Fig 9.26). In the investigated range of OHNESORGE numbers $1.456 \times 10^{-3} <$ Oh $< 44 \times 10^{-3}$ the dimensionless frequency varies between 7 and 10. The extrapolation to a static contact angle $\gamma_s = 0$ yields a dimensionless frequency of 7 for the lowest Oh number considered. In general the frequencies for the free contact line condition are lower than the frequencies for the fixed contact line.

9.4 Summary Center Point Observations

9.4.1 Frequency of Oscillation

The influence of the static contact angle γ_s on the dimensionless frequency $\Omega_0 = \omega_0 t_{pu}$ is presented in Fig. 9.25, where experimental and numerical results

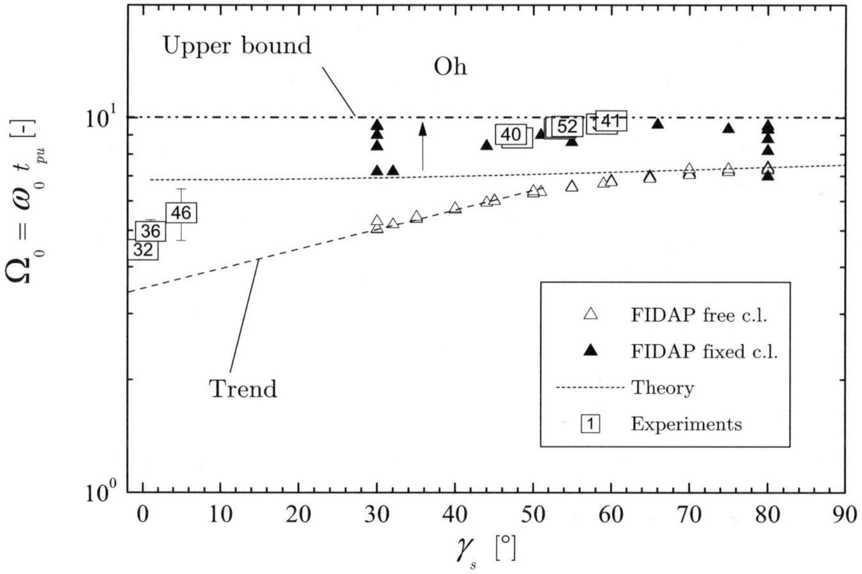

Fig. 9.25. Dimensionless natural frequency versus γ_s and varying OHNESORGE number. Indication of trend for small contact angles (*dashed line*), upper bound (*dash-dot-dotted line*), (9.9) with $m_s = 1$ (*dotted line*)

are compared. Furthermore the correlation

$$\Omega_0 = \tanh\left(3.83\frac{h_0}{R}\right)[56.2 + 3.83\text{Bo} - 2m_s 4.76\cos(\gamma_s)] \ , \qquad (9.9)$$

(referring to [3]) is included in Fig. 9.26. Equation 9.9 is valid for $m_s = 1$ only. Other modes require other constants. For constant OHNESORGE numbers, the frequency Ω_0 decreases if the static contact angle is varied to lower values. This trend is revealed by both the experimental data and the numerical data. For comparison of the experimental data with the numerical results one must distinguish between experiments with $\gamma_s < 10°$ and $\gamma_s > 10°$. For the latter experiments the contact point obeys a fixed contact line behavior, where the agreement with the corresponding FIDAP results (▲ for fixed contact line) is remarkable. Experiments with low static contact angles seem to follow the free contact line condition (△), in particular when the contact point exceeds the equilibrium position during the initial capillary rise which causes a remaining liquid layer at the cylinder wall.

This overshoot depends on the OHNESORGE number. Consequently the free contact line condition, which yields lower frequencies Ω_0 (as indicated in Fig. 9.25), is preferred to be applicable for experiments with rather low OHNESORGE numbers Oh $\lesssim 2 \times 10^{-3}$ (compare Sect. 9.1.1). Extrapolation of the trend given by the numerical calculation based on the free contact line (△) leads to a frequency of $\Omega_0 \approx 3$ for $\gamma_s = 0$ for small OHNESORGE numbers. This

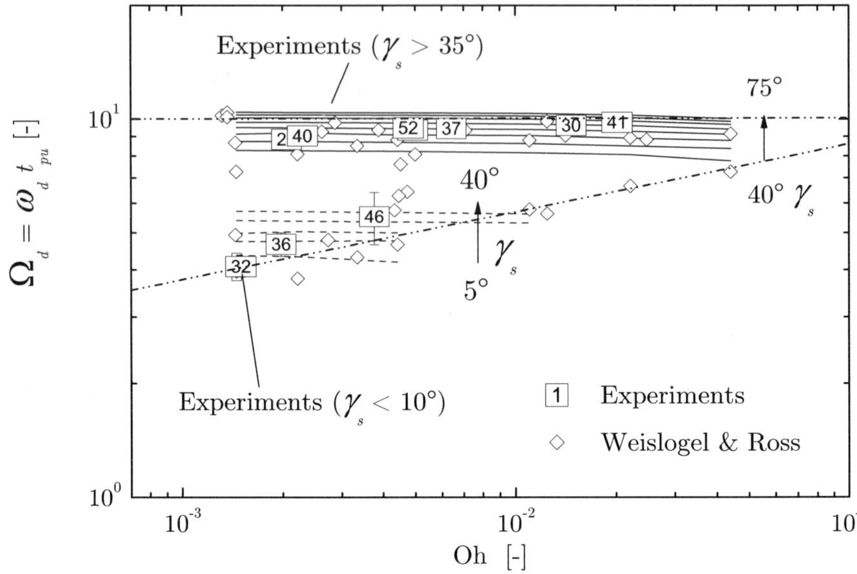

Fig. 9.26. Dimensionless natural frequency versus OHNESORGE number with variation of γ_s. Upper and lower bound (*dash-dot-dotted line*), experimental results, numerical results FIDAP with fixed contact line (*solid line*), FIDAP with free contact line (*dashed line*)

influence of the OHNESORGE number on the frequency is presented in more detail in Fig. 9.26, where $\Omega_d = \omega_d t_{pu}$ for the present experimental data (□) and experimental data from [2] (◊) are compared with numerical calculations.

The experimental data for high static contact angles, which lie beneath the upper bound show only a slight influence on the OHNESORGE number in good agreement with FIDAP (▲ for the fixed contact line). Experiments with a very small static contact angle ($\gamma_s < 10°$) show a distinct dependence on the OHNESORGE number (denoted in Fig. 9.26 with the lower bound). This dependence is caused by the behavior of the contact point governed by the OHNESORGE number. While higher OHNESORGE numbers cause the contact point to rest below the equilibrium position during the initial capillary rise, rather low OHNESORGE numbers lead to the mentioned overshoot and the subsequent formation of a liquid layer at the wall. Thus higher OHNESORGE numbers lead to an increase of the frequency since the configuration is closer to a pinned contact line condition. Lower frequencies can be observed for low OHNESORGE numbers, where the contact line obeys a fixed contact line condition. However for extrapolation to $\mathrm{Oh} = 10^{-4} - 10^{-5}$ one must consider, that calculations with the free contact line condition (△) do not significantly depend on the Ohnesorge number. Thus it can not be expected that the trend given by the lower bound continues to $\mathrm{Oh} = 10^{-5}$.

The correlation (9.9) (with $h_0/R = 2$ and $m_s = 1$) does not depend on viscous effects and implies a free boundary condition. It coincides well with the FIDAP calculations (free contact line condition $\triangle$) for contact angles $\gamma_s > 60°$. For experiments with small static contact angles the effect of a decreasing frequency with decreasing OHNESORGE number can not be described by (9.9). While experimental data give a trend for very low OHNESORGE numbers (Oh $\approx 10^{-3}$) to dimensionless frequencies of $\Omega_0 \approx 4 - 6$, (9.9) leads to a values of $\Omega_0 \approx 7$.

9.4.2 Damping Behavior

Experiments with a high static contact angle show a short nonlinear period with a following linear oscillation after the contact line is fixed (compare time t_{wf} in Sect. 9.1.1). In the initial nonlinear period these experiments exhibit a lower frequency and a higher damping of the system. After pinning of the contact point has occurred the surface oscillation shows a linear behavior with a higher, but almost constant frequency and a lower damping. A comparison of experimental data (□) with numerical results is shown in Fig. 9.27. For high static contact angles the damping ratio D increases for higher OHNESORGE number with a relation on the order of $D \sim \mathrm{Oh}^{0.5}$.

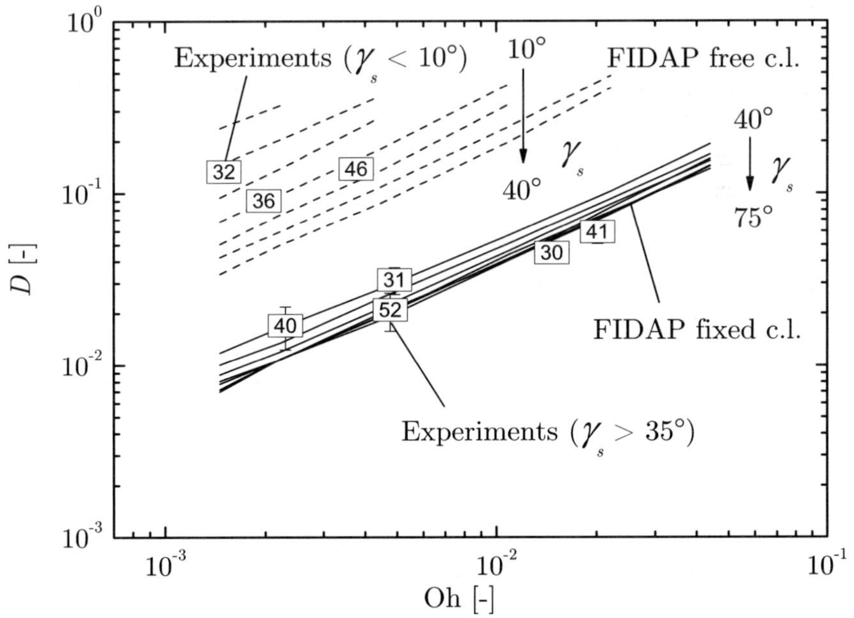

Fig. 9.27. Damping ratio D versus OHNESORGE number and varying static contact angle. Experimental results, numerical results FIDAP with fixed contact line (*solid lines*), FIDAP with free contact line (*dashed line*)

For small static contact angles the damping ratio D shows higher values in comparison to experiments with higher static contact angles. A clear trend for the dependence on the OHNESORGE number can not be identified. However it can be assumed that the behavior of the contact point must determine the damping behavior of the flow. As described above, the initial behavior of the contact point depends on the OHNESORGE number, whether it exceeds the equilibrium position z_{we} or it rests below. If for low OHNESORGE numbers z_{we} is overshoot during the initial capillary rise the contact point may be characterized by a free contact line condition with a certain damping behavior. On the other hand, for higher OHNESORGE numbers the contact point does not exceed the equilibrium position during the initial rise and a fixed contact line condition may be considered. In this case the frequency increases as mentioned above. Viscous forces become dominant compared to inertia forces if the OHNESORGE number is increased and thus a creeping motion could be observed for very high OHNESORGE numbers. The assumption $\delta \sim 2\nu K^2$ for the exponential decay of a high BOND number oscillation and the assumption of the wave number $K = 3.83/R$ (i.e. mode $m_s = 1$) leads to

$$z(t) = z_0 e^{-C_1 \mathrm{Oh} \omega_0 t} . \tag{9.10}$$

The linear oscillation theory (with $\tau = \omega_d t$ and $\omega_d = \omega_0 \left(1 - D^2\right)^{1/2}$, compare Sect. 7.3.3) yields $z(t) = z_0 e^{-D\omega_0 t}$ and thus a linear dependence of the damping ratio D on the OHNESORGE number is expected. Introducing the STOKES boundary layer thickness $\delta_w \sim (\nu/\omega_0)^{1/2}$ [5] of an oscillating plate yields a weighted ansatz $\delta_1 = \delta R/\delta_w$ for the exponential decay. The same assumption for δ and K leads to

$$z(t) = z_0 e^{-C_2 \mathrm{Oh}^{0.5} \omega_0 t} . \tag{9.11}$$

The dependence $D \sim \mathrm{Oh}^{1/2}$ seems to predict the fixed contact line behavior, thus high contact angles, with a good accuracy. The relation $D \sim \mathrm{Oh}$ is better suitable for the free contact line condition given by the computations with FIDAP. The experimental data for low contact angles cannot be used to corroborate either of the above mentioned hypotheses. This justifies the need for more data in the lower OHNESORGE number and low contact angle range.

The damping behavior of the flow with respect to the influence of the static contact angle is depicted in Fig. 9.28. The numerical calculations exhibit a difference between the free contact line (solid line) and the fixed contact line (dashed line) condition, if the static contact angle is varied. While the fixed contact line condition shows a weaker influence of the static contact angle on D, the damping ratio for the free contact line increases with decreasing γ_s. Again, numerical calculations with the fixed contact line condition show a better agreement for experiments with higher static contact angles, while the free contact line condition seems to be appropriate to simulate experiments with low static contact angles. In the case of low contact angles and low

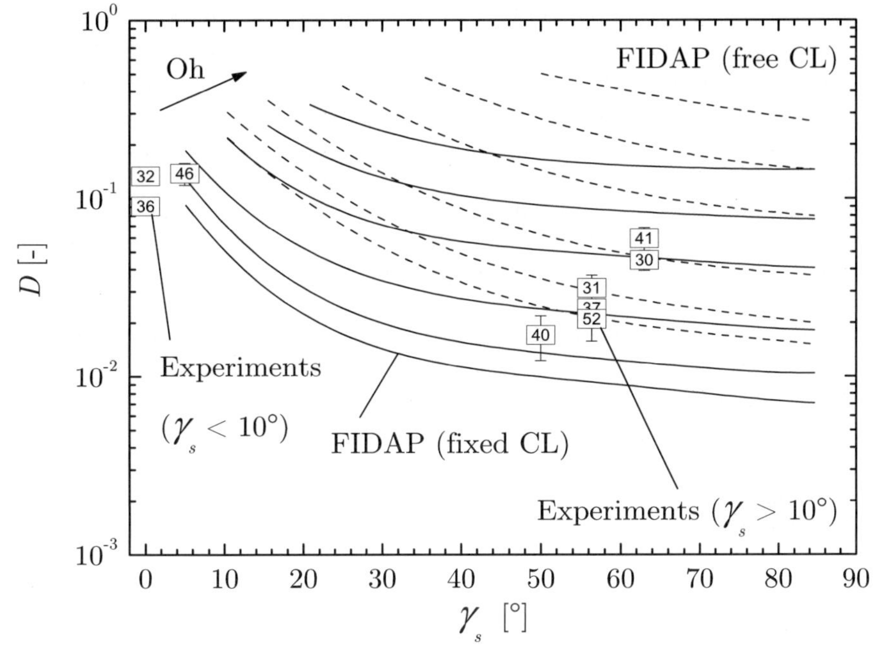

Fig. 9.28. Damping ratio D versus static contact angle and varying OHNESORGE number. Experimental results (*squares*), numerical results FIDAP with fixed contact line (*solid line*), FIDAP with free contact line (*dashed line*)

OHNESORGE numbers a liquid layer appears in front of the bulk at the wall. The reorientation of the bulk is decoupled from the behavior of the contact point on the end of the liquid layer. The apparent contact point, the transition point between liquid layer and bulk, is free to move on the liquid layer. The behavior of the apparent contact point is comparable to the behavior of the real contact point in the numerical simulation with FIDAP and the free contact line condition, where the contact point is also free to move. High OHNESORGE numbers prevent the film formation in front of the bulk. The contact point creeps slowly to the new equilibrium position and the contact line follows a quasi fixed condition. The reorientation behavior changes towards the fixed contact line behavior with increasing OHNESORGE number.

9.4.3 First Amplitude

The dimensionless deflection of the center point at the first peak describes the initial amplitude of the system. Figure 9.29 shows a comparison of experimental data with numerical results from FIDAP (△) and FLOW-3D (▶ for $\gamma_s = 0°$). The data indicate a trend of an increasing deflection z_{cp}/R with decreasing γ_s. The dependence of the OHNESORGE number on the first amplitude is stronger for small γ_s, where higher OHNESORGE numbers lead to a reduction

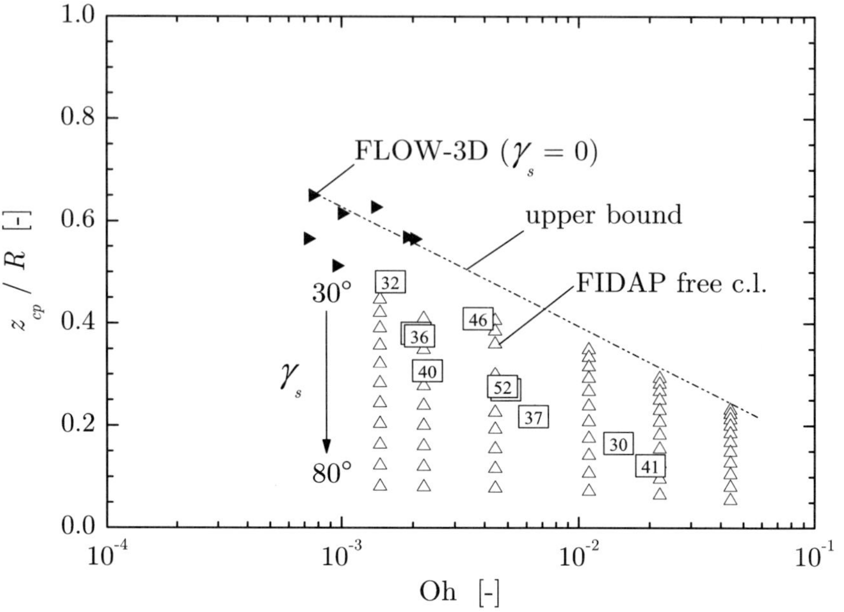

Fig. 9.29. Dimensionless maximum deflection of the center point versus OHNESORGE number and varying γ_s. Experimental results (*squares*), numerical results FIDAP with free contact line (*white triangles*), FLOW-3D (*black triangles*)

of the first maximum deflection of the center point. The numerical calculations with FIDAP are in good agreement with the experimental results, but are slightly high due to the free contact line condition. However, calculations with FLOW-3D show too high values in comparison with the experimental data. Extrapolation of numerical data from FIDAP (calculated with the free contact line condition) to static contact angles of $\gamma_s = 0°$ would likely lead to values obtained in the experiments, as shown in Fig. 9.30.

The time of the first peak t_{cp}, scaled with t_{pu} is plotted versus OHNESORGE number in Fig. 9.31. A clear influence on t_{cp}/t_{pu} is given by the static contact angle which leads to an increasing time with decreasing γ_s. This characteristic should be mainly caused by a higher equilibrium position z_{ce} with increasing γ_s. The variation of the OHNESORGE number shows no obvious and clear dependence on the data points. However it may be expected, that an increasing OHNESORGE number leads to longer times caused by viscous effects. Computational results with FIDAP ($\triangle$ for free contact line) lie slightly above, while FLOW-3D (▶) calculations lie clearly above the experimental data points. The discrepancy of the results with FIDAP ($\triangle$ for free contact line) is caused by the free contact line condition, which does not consider the variation of the dynamic contact angle. The free contact line condition causes a higher wall peak associated with a higher center peak. The overestimation of the results with FLOW-3D (▶) is also caused by the behavior of the contact line.

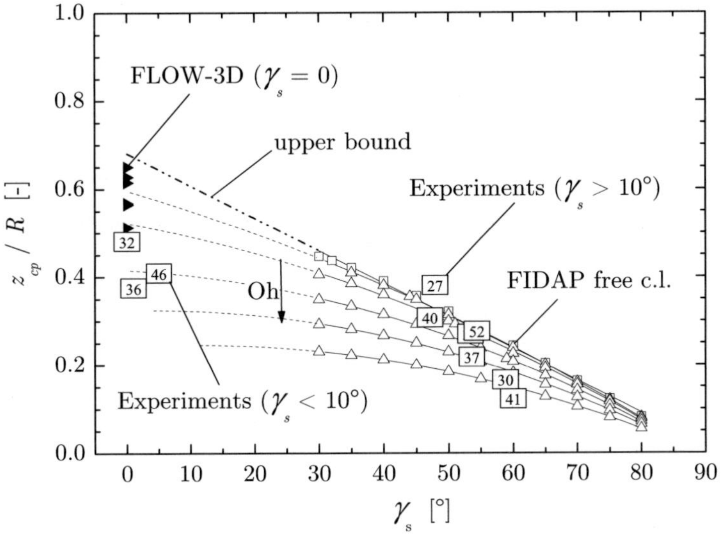

Fig. 9.30. Dimensionless maximum deflection of the center point versus γ_s and varying OHNESORGE number. Upper bound (*dash-dot-dotted line*), connected symbols for visualization only, trend from numerical calculations (*dashed line*), numerical calculations FIDAP with free contact line (*white triangles*), FLOW-3D calculations (*black triangle*)

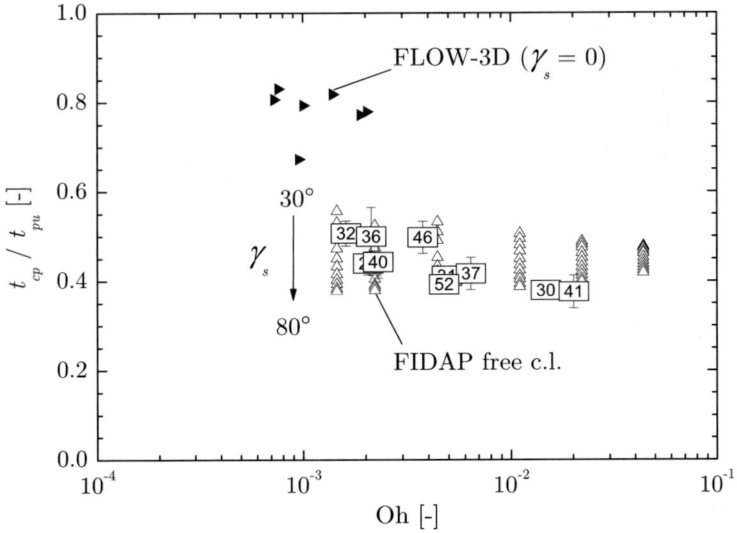

Fig. 9.31. Dimensionless time at the first maximum of the center point versus OHNESORGE number with variation of γ_s. Experimental results (*squares*), numerical results FIDAP with free contact line (*white triangles*), FLOW-3D calculations with $\gamma_s = 0$ (*black triangels*)

9.5 General Summary

The main conclusions from the contact point and the center point observations are repeated here for convenience. The main findings for the **contact point** are:

- Maximum rise velocity $0.5 > u^*_{wmax} > 0.09$ for range $10^{-3} < \mathrm{Mo} < 10^{-1}$.
- Mean rise velocity $0.2 > u^*_{wr} > 0.04$ for the same range of MORTON numbers.
- Both velocities decrease with increasing static contact angle.
- Liquids with low contact angles $\gamma_s < 10°$ show an overshoot over the equilibrium position z_{we} for $\mathrm{Oh} < 2 \times 10^{-3}$. A layer formation could be observed.
- Liquids with low contact angles and higher OHNESORGE numbers show a creeping behavior of the contact point towards z_{we}.
- Liquids with high static contact angles $48° < \gamma_s < 60°$ always show an overshoot in the investigated range of OHNESORGE numbers $1.6 \times 10^{-3} < \mathrm{Oh} < 2 \times 10^{-2}$.

The corresponding items for the **center point** are:

- The time the disturbance needs to reach the center is
$$t_{RLc} \approx R \left(\frac{\rho}{k_{ki}\sigma} \right)^{1/4} . \tag{9.12}$$
After that time the whole system is affected by the surface oscillation.
- The frequency of the oscillation is $9 < \Omega_d < 10$ for liquids with high static contact angle. These frequencies could be computed using a fixed contact line condition, they are mainly independent of γ_s and Oh.
- The damping ratio is $0.01 < D < 0.05$ with a $D \sim \mathrm{Oh}^{1/2}$ dependence.
- The frequency for low contact angle liquids is $4 < \Omega_d < 6$ with a damping ratio of $D \approx 0.1$. A free contact line condition is observed and the frequency depends on the OHNESORGE number.
- No clear trend could be identified for the damping behavior in dependence of the OHNESORGE number but the damping behavior differs with the existence of a liquid layer on the wall.

References

1. C.E. Siegert, D.A. Petrash, E.W. Otto: NASA TN D-2458 (1964)
2. M.M. Weislogel, H.D. Ross: Microgravity Sci. Technol. **3**, 24 (1990)
3. W.C. Reynolds, H.M. Satterlee: Liquid Propellant Behavior at Low and Zero Gravity. In: *The Dynamics Behavior of Liquids in Moving Containers*, NASA SP-106, ed by H.N. Abramson (NASA, Washington 1966) pp 417
4. M. Michaelis: *Kapillarinduzierte Schwingungen freier Flüssigkeitsoberflächen*, Nr. 454 (VDI Verlag, Düsseldorf 2003)
5. H. Schlichting, K. Gersten: *Grenzschichttheorie*, (Springer, Berlin 1997)

Part III

Capillary Rise in Cylindrical Tubes

In this part the capillary-driven flow of a perfectly wetting liquid ($\gamma_s = 0$) into circular cylindrical tubes is studied. On earth, capillary forces are masked by the hydrostatic pressure and flows can be observed only if the pore size is sufficiently smaller than the capillary length. In the compensated gravity environment of a coasting spacecraft, capillary forces become dominant and may lead to the displacement of large amounts of liquids. The capillary length is enlarged and tube diameters in the range of centimeters to decimeters cause capillary flows. Inertia and convective forces come into play and the simple viscous dominated approach of capillary flows in porous media must be abandoned.

Based on an analysis of previous approaches, a comprehensive theoretical model is derived which is not limited to certain special cases. This model considers the meniscus reorientation, the dynamic contact angle as well as viscous, inertia, and convective losses inside the tube and the reservoir. The dimensionless formulation of the second order nonlinear differential equation for the meniscus rise shows the dominating influence of the OHNESORGE number and the initial liquid height. The differential equation is corroborated by numerous experiments in the Bremen drop tower, which cover a wide range of the governing parameters. Writing the equation in terms of forces enables the discussion of the major contribution in the different time regimes.

The capillary-driven flow is divided into three successive phases. At the beginning the driving capillary force is counteracted by the inertia force and the meniscus height is proportional to the square of time. The second phase is controlled by convective accelerations and the meniscus height increases linearly with time. Finally the flow enters the third region which is dominated by the viscous force and the meniscus height is proportional to the square root of time: the so-called Lucas-Washburn behavior. The three phases are separated by two characteristic transition times. The extension of the three regions is determined by the OHNESORGE number and the initial liquid height.

Experiments were carried out under microgravity conditions in a wide range of OHNESORGE numbers and initial liquid heights to verify the existence of all three flow regimes predicted by the theory. Tube diameter range from 4 to 70 mm and three low viscosity test liquids have been used. Good agreement of theoretical and experimental data is found throughout the complete range of experiment parameters. Surface reorientations and oscillations could be observed for tube diameters larger than 36 mm. This is basically the same phenomenon as reported in part II of this book but overlayed by the capillary rise.

Approximate solution are derived with some simplifications of the governing equation, which yield simple and easy to compute relations for the meniscus height and velocity in the different time regimes.

10

Introduction

Under terrestrial conditions capillary-driven flows of liquids are counteracted by the hydrostatic pressure and thus restricted to small geometries. Familiar applications are the penetration of porous media by liquids in paper printing, agronomy, textiles, and oil recovery. In space, gravitational effects are negligibly small and surface tension effects dominate the liquid behavior. This fact is exploited in surface tension tanks, where propellants are pumped and positioned through different geometries with lengths of some centimeters to meters exclusively by capillary forces [1]. Since porous media models often consist of a single cylindrical capillary [2, 3, 4], a single non-circular tube [5, 6] or an assembly of parallel capillaries [7, 8, 9, 10], knowledge of the dynamics of capillary-driven flow in a single circular-cylindrical tube is still of importance. A general theory which is not limited to certain special cases is still missing.

Under terrestrial conditions the capillary rise into a vertical tube is determined by the tube radius $R(= d/2)$, density ρ, dynamic viscosity μ, surface tension σ of the liquid, the static contact angle of the liquid with the solid γ_s, the gravitational acceleration g_0, and the time t. From dimensional considerations (compare Chap. 2) two characteristic time scales for this problem are derived. The time

$$t_{pu} \sim \sqrt{\frac{\rho R^3}{\sigma}} \tag{10.1}$$

is known as a characteristic response time for surface oscillations [11]. It will be shown later that for the capillary rise an inertia time comparable to t_{pu} can be formulated which marks the time until inertia effects cease. The second time

$$t_{vu} \sim \frac{R^2}{\nu} \tag{10.2}$$

is the well-known viscous time scale.

Furthermore two dimensionless groups are found. The OHNESORGE number can be derived from the ratio of the two time scales $\mathrm{Oh} \sim t_{pu}/t_{vu}$. However, the OHNESORGE number is preferably formed with the tube diameter

Michael E. Dreyer: *Free Surface Flows under Compensated Gravity Conditions*
STMP **221**, 135–137 (2007)
DOI 10.1007/3-540-44628-1_10

$$\mathrm{Oh} = \sqrt{\frac{\nu^2 \rho}{\sigma d}} . \tag{10.3}$$

The OHNESORGE number is known as a damping measure for capillary dominated oscillations [12]. It can also be written as $\mathrm{Oh} = \sqrt{\mathrm{Ca}/\mathrm{Re}_d}$, with the capillary number $\mathrm{Ca} = \mu U/\sigma$ and the Reynolds number based on the tube diameter $\mathrm{Re}_d = dU/\nu$ (with the kinematic viscosity ν). But the capillary number and the Reynolds number do not completely classify the flow since the mean velocity changes during the unsteady capillary rise process.

The BOND number

$$\mathrm{Bo} = \frac{\rho k R^2}{\sigma} = \frac{R^2}{L_c^2} \tag{10.4}$$

relates body forces to capillary forces and represents the influence of gravity. It can also be formed by the ratio of R^2 and the square of the capillary length $L_c = \sqrt{\sigma/\rho k}$.

The static contact angle γ_s is a dimensionless group of its own. It decides whether a capillary ascent or decent occurs. For $0 < \gamma_s < 90°$, the liquid is wetting the tube wall and a capillary rise takes place (ascent). For $90° < \gamma_s < 180°$, the liquid does not wet the wall and the liquid is pushed out of the capillary (decent). For the case $\gamma_s = 90°$, a flat surface in the tube is formed, and the capillary force is zero. In this chapter only perfectly wetting liquids with $\gamma_s = 0$ are used. This leads to the largest capillary force.

A further criterion for a significant liquid rise height is a small BOND ($\mathrm{Bo} \ll 1$) number, which can be achieved either if the tube radius is small compared to the capillary length ($R \ll L_c$) or if the residual acceleration k is low or compensated. The use of narrow tubes leads to the drawback of small time scales t_{pu} (typically in the order 10^{-3}s) and t_{vu} (order 0.1 s), thus changes in the behavior of the liquid movement are difficult to observe. Under microgravity conditions, wide tubes can be used to increase the time scales t_{pu} and t_{vu}, which makes the changes in the flow dynamics observable. However, if short time microgravity facilities are used, the experiment duration may be smaller than t_{vu} and the final viscous dominated stage of the flow cannot be reached within the experiment time.

Numerous theoretical and experimental studies concerning the capillary rise phenomenon are found in literature and different liquid height-time dependencies are described therein. Either the height h of the meniscus (i.e. the free liquid surface inside the tube) is proportional to the square of time $h \sim t^2$, or h increases linear with time $h \sim t$, or a $h \sim \sqrt{t}$ behavior is found. This holds for the capillary rise from an infinite reservoir where the wetted length increases with time. Steady capillary driven flows have been observed by Weislogel [13] in partially coated tubes. A liquid slug of constant length is confined in a tube and exposed to different wetting conditions on either sides. As long as one side of the slug experiences a different wetting condition, the liquid slug moves with a constant velocity. A similar behavior is expected if one meniscus would be exposed to a different temperature causing a different surface tension there. Constant velocities in capillary driven flows were also reported by Bico & Quéré [14] using bi-slugs consisting of different immiscible liquids in untreated horizontal tubes. The different capillary pressures on both sides are caused here by the different liquid properties. From that one may conclude that a confined amount of liquid will be always set in motion if the capillary pressures on both sides are different, no matter how that difference has been created.

In this chapter we present an improved mathematical model for the capillary rise with all relevant effects including the dynamic contact angle and the meniscus reorientation at the start of the flow process as well as inertial forces, viscous forces, and convective losses inside the tube and the reservoir. The equation is scaled with the driving capillary force and the dimensionless equation is found to depend on the OHNESORGE number Oh and the initial liquid height Λ_1. The model predicts that all three $h(t)$ dependencies described in literature are established consecutively if the process is observed long enough. The three regions are subdivided by two time scales comparable to t_1 and t_2, which will be determined in detail for the tube geometry. To verify the theory and to prove the existence of the three domains, experiments were performed under microgravity condition ($\mathrm{Bo} \rightarrow 0$) in the 4.7 s drop tower Bremen covering a wide range of the characteristic parameters. Tubes with internal diameters from 4 to 70 mm were tested with three liquids. The OHNESORGE number range of the experiments is $5.8 \times 10^{-4} < \mathrm{Oh} < 4.7 \times 10^{-3}$.

References

1. M.E. Dreyer, J. Gerstmann, M. Stange et al.: Space Forum **3**, 87 (1998)
2. S. Levine, P. Reed, G. Shutts et al: Powder Technology **17**, 163 (1977)
3. J.O. Carnali, C.A. Kotkin: J. Colloid Interface Sci. **159**, 319 (1993)
4. A. Marmur, R.D. Cohen: J. Colloid Interface Sci. **189**, 299 (1997)
5. G. Mason, N.R. Morrow: J. Colloid Interface Sci. **141**, 262 (1991)
6. R.M. Turian, F.D. Kessler: AIChE Journal **46**, 695 (2000)
7. C.S. Kao, J.R. Hunt: Water Resour. Res. **31.1**, 55 (1996)
8. H.M. Princen: J. Colloid Interface Sci. **30**, 69 (1969)
9. H.M. Princen: J. Colloid Interface Sci. **30**, 359 (1969)
10. H.M. Princen: J. Colloid Interface Sci. **31.2**, 171 (1970)
11. M.M. Weislogel, H.D. Ross: Microgravity Sci. Technol. **3**, 24 (1990)
12. G. Wölk, M.E. Dreyer, H.J. Rath et al: J. Spacecr. Rockets **34.1**, 110 (1997)
13. M.M. Weislogel: AIChE Journal **43.3**, 645 (1997)
14. J. Bico, D. Quéré: Europhys. Lett. **51.5**, 546 (2000)

Standard International **Standard International** **Standard International**

Books From California
51 W. Easy St.
Simi Valley, CA 93065
UNITED STATES
amazonuksales@booksfromca.com

Books From California
51 W. Easy St.
Simi Valley, CA 93065
UNITED STATES

To: F.Javier Garrido-López
Cavanilles 60 -3A
San Cipriano 53 -8B
Madrid, Madrid 28007
SPAIN

Marketplace:	Amazon UK
Order Number:	1650901
Ship Method:	Standard International
Customer Name:	F.Javier Garrido-López
Order Date:	28/09/2010
Marketplace Order #:	026-8364055-5765905
Email:	5z04rnkx1y52nyc@marketplace.amazon.co.uk

Items:

11

Experiments on Capillary Rise

A cylindrical acrylic glass (PMMA, plexiglass) container with internal diameter 230 mm and total height 420 mm was used for the experiments. The container was filled with liquid up to a level of 100 mm. The tubes, open on bottom and top, were immersed into the liquid from above. Internal diameters of $d = 4, 9, 11, 16, 21, 36, 54$, and 70 mm were used. The immersion depth of the tubes (= initial liquid height h_0) was varied from $h_0 = 10$ to 50 mm and the dimensionless initial liquid height $\Lambda_1 = h_0/R$ covered a range from 0.3 to 25. Measures were attached at the outer sides of the tubes to evaluate the meniscus height. Wetting barriers were used to prevent the liquid from climbing up at the outer side of the tubes and the inner container wall (Figs. 11.1 and 12.1). Thus an undisturbed observation of the meniscus movement inside the tube was possible. The experiments were observed using a CCD camera and a Hi8 video system.

Three liquids were used: two Dow Corning 200 Silicone Fluids that are labeled here by their kinematic viscosities (SF 0.65 with $\nu = 0.65$ and SF 1.00 with $\nu = 1.00$ cSt) and 3M Fluorinert Dielektrica FC-77 (Table 3.1). The surface tension of the liquids was measured with a Du Noüy ring tensiometer (model K10T, Krüss GmbH, Hamburg, Germany). The values of the density and the viscosity were taken from the data sheets of the manufacturer of the liquids. A small amount of dye was added to the liquids with no measurable effect on the surface tension. The temperature of the liquids was measured during the experiment.

All three liquids perfectly wet the Plexiglass tubes (static contact angle $\cos\gamma_s = 0$), thus the tube surface can be assumed to be prewetted by a thin liquid film (see Kistler [1]). Additionally, the tubes were cleaned and rinsed with the respective test liquid before each experiment. This avoided any influence of the tube surface condition on the meniscus movement reported by other authors [2, 3].

The experiments were performed in the drop tower Bremen which has a free fall height of 110 m providing a microgravity time of 4.7 s. After the release of the drop capsule the weightlessness condition is reached within a few

Michael E. Dreyer: *Free Surface Flows under Compensated Gravity Conditions*
STMP **221**, 139–143 (2007)
DOI 10.1007/3-540-44628-1_11

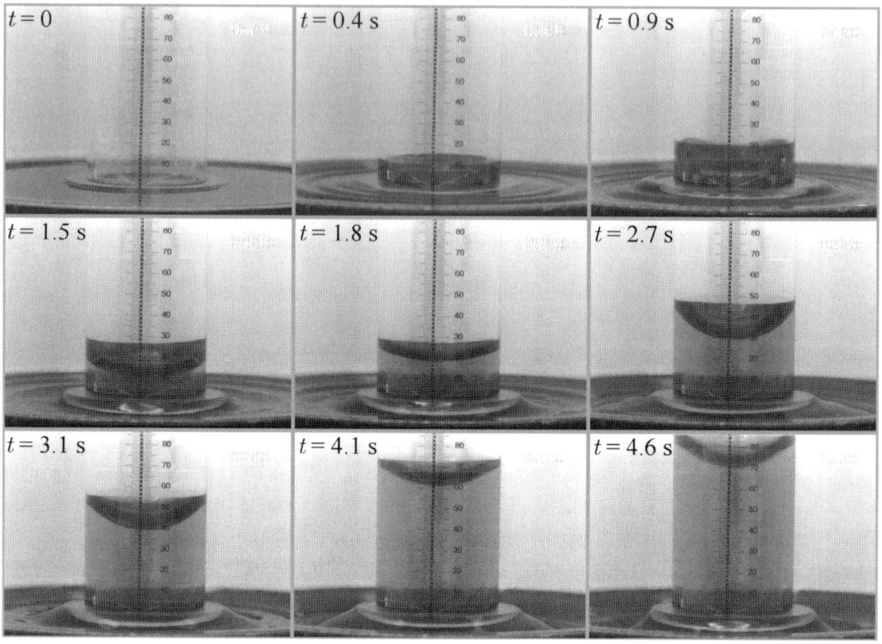

Fig. 11.1. Capillary driven flow of FC-77 liquid into a tube with $d = 54$ mm internal diameter and $h_0 = 10$ mm initial liquid height (experiment #38) under microgravity condition in the drop tower Bremen. The first picture ($t = 0$) shows the $1g_0$ free surface configuration before the drop of the capsule. With the release of the drop capsule the surface reorientation begins and the liquid starts to flow into the tube. Due to the initial reorientation the meniscus performs a damped oscillation during the liquid rise. The increasing curvature of the free surface in the reservoir is visible

milliseconds. During this time the mechanical oscillations (with high frequency and small amplitude) of the capsule structure decay. No disturbance of the liquid surface due to the release mechanism was noticeable. The liquid surface inside the tube changes from the $1g_0$ shape (flat at the tube center and curved in the vicinity of the liquid/solid contact line, see Fig. 11.1) to the spherical microgravity shape and at the same time the liquid starts to flow into the tube. In ground experiments, small capillaries have to be used to reach small Bond numbers and the liquid flow begins immediately as soon as the tube gets in contact with the reservoir liquid. The use of a drop tower provides the unique possibility to adjust the initial liquid height h_0 before the experiment starts since the capillary-driven liquid flow does not start until the capsule is released into free fall. Under microgravity it is also possible to use wide tubes and by this to extend the range of the characteristic experiment parameters.

Figure 11.1 shows a sequence of still frames taken from an experiment video. The internal diameter of the tube was $d = 54$ mm, the initial liquid height was $h_0 = 10$ mm and the liquid was FC-77 (experiment #38). The

Table 11.1. Experiment parameters for capillary rise in tubes. $L_e = 0.06d\,\mathrm{Re}_{d\,max}$, $\mathrm{Re}_{d\,max} = dv_{max}/\nu$, $v_c = \sqrt{2\sigma/(\rho R)}$, $t_c = R^2/(8\nu)$, $t_i = (h_0 + R)/v_c$ (13.13)

No	Liquid	d (mm)	h_0 (mm)	v_{max} (mm/s)	L_e (mm)	t_c (s)	v_c (mm/s)	t_i (s)
#45	SF 0.65	4	10	103	143	0.7	144	0.1
#37	SF 0.65	4	50	80	111	0.7	144	0.4
#51	SF 1.00	4	10	100	87	0.5	15	0.1
#49	SF 1.00	4	50	67	58	0.5	15	0.4
#43	FC-77	4	10	60	67	0.6	92	0.1
#40	FC-77	4	50	35	39	0.6	92	0.6
#25	SF 0.65	9	10	86	604	3.7	96	0.2
#21	SF 0.65	9	50	73	513	3.7	96	0.6
#67	SF 1.00	9	50	61	270	2.3	96	0.6
#26	FC-77	11	10	45	378	4.4	55	0.3
#22	FC-77	11	50	37	310	4.4	55	1.0
#39	SF 0.65	16	10	67	1487	11.6	72	0.3
#31	SF 0.65	16	50	62	1376	11.6	72	0.8
#53	SF 1.00	16	10	65	908	7.3	72	0.3
#57	SF 1.00	16	50	57	796	7.3	72	0.8
#41	FC-77	16	10	39	692	9.3	46	0.4
#30	FC-77	16	50	33	586	9.3	46	1.3
#47	SF 0.65	36	10	46	5170	58.5	48	0.6
#61	SF 0.65	36	50	42	4720	58.5	48	1.4
#65	SF 1.00	36	10	45	3182	36.8	48	0.6
#60	SF 1.00	36	50	39	2758	36.8	48	1.4
#34	FC-77	36	10	28	2516	46.8	31	0.9
#32	FC-77	36	50	22	1977	46.8	31	2.2
#59	SF 0.65	54	10	34	8597	131.7	39	1.0
#36	SF 0.65	54	36	31	7839	131.7	39	2.0
#62	SF 1.00	54	62	34	5410	82.9	39	0.9
#63	SF 1.00	54	63	30	4774	82.9	39	2.0
#38	FC-77	54	10	20	4044	105.3	25	1.5
#50	FC-77	54	50	14	2831	105.3	25	3.1
#44	SF 0.65	70	10	26	11048	221.3	34	1.3
#42	SF 0.65	70	50	23	9773	221.3	34	2.5
#18	SF 1.00	70	32	21	5615	139.3	35	1.9
#46	FC-77	70	10	13	4417	177.0	22	2.1
#54	FC-77	70	50	9	3058	177.0	22	3.9

Table 11.2. Experiment parameters for capillary rise in tubes. $\mathrm{Oh} = \sqrt{\rho\nu^2/(\sigma d)}$, $\mathrm{Ca}_{max} = \mu v_{max}/\sigma$, $t_{i*} = 8(\Lambda_1 + 1)\mathrm{Oh}$ (13.11)

No	Λ_1	Oh 10^{-3}	Ca_{max} 10^{-3}	$\mathrm{Re}_{d\ max}$	t_{i*}
#45	5.0	2.41	3.46	595	0.112
#37	25.0	2.41	2.68	462	0.480
#51	5.0	3.81	5.27	364	0.174
#49	25.0	3.81	3.53	244	0.745
#43	5.0	4.73	6.19	277	0.228
#40	25.0	4.73	3.61	162	0.914
#25	2.2	1.61	2.89	1119	0.040
#21	11.1	1.61	2.45	950	0.148
#67	11.1	2.54	3.21	499	0.243
#26	1.8	2.85	4.64	572	0.063
#22	9.1	2.85	3.82	470	0.219
#39	1.3	1.20	2.25	1549	0.021
#31	6.3	1.20	2.08	1434	0.068
#53	1.3	2.69	3.42	946	0.032
#57	6.3	2.69	3.00	829	0.105
#41	1.3	2.36	4.03	721	0.042
#30	6.3	2.36	3.41	610	0.135
#47	0.6	0.80	1.54	2393	0.009
#61	2.8	0.80	1.41	2185	0.024
#65	0.6	1.27	2.37	1473	0.015
#60	2.8	1.27	20.5	1277	0.037
#34	0.6	1.58	2.89	1165	0.018
#32	2.8	1.58	2.27	915	0.046
#59	0.4	0.66	1.14	2654	0.007
#36	1.9	0.66	1.04	2419	0.014
#62	0.4	1.04	1.79	1670	0.011
#63	1.9	1.04	1.58	1473	0.023
#38	0.4	1.29	2.06	1248	0.013
#50	1.9	1.29	1.45	874	0.028
#44	0.3	0.58	0.87	2630	0.006
#42	1.4	0.58	0.77	2327	0.011
#18	0.9	0.91	1.11	1337	0.014
#46	0.3	1.13	1.34	1052	0.011
#54	1.4	1.13	0.93	728	0.021

first picture ($t = 0$) shows the $1g_0$ fluid configuration before the drop of the capsule. With the release of the drop capsule the reorientation of the liquid surface begins and the liquid starts to flow into the tube. The meniscus shape changes during the liquid movement due to an oscillation, which is induced by the initial reorientation of the surface and is damped during the flow. For tube diameters $d < 36$ mm no oscillation of the meniscus was observed. In all experiments a smooth meniscus movement without disturbances like slip-stick motion was observed. The progression of the meniscus height h with experiment time t was evaluated from the video tapes. To detect the start of the experiment, the videos were analyzed frame by frame. The first frame showing a slight change of the liquid surface was defined as the start of the experiment. Consequently the foregoing frame marks the time $t = 0$ which is the origin of Figs. 13.1–13.3. The lowest point of the meniscus (at the tube center) was read off the measure in time steps of 0.1 s and was taken as the liquid height h. For the wider tubes with $d > 36$ mm the meniscus position was not observable at the beginning of the liquid rise due to the initial surface oscillation (see Fig. 11.1). This explains the lack of data points in Fig. 13.3.

Errors of the meniscus height measurement due to refraction at the tube walls were corrected. The accuracy of the meniscus height measurement is ± 0.5 mm to ± 1.5 mm depending on the area of observation. The time measurement accuracy is ± 0.02 s due to the 50 Hz frame frequency of the video system. Since both measurement errors are only on the order of 1 % no error bars are printed in Figs. 13.1–13.3. It has been verified with several tests that the experimental measurements of the meniscus height h vs. time t are reproducible within the accuracy of the meniscus height measurement.

The parameters of all experiments presented here are listed in Table 11.1 and 11.2 together with the experiment numbers (first column), which are also given in Figs. 13.1–13.7 to identify the single experiments. The measured maximum meniscus velocities ranged from 9 mm/s $< v_c <$ 103 mm/s and maximum capillary numbers $7.7 \times 10^{-4} < \mathrm{Ca} < 6.2 \times 10^{-3}$ (with $\mathrm{Ca}_{max} = \mu v_{max}/\sigma$) were achieved. The maximum Reynolds numbers $\mathrm{Re}_{d\,max} = d v_{max}/\nu$ ranged $162 < \mathrm{Re}_{d\,max} < 2654$, indicating possible turbulent flows for the $d = 54$ mm and $d = 70$ mm tubes with SF 0.65. The entrance lengths [4] calculated with $L_e = 0.06\ d\ \mathrm{Re}_{d\,max}$ ranged 39 mm $< L_e < 11 \times 10^3$ mm. Thus the meniscus height $h + h_0$ exceeded the entrance length L_e only in the case of $d = 4$ mm tube diameter.

References

1. S.F. Kistler: Hydrodynamics of Wetting. In: *Wettability*, ed by J.C. Berg (Marcel Dekker, New York 1993) pp 311-429
2. N. Ichikawa, M. Misawa, K. Kawasaki et al: AIChe Journal **92**, 186 (1996)
3. T.E. Mumley, C.J. Radke, M.C. Williams: J. Colloid Interface Sci. **109.2**, 398 (1986)
4. F.M. White: *Viscous Fluid Flow*, (McGraw-Hill, New York 1974)

12

Mathematical Model for the Capillary Rise

To derive the equation for the meniscus acceleration we start with an integral relation for the linear momentum of an arbitrary moving and deformable control volume within a non-inertial reference frame [1]. Cylindrical coordinates r, φ, z are introduced with the origin on the tube axis at the liquid filling level under normal gravity (Fig. 12.1). The tube's immersion depth into the liquid is labeled h_0, the position of the center point of the meniscus is $h(t)$.

The liquid volume V_m located between the height $h(t)$ and the meniscus (see Fig. 12.1) is considered as a cylinder with identical volume $V_m = \pi R^2 l_m$, with the cylinder height l_m. The volume V_m can be calculated exactly in consideration of the dynamic contact angle γ_d. The maximum volume with $l_m = R/3$ is found for a zero degree contact angle, the minimum volume with $l_m = 0.2R$ is reached for the maximum dynamic contact angle of 44°. Since V_m has only a minor influence on the liquid rise, we neglect the dependence of l_m from the dynamic contact angle γ_d and define a constant volume with $l_m = 0.3R$. Thus the total length of the liquid column inside the tube at time t is represented by $h(t) + h_0 + 0.3R$, in the following denoted by $h_t = h(t) + h_0 + 0.3R$.

The meniscus velocity is labeled $\dot{h}(t)$ and the meniscus acceleration is $\ddot{h}(t)$. For convenience we write h, $\dot{h}$, and $\ddot{h}$ instead of $h(t)$, $\dot{h}(t)$, and $\ddot{h}(t)$. An incompressible, homogeneous, Newtonian liquid under isothermal conditions is assumed. The influence of the displaced air on the meniscus movement is neglected since density and viscosity of air are much smaller than the ones of the advancing liquid. The equation for the conservation of linear momentum in z-direction for the control volume CV_a as shown in Fig. 12.1 reads

Michael E. Dreyer: *Free Surface Flows under Compensated Gravity Conditions*
STMP **221**, 145–154 (2007)
DOI 10.1007/3-540-44628-1_12

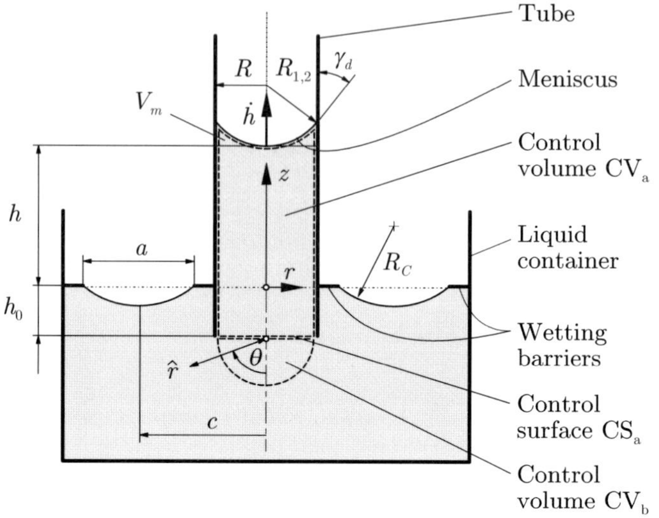

Fig. 12.1. Schematic representation of the geometry and the control volumes (*dashed lines*). The initial liquid height is named h_0 (= depth of immersion), the meniscus height h and the meniscus velocity $\dot{h}$. The tube radius is R, the principal radii of the meniscus curvature are R_1 and R_2, the dynamic contact angle is γ_d. The radius of the free surface inside the container R_C is calculated with the distance a between the wetting barriers and the radius c of the centerline of the free surface inside the reservoir. The liquid volume above the measured liquid height h and the meniscus is named V_m

$$F_{pu} + F_{pl} + F_f + F_w - \int_{-h_0}^{h+0.3R}\int_0^{2\pi}\int_0^R \rho k_z r\mathrm{d}r\mathrm{d}\varphi\mathrm{d}z =$$

$$\frac{d}{dt}\left\{\int_{-h_0}^{h+0.3R}\int_0^{2\pi}\int_0^R \rho v_z r\mathrm{d}r\mathrm{d}\varphi\mathrm{d}z\right\} - \int_0^{2\pi}\int_0^R \rho v_z^2 r\mathrm{d}r\mathrm{d}\varphi. \qquad (12.1)$$

Equation (12.1) compares the sum of forces acting on the control volume CV_a minus the inertia force due to the acceleration k_z of the reference system with the rate of change of linear momentum inside the control volume CV_a minus the inlet momentum flux entering across the control surface CS_a. No outlet momentum flux occurs, since the upper boundary of the control volume moves with the meniscus velocity $\dot{h}$. The forces acting on the control volume are the pressure forces F_{pu} at the upper and F_{pl} at the lower boundary of the control volume CV_a, the friction force F_f along the tube walls, and the weight F_w of the liquid inside the control volume CV_a.

For the determination of the momentum balance we assume a uniform velocity distribution with

$$v_z(r,t) = \dot{h} \qquad (12.2)$$

for the following reasons. The usually assumed parabolic velocity profile does not apply at the tube entrance and at the meniscus where the velocity is $v_z = \dot{h}$. Furthermore the flow needs a certain flow length (i.e. the entrance length L_e) to change from the uniform profile at the tube entrance to the fully-developed parabolic profile. In our experiments the flow length h_t exceeded the entrance length only for the d = 4 mm tubes (see Table 11.1), thus a fully-developed flow field was hardly ever reached. The velocity distribution of (12.2) affects only the right hand side of (12.3) which determines the inertia of the liquid located inside the tube. The molecular (or viscous) momentum transport yielding a pressure gradient in the tube (which is hidden in F_f) will be discussed in greater detail later.

Substituting (12.2) and solving the integrals, (12.1) yields

$$F_{pu} + F_{pl} + F_f + F_w - \rho\pi R^2 h_t k_z = \rho\pi R^2 h_t \ddot{h} \ . \tag{12.3}$$

12.1 Pressure Force at the Upper Boundary of the Control Volume CV$_a$

The pressure force F_{pu} at the upper boundary of the control volume CV$_a$ reads

$$F_{pu} = -\int_0^{2\pi}\int_0^{R} p_u \ r\mathrm{d}r\mathrm{d}\varphi \ . \tag{12.4}$$

The pressure p_u in the liquid is related to the curvature of the meniscus via the GAUSS-LAPLACE equation (2.7)

$$\Delta p_c = p_u - p_a = -\sigma\left(\frac{1}{R_1} + \frac{1}{R_2}\right) , \tag{12.5}$$

where p_a is the ambient pressure of the gas phase above the meniscus. The meniscus inside the tube is assumed to form a spherical cap under microgravity condition. Thus the principal radii of curvature are

$$R_1 = R_2 = \frac{R}{\cos\gamma_d} \tag{12.6}$$

where γ_d denotes the dynamic contact angle, which reduces the capillary pressure depending on the meniscus velocity $\dot{h}$. The dynamic contact angle γ_d is calculated using a commonly accepted correlation by [2] based on experimental data given by [3] (see [4]). For liquids with static contact angles $\gamma_s = 0$ this correlation reads

$$\cos\gamma_d = 1 - 2\tanh\left(4.96 \ \mathrm{Ca}^{0.7}\right) , \tag{12.7}$$

with the capillary number $\mathrm{Ca} = \mu\dot{h}/\sigma$. The capillary numbers for our experiments ranged $7.7 \times 10^{-4} < \mathrm{Ca} < 6.2 \times 10^{-3}$. Thus during the meniscus rise the dynamic contact angle increased from the static value of $\gamma_s = 0$ to $21° < \gamma_s < 44°$ reducing the driving capillary pressure by 6 % ($d = 70$ mm tubes) to 28 % ($d = 4$ mm tubes). This demonstrates that the velocity dependence of the contact angle cannot be neglected as in previous approaches.

After the step-reduction in gravity due to the release of the experiment capsule into free fall, the meniscus needs a certain time to change from the initial flat $1g_0$ configuration to the final spherical microgravity shape. After an initial overshoot, the surface experiences a damped oscillation around its final low-gravity equilibrium shape while the liquid continue to flow into the tube. In our experiments, this effect was noticeable for all tube diameters $d \geq 36$ mm. Siegert et al. [5] performed drop tower experiments to study the surface reorientation (see Sect. 4.5). Equation (4.24) was confirmed by recent experiments of [6].

For our model we approximate the meniscus reorientation with an exponential function to remove the non-physical jump from a flat meniscus to the final meniscus shape after the step reduction in gravity. The function

$$s(t) = 1 - e^{-4.6t/t_{cr}} \tag{12.8}$$

describes the transition from the flat initial surface to 99 % of the final equilibrium curvature within the reorientation time t_{cr} (4.24). The further meniscus oscillation is not considered for our model since the surface is assumed to oscillate around its final equilibrium shape. It has to be noted that t_{cr} is used here although it has been derived for the pure surface reorientation, which is different to our experiments where the surface reorientation is overlaid by the liquid flow into the tube which increases the speed of the liquid/solid contact line. This assumption is confirmed by Weislogel [7] who observed the superposition of the surface oscillations on the bulk flow and concluded that they damped out faster than in the case of a pure reorientation without net flow.

Using (2.7), (12.6)–(12.8) the pressure force at the upper boundary of the control volume CV_a (12.4) reads

$$F_{pu} = -\pi R^2 \left(p_a - 2\sigma s \cos\gamma_d / R\right) \,. \tag{12.9}$$

12.2 Pressure Force at the Lower Boundary of the Control Volume CV_a

To determine the pressure force

$$F_{pl} = \int_0^{2\pi} \int_0^R p_l \, r \mathrm{d}r \mathrm{d}\varphi \tag{12.10}$$

at the lower boundary of control volume CV_a, a second momentum balance for a hemispherical control volume CVb (Fig. 12.1) of radius R with spherical polar coordinates $\hat{r}, \theta, \phi$ is applied. To derive F_{pl} we follow an approach of [8] with some improvements. The derivation is given in the appendix (see (A.32)). The result reads

$$F_{pl} = \int_0^{2\pi} \int_0^R p_l \; r\mathrm{d}r\mathrm{d}\varphi$$
$$= \pi R^2 \left[p_\infty + \rho \left(g_0 + k_z \right) h_0 - \rho \left(\frac{11}{12} R\ddot{h} + \frac{2\nu}{R}\dot{h} + \frac{23}{24}\dot{h}^2 \right) \right] . \quad (12.11)$$

This equation describes the retarding capillary pressure of the free liquid surface inside the container and the total flow losses inside the container due to convective and molecular momentum transports. The pressure p_∞ of the reservoir may be different from ambient pressure due to the curvature of the free surface

$$p_\infty - p_a = \Delta p_c = -2H\sigma = -\frac{\sigma}{R_c} . \quad (12.12)$$

The capillary pressure σ/R_C due to the curvature $1/R_C$ of the free surface inside the reservoir is considered neglecting the second principal radius of curvature. This capillary pressure retards the liquid rise if wide tubes (with small a/R ratios) are used. Under microgravity condition, the assumption of a flat liquid surface inside the reservoir is valid only for small tube diameters (high a/R ratios). The radius R_C of the free surface inside the reservoir results from the decreasing liquid volume in the reservoir due to the liquid rise h into the tube and depends on the geometry parameters R, a, and c (Fig. 12.1) for the respective tube/container setup. For the experiment setup used in this study the approximation for R_C reads

$$R_C = \frac{C}{R^2 h} = \frac{a^3 c}{6 R^2 h} , \quad (12.13)$$

with the constant for the container geometry $C = a^3 c/6$. This approximation is only applicable within the experiment duration of 4.7 s. The geometrical parameters of (12.13) are listed in Table 12.1.

12.3 Friction Force inside the Tube

In general the total pressure drop from the tube entrance with pressure p_i to an axial position z with pressure p_z reads [9]

$$\frac{p_i - p_z}{\frac{1}{2}\rho\dot{h}^2} = \frac{K_{tl} z}{\mathrm{Re}_d \; d} + K_{ts}(z) , \quad (12.14)$$

with the mean velocity $\dot{h}$ and $\mathrm{Re}_d = \dot{h}d/\nu$. The total pressure drop consists of the pressure drop for the fully developed laminar Poiseuille flow and an excess

Table 12.1. Geometrical parameters of the tube/container setup

R (mm)	a (mm)	c (mm)	$C = a^3c/6$ 10^6 (mm^4)	$\Lambda_2 = R^4/C$ (-)
2.0	91.5	59.3	7.6	2.1×10^{-6}
4.5	85.0	62.5	6.4	6.4×10^{-5}
5.5	83.0	63.0	6.0	1.5×10^{-4}
8.0	80.0	65.0	5.6	7.4×10^{-4}
18.0	70.5	69.8	4.1	2.6×10^{-2}
27.0	65.0	72.5	3.3	1.6×10^{-1}
35.0	55.0	77.5	2.2	7.0×10^{-1}

pressure drop $K_{ts}(z)$ due to the transition from the inlet velocity profile to the fully developed profile. For a fully-developed parabolic velocity profile the friction coefficient reads $K_{tl} = 64$. The excess pressure drop $K_{ts}(z)$ is due to the velocity profile development within the entrance length L_e, which is commonly accepted [1] to be $L_e \approx 0.06\ d\mathrm{Re}_d$ for laminar flow. In this study, the flow length exceeded the entrance length L_e only for the 4 mm tubes, thus the excess pressure $K_{ts}(z)$ drop cannot be neglected a priori as in previous studies.

A useful relation for the excess pressure drop was given by [10] based on the linearization of the inertia terms of the momentum equation assuming a uniform velocity profile across the inlet section

$$K_{ts2}(\hat{z}) = \frac{4}{3} + \sum_{i=1}^{25} \frac{8}{\alpha_i^2} \left(e^{-4\alpha_i^2 \hat{z}} - 3 \right) e^{-4\alpha_i^2 \hat{z}} , \tag{12.15}$$

with $\hat{z} = z/(d\ \mathrm{Re}_d)$. The first 25 values of the eigenvalues α_i were computed by [10] (and are given in the appendix A). Within the entrance length L_e the excess pressure drop $K_{ts}(z)$ depends on the flow length z. For $z > L_e$ (12.15) yields $K_{ts2} = 4/3$ which corresponds to the results found by other authors listed by [9]. The approach of [10] seems applicable for this practical purpose. The comparison with our own CFD computations is shown in the appendix A.

The total friction force for the capillary rise problem then reads (setting $K_{ts} = K_{ts2}$)

$$F_f = -\pi R^2 \frac{\rho \dot{h}^2}{2} \left[\frac{64}{\mathrm{Re}_d} \frac{h_t}{d} + \frac{4}{3} + \sum_{i=1}^{25} \frac{8}{\alpha_i^2} \left(e^{-4\alpha_i^2 h_t/(d\ Re_d)} - 3 \right) e^{-4\alpha_i^2 h_t/(d\ Re_d)} \right] \tag{12.16}$$

The weight of the liquid volume inside the tube is

$$F_w = -\rho \pi R^2 h_t g_0 . \tag{12.17}$$

12.4 Equation for the Capillary Rise and Variation of the Coefficients

Substituting (12.9), (12.11), (12.16), and (12.17) into (12.3) the equation for the capillary rise in terms of forces reads

$$\begin{aligned}\rho\pi R^2\left(h_t+\frac{11}{12}R\right)\ddot{h} &= \sigma\pi R^2\left(\frac{2s\cos\gamma_d}{R}-\frac{1}{R_C}\right)\\ &\quad -\rho\pi R^2\left(\frac{K_{tl}}{\mathrm{Re}_d}\frac{h_t}{2R}+K_{ts2}\right)\frac{\dot{h}^2}{2}\\ &\quad -\rho\pi R^2\left(\frac{23}{12}+\frac{8}{\mathrm{Re}_d}\right)\frac{\dot{h}^2}{2}\\ &\quad -\rho\pi R^2\left(h+0.3R\right)\left(g_0+k_z\right)\ . \end{aligned} \tag{12.18}$$

The first term on the right hand side of (12.18) describes the driving capillary force inside the tube minus the retarding capillary force of the curved free surface inside the liquid container. The second term denotes the total friction force inside the tube. The third term represents convective and viscous losses inside the reservoir. The fourth term compares the weight of the liquid inside the tube with the inertia force due to the acceleration k_z of the capsule. For the drop tower Bremen the residual acceleration during the experiment is $g_0 + k_z < 10^{-5} g_0$, thus weight and inertia compensate each other and this term may be neglected. If this term is retained, (12.18) is also applicable to predict the capillary rise under normal gravity conditions within a fixed frame of reference with $k_z = 0$. The left hand side of (12.18) describes the inertia force of the liquid volume inside the tube h_t and the liquid volume below the tube entrance $11/(12R)$.

Since the single terms of (12.18) are based on different assumptions, the corresponding coefficients were varied to evaluate their influence on the liquid rise and to improve the agreement of the model with the experimental data. In the expression for the driving capillary force, the coefficient 0.413 for the reorientation time (4.24) was varied. In the complete range of our experiment data the variation did not show a significant influence on the total meniscus height calculation, thus the original value of 0.413 from [5] is retained. The total friction force inside the tube was not varied since both, the Poiseuille friction coefficient $K_{tl} = 64$ and the excess pressure drop K_{ts2} (12.15), are reliable solutions from literature. The convective and viscous losses inside the reservoir result from the second momentum balance which is based on several assumptions [8] and therefore should be used with care. We found the best agreement between the theoretical data and our entire experimental data if the coefficient 1 instead of 23/12 (calculated with (12.11)) is used for the convective losses inside the reservoir. The viscous loss represented by $8/\mathrm{Re}_d$ has only a negligible influence in the complete range of our experiment parameters where $162 < \mathrm{Re}_d < 2654$ (well above creeping flow) and is thus not modified. The inertia coefficient $11/(12R)$ has only a minor influence and is also retained as derived from the momentum balance for control volume CVb.

Finally the improved equation for the capillary rise reads

$$\begin{aligned}\rho\left(h_t+\frac{11}{12}R\right)\ddot{h} = \sigma\left(\frac{2s\cos\gamma_d}{R}-\frac{1}{R_C}\right)\\ -\frac{\rho\dot{h}^2}{2}\left(\frac{K_{tl}}{\mathrm{Re}_d}\frac{h_t}{2R}+K_{ts2}\right)\\ -\frac{\rho\dot{h}^2}{2}\left(1+\frac{8}{\mathrm{Re}_d}\right) .\end{aligned} \tag{12.19}$$

Equation (12.19) is rearranged to get the second order nonlinear differential equation for the liquid flow into the tube

$$\begin{aligned}\ddot{h} = \frac{1}{h_t+11/(12R)}\left[\frac{\sigma}{\rho}\left(\frac{2s\cos\gamma_d}{R}-\frac{1}{R_C}\right)\right.\\ \left.-\left(\frac{8h_t}{R^2}+\frac{2}{R}\right)\nu\dot{h}-(1+K_{ts2})\frac{1}{2}\dot{h}^2\right] .\end{aligned} \tag{12.20}$$

We compare our (12.20) only with the most detailed theory up to now given by [8]. For the comparison (4.32) is rewritten as

$$\ddot{h} = \frac{1}{h+37/(36R)}\left[\frac{2\sigma}{\rho R}-\left(\frac{8h}{R^2}+\frac{2}{R}\right)\nu\dot{h}-\frac{7}{6}\dot{h}^2-g_0 h\right] . \tag{12.21}$$

Due to the different geometry [8] did not need to consider an initial liquid height h_0. Their length $37/(36R)$ for the inertial mass below the tube entrance is comparable to $11/(12R)$ in (12.20). The initial meniscus development, the dynamic contact angle, and the capillary counter pressure in the reservoir were not considered by [8]. Apart from the initial liquid height h_0 the terms for the viscous losses are identical. For the convective losses Levine et al. found 7/6. If the fully-developed value of $K_{ts} = 4/3$ is used in (12.20) we also find 7/6 as the coefficient for the sum of the convective losses if the flow length exceeds the entrance length L_e. The term $g_0 h$ for the hydrostatic pressure in (12.21) can be neglected in the case of microgravity conditions.

12.5 Dimensionless Equation for the Capillary Rise

Equation (12.20) is scaled using the characteristic viscous time $t_c = R^2/(8\nu)$ (with $t_c \sim t_{vu}$, (10.2)) and the characteristic velocity $v_c = \sqrt{2\sigma/(\rho R)}$ (compare u_{pu} (2.19)). The time t_c was used before by [11, 12, 8, 13]. It also appears in (12.20) as a coefficient for the laminar pressure loss inside the tube. The velocity v_c was used before by [8] and [14]. Introducing dimensionless variables based on t_c and v_c

$$t_* = \frac{t}{t_c}, h_* = \frac{h + 0.3R}{t_c v_c} = 8\mathrm{Oh}\frac{h + 0.3R}{R}, \dot{h}_* = \frac{\dot{h}}{v_c}, \ddot{h}_* = \frac{t_c}{v_c}\ddot{h} \tag{12.22}$$

leads to the dimensionless form of (12.20)

$$\ddot{h}_* = \frac{1}{h_* + 8\left(\Lambda_1 + 73/60\right)\mathrm{Oh}}\left\{s_* \cos\gamma_d - \frac{\Lambda_2 h_*}{16\mathrm{Oh}} - \left[h_* + \left(8\Lambda_1 + 2\right)\mathrm{Oh}\right]\dot{h}_* - \left(1 + K_{ts2*}\right)\frac{1}{2}\dot{h}_*^2\right\} \tag{12.23}$$

with the OHNESORGE number $\mathrm{Oh} = \sqrt{\nu^2\rho/(\sigma d)} = R/(8t_c v_c)$ and the geometric ratios $\Lambda_1 = h_0/R$ (dimensionless initial liquid height, Table 11.2) and $\Lambda_2 = R^4/C$ (dimensionless tube/container ratio, Table 12.1). The function s_* for the initial meniscus development, the dynamic contact angle γ_d, and the excess pressure loss K_* expressed with dimensionless quantities read

$$s_* = 1 - e^{-t_*/\mathrm{Oh}}, \tag{12.24}$$

$$\cos\gamma_{d*} = 1 - 2\tanh\left[4.96\left(2\mathrm{Oh}\dot{h}_*\right)^{0.7}\right], \tag{12.25}$$

$$K_{ts2*} = \frac{4}{3} + \sum_{i=1}^{25}\frac{8}{\alpha_i^2}\left[e^{-4\alpha_i^2(h_* + 8\Lambda_1\mathrm{Oh})/(8\dot{h}_*)} - 3\right]e^{-4\alpha_i^2(h_* + 8\Lambda_1\mathrm{Oh})/(8\dot{h}_*)}. \tag{12.26}$$

The dimensionless equation (12.23) for the meniscus rise depends only on the OHNESORGE number Oh and the geometric ratios Λ_1 and Λ_2. The dimensionless capillary force $s_* \cos\gamma_{d*}$ tends to unity after the meniscus development is completed ($s_* = 1$) and for low flow velocities where $\cos\gamma_{d*} \approx 1$. This means that by using t_c and v_c, (12.20) is scaled with the driving capillary force to get the dimensionless formulation of (12.23).

Equations (12.20) and (12.23) are solved numerically using a fourth order Runge-Kutta algorithm. The initial conditions are $h = \dot{h} = 0$ at $t = 0$ or $h_* = \dot{h}_* = 0$ at $t_* = 0$, respectively. The solution of (12.20) and (12.23) yields the meniscus height h (or h_*) and the meniscus velocity $\dot{h}$ (or $\dot{h}_*$) as a function of time t (or t_*), which are compared with our experimental data. For the comparisons, numerical calculations are performed for each experiment configuration using the fluid properties at the temperature, which was measured during the respective experiment run.

References

1. F.M. White: *Viscous Fluid Flow*, (McGraw-Hill, New York 1974)

2. T.S. Jiang, S.O. Oh, J.C. Slattery: J. Colloid Interface Sci. **69.1**, 74 (1979)
3. R.L. Hoffman: J. Colloid Interface Sci. **50.2**, 228 (1975)
4. S.F. Kistler: Hydrodynamics of wetting. In: *Wettability*, ed by J.C. Berg (Marcel Dekker, New York 1993) pp. 311–429
5. C.E. Siegert, D.A. Petrash, E.W. Otto: NASA TN D-2458, (1964)
6. M. Michaelis, M.E. Dreyer, H.J. Rath: Ann. N.Y. Acad. Sci. **974**, 246 (2002)
7. M.M. Weislogel: AIChE Journal **43.3**, 645 (1997)
8. S. Levine, P. Reed, E.J. Watson, G. Neale: A Theory of the Rate of Rise of a Liquid in a Capillary. In: *Colloid and Interface Science*, Chap. 3, ed by M. Kerker (Academic Press, New York 1976) pp. 403–419
9. F.W. Schmidt, B. Zeldin: AIChe Journal **15.4**, 612 (1969)
10. E.M. Sparrow, S.H. Lin, T.S. Lundgren: Phys. Fluids **7.3**, 338 (1964)
11. C.H. Bosanquet: Philos. Mag. Ser. 6 **45**, 525 (1923)
12. J. Szekely, A.W. Neumann, Y.K. Chuang: J. Colloid Interface Sci. **35.2**, 273 (1971)
13. N. Ichikawa, Y. Satoda: J. Colloid Interface Sci. **162**, 350 (1994)
14. D. Quéré: Europhys. Lett. **39**, 533 (1997)

13

Results and Discussion

13.1 Comparison of Experimental and Theoretical Results

The results of the drop tower tests are presented in Figs. 13.1–13.3 in comparison with the corresponding numerical solutions of (12.20). The meniscus height h is plotted versus time t. For comparison all plots have the same time range of 5 s and the same maximum liquid height of 250 mm. In the complete range of experiment parameters of our study (Table 11.1 and 11.2) good agreement of experimental and theoretical data is found.

In the case of a tube with $d = 4$ mm and FC-77 test liquid (Fig. 13.1, #43, #40) the highest OHNESORGE number of $\mathrm{Oh} = 4.7 \times 10^{-3}$ of this study occurs. The slope of all 4 mm curves decreases almost immediately after the start of the meniscus rise and the curves follow the Lucas-Washburn behavior with $h \sim \sqrt{t}$. For the $d = 9$ mm and 11 mm tubes (Fig. 13.1, right column) the slope of the curves increases within the first 0.6 s of the experiments followed by a region with constant slope ($h \sim t$). About the end of the experiments the curve's slope decreases indicating a beginning Lucas-Washburn behavior.

In the case of a tube with $d = 16$ mm (Fig. 13.2, left column) the initial region with increasing slope is enlarged compared to the $d = 9$ mm tubes. If the tube diameter is increased furthermore (Fig. 13.2, right column), the first region with increasing slope is enlarged further and Lucas-Washburn flow is not reached within the experiment time of 4.7 s.

For tube diameters $d \geq 36$ mm the experiment data of the meniscus height h oscillate around the calculated curves. This oscillation is visible because the lowest point of the meniscus at the tube center is taken for the evaluation of the liquid height. The curves for the $d = 36$ mm tubes show that the oscillation decays during the meniscus rise. For the $d = 54$ mm and 70 mm tubes (Fig. 13.3) the initial oscillation amplitude is higher and the meniscus oscillation is damped but does not vanish completely during the observed experiment time. The lowest OHNESORGE number of this study $\mathrm{Oh} = 5.8 \times 10^{-4}$ occurs with $d = 70$ mm and SF 0.65 test liquid (Fig. 13.3, #44, #42).

Michael E. Dreyer: *Free Surface Flows under Compensated Gravity Conditions*
STMP **221**, 155–167 (2007)
DOI 10.1007/3-540-44628-1_13

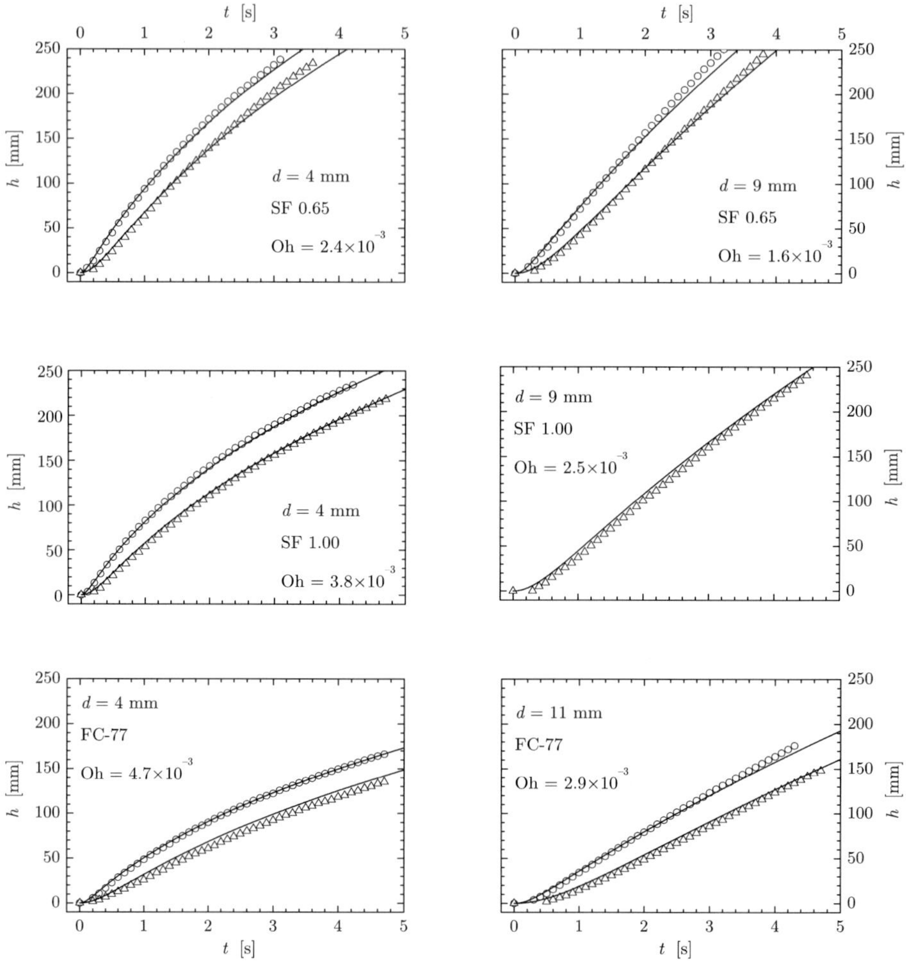

Fig. 13.1. Meniscus height h versus time t for tube diameters $d = 4, 9$, and 11 mm, initial liquid heights $h_0 = 10$ (*open circles*) and 50 mm (*open triangles*) and three test liquids. Experimental data (*symbols*) from drop tower tests compared with numerical solutions of (12.20) (*lines*)

13.2 Dimensionless Representation of the Results

The comparison of experimental and numerical data of the meniscus height h versus time t (Figs. 13.1–13.3) yields only information about the total effect of the sum of forces acting on the meniscus rise. The inertia and the viscous time cannot be determined exactly in this representation. More details become visible if the dimensionless meniscus velocity $\dot{h}_*$ is plotted versus the

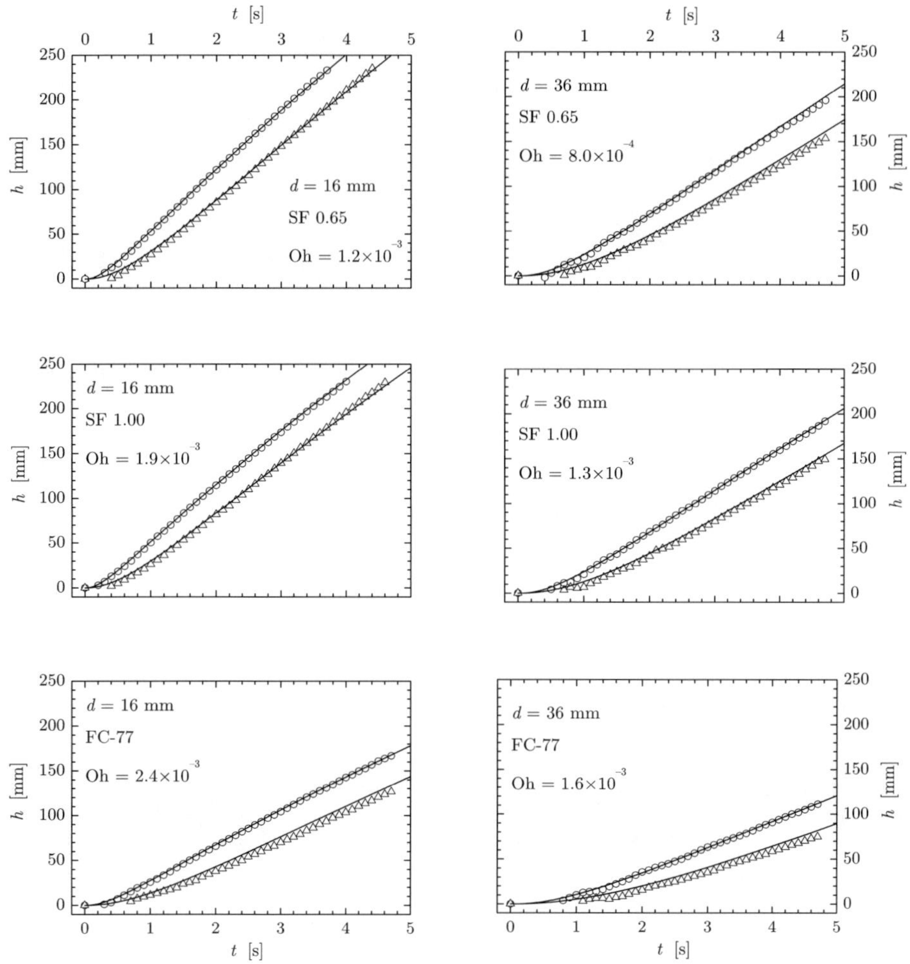

Fig. 13.2. Meniscus height h vs. time t for tube diameters d = 16 and 36 mm, initial liquid heights h_0 = 10 (*open circles*) and 50 mm (*open triangles*) and three test liquids. Experimental data (*symbols*) from drop tower tests compared with numerical solutions of (12.20) (*lines*)

dimensionless time t_*. The theoretical velocity data are given by the numerical solution of (12.23).

The meniscus velocity $\dot{h}$ in the experiments is calculated from the data of the meniscus height h using a fourth order centered approximation method [1] where five adjacent data points are used with different weights to calculate the slope at the center point.

The dimensionless velocity plots are discussed below (Figs. 13.4–13.7) together with the plots of the single forces, which are calculated with (12.23) but

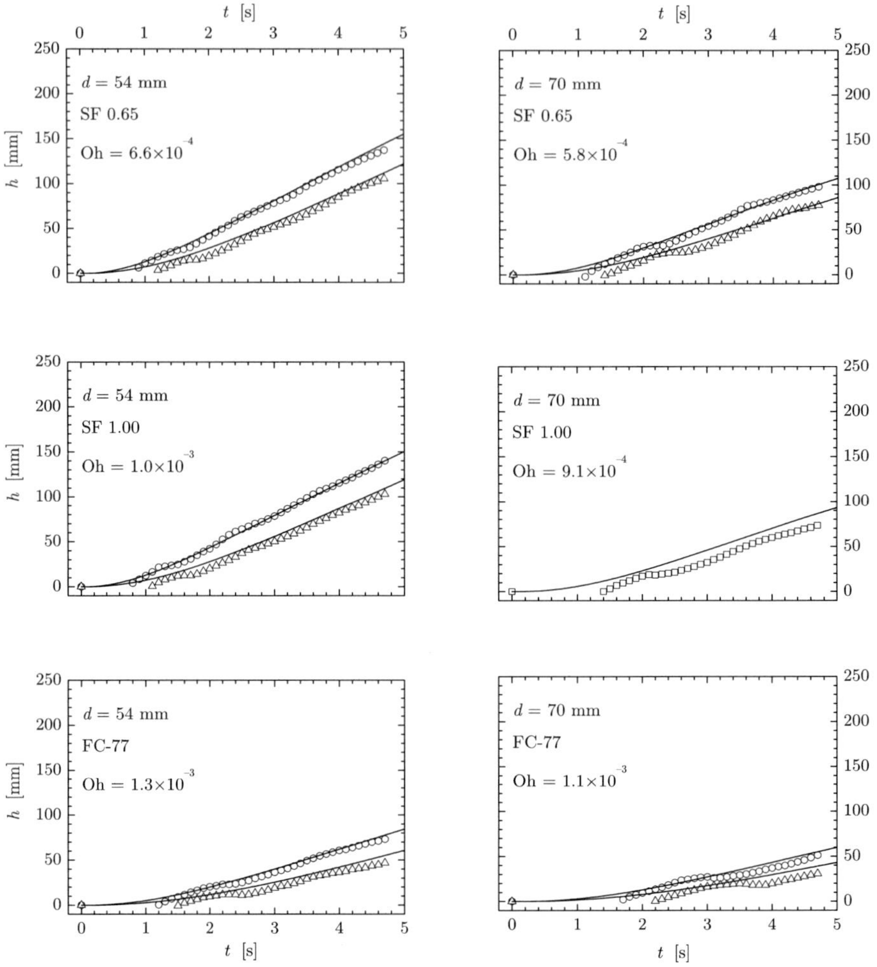

Fig. 13.3. Meniscus height h vs. time t for tube diameters $d = 54$ and 70 mm, initial liquid heights $h_0 = 10$ (*open circles*), 32 mm (*open squares*) and 50 mm (*open triangles*) and three test liquids. Experimental data (*symbols*) from drop tower tests compared with numerical solutions of (12.20) (*lines*)

cannot be measured during the experiment. The single dimensionless forces in details are:

1. The capillary force of the meniscus inside the tube $F_{ct*} = s_* \cos \gamma_{d*}$.
2. The capillary force of the free surface inside the reservoir $F_{cr*} = \Lambda_2 h_* / (16\text{Oh})$ is set to $F_{cr*} = 0$ (representing a flat liquid surface in the reservoir) since (12.13) is valid only within the experiment duration of 4.7 s but the forces are plotted here up to $100 \times t_c$ which is up to 6 hours for the wide tubes.

3. The friction force (viscous) inside the tube according to Poiseuille $F_{ftP*} = (h_* + 8\Lambda_1 \mathrm{Oh})\dot{h}_*$.
4. The friction force (viscous) inside the reservoir $F_{fr*} = 2\ \mathrm{Oh}\dot{h}_*$.
5. The excess pressure loss (convective) inside the tube according to Sparrow et al. [2] $F_{ftS*} = 0.5K_{ts2*}\dot{h}_*$.
6. The convective pressure loss inside the reservoir $F_{pr*} = 0.5\dot{h}_*^2$.
7. The inertia force inside the tube (liquid mass entering the tube during the meniscus rise plus the liquid mass in the vicinity of the meniscus) $F_{it*} = (h_* + 12\ \mathrm{Oh}/5)\ddot{h}_*$.
8. The inertia force inside the reservoir (initial liquid height inside the tube plus the liquid mass below the tube entrance) $F_{ir*} = 8\,(\Lambda_1 + 11/12)\,\mathrm{Oh}\ddot{h}_*$.

The top diagram of Fig. 13.4 compares the experimental and theoretical (12.23) dimensionless meniscus velocity $\dot{h}_*$ plotted over the dimensionless time t_* for $d = 4$ mm tube diameter with FC-77 liquid and two initial liquid heights $h_0 = 10$ (#43) and 50 mm (#40). These are the configurations with the highest Ohnesorge number $\mathrm{Oh} = 4.7 \times 10^{-3}$ of this study. For $t_* < 1$ the meniscus velocity increases and decreases for $t_* > 1$ with a slope of $-1/2$. Both curves do not reach the velocity $\dot{h}_* = 1$ (representing the characteristic velocity of $v_c = 92$ mm/s). The maximum velocity of the curve with higher initial liquid height $\Lambda_1 = 25$ (#40) is smaller than for $\Lambda_1 = 5$ (#43). For $t_* > 1$ both curves merge into a single curve. The characteristic time $t_c = 0.58$ s is smaller than the experiment duration of 4.7 s and most of the experimental data points are located in the range of $t_* \geq 1$.

The middle and the bottom diagrams of Fig. 13.4 show the progression of the single dimensionless forces F_* plotted over the dimensionless time t_* for both configurations of the top diagram. Both force plots show an initial increase of the capillary force of the meniscus inside the tube F_{ct*} due to the development of the meniscus after the step-reduction in gravity. After the maximum, F_{ct*} decreases due to the increasing meniscus velocity, which increases the dynamic contact angle γ_d and thus reduces the capillary force. Later F_{ct*} increases again due to the decreasing meniscus velocity for $t_* > 1$.

The first force acting against the driving capillary force F_{ct*} is the inertia force F_{ir*} inside the reservoir. In the case of the smaller dimensionless initial liquid height $\Lambda_1 = 5$ (#43) the next significant counter force is the sum of the convective losses $F_{ftS*} + F_{pr*}$ which reach 60 % of the capillary force F_{ct*} and are followed by the friction force inside the tube F_{ftP*} after the intersection at $t_* = 1$. For $\Lambda_1 = 25$ (#40) the friction force F_{ftP*} follows already after the inertia force F_{ir*} since the dominating counter force and the convective losses have a smaller maximum after the intersection of F_{ir*} and F_{ft*}. The inertia force of the liquid mass, which enters the tube F_{it*}, is small compared to the capillary force F_{ct*} and the friction force inside the reservoir F_{fr*} is close to zero during the complete time range. In Fig. 13.4 most of the experimental data are located in the region $t_* > 1$ which is dominated by the friction force

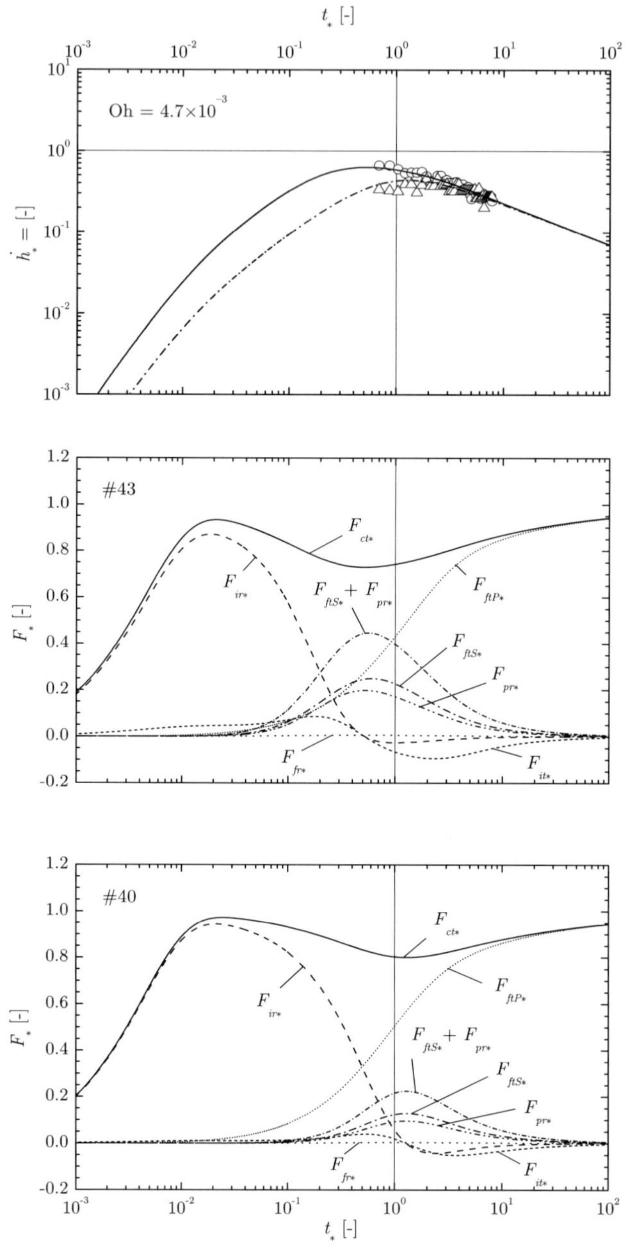

Fig. 13.4. Dimensionless representation of experimental and numerical results for tube diameter $d = 4$ mm and FC-77 liquid with initial liquid heights $h_0 = 10$ mm (#43, *open circles*) and $h_0 = 50$ mm (#40, *open triangles*). *Top*: experimental meniscus velocity $\dot{h}_*$ (*symbols*) versus time t_* compared with the corresponding numerical solutions of Eq. (12.23) (*lines*). *Middle* and *bottom*: force progressions F_* versus time t_* for #43 and #40 calculated with Eq. (12.23)

F_{ftP*}. This explains why a variation of the other forces does not change the calculated meniscus rise significantly for this configurations.

Figure 13.5 shows the same plots as Fig. 13.4 but for $d = 11$ mm, FC-77, and Oh $= 2.9 \times 10^{-3}$. Here the characteristic time $t_c = 4.37$ s corresponds to the experiment duration of 4.7 s and most experimental data points are located in the range of $t_* < 1$. In contrast to Fig. 13.4 the regions with increasing and decreasing velocities are connected here by a region with constant velocity which is close to $\dot{h}_* = 1$ (representing $v_c = 55$ mm/s). The time with constant velocity is longer in case of the smaller initial height $\Lambda_1 = 1.8$ (#26) and shorter for the higher initial height $\Lambda_1 = 9.1$ (#22). Again the two velocity curves merge into one curve with slope $-1/2$ for $t_* > 1$.

The force plot for $\Lambda_1 = 1.8$ (Fig. 13.5, middle) shows that the sum of the convective losses $F_{ftS*}+F_{pr*}$ is the dominating counter force during the phase with constant meniscus velocity. The convective losses $F_{ftS*} + F_{pr*}$ reach a maximum of 80% of the capillary force F_{ct*}. For $\Lambda_1 = 9.1$ the convective losses $F_{ftS*} + F_{pr*}$ are less significant and reach only 60 % of F_{ct*}. Like before F_{ir*} is the dominating counter force during the phase with increasing velocity and F_{ftP*} remains the only counter force in the final phase with decreasing meniscus velocity. The influence of F_{it*} and F_{it*} can be neglected compared to the other forces. For the configurations shown in Fig. 13.5 the experimental data points are located in the region where the convective losses dominate the flow. For a good agreement of the calculated meniscus heights and the experimental data, especially the convective losses have to be calculated accurately.

Figure 13.6 shows the corresponding plots for $d = 36$ mm, SF 0.65, and Oh $= 8 \times 10^{-4}$. The characteristic time $t_c = 58.5$ s is one order of magnitude higher than the experiment duration, thus the experimental data are located in the range of $t_* = 0.01$ to 0.08 and are shifted towards the region with increasing velocity. The scatter of the experimental velocity data around the calculated curves is due to the initial meniscus oscillation during the meniscus rise, which also affects the measured meniscus velocity. Compared to Fig. 13.4 and Fig. 13.5 the region with constant velocity is enlarged furthermore and the convective losses $F_{ftS*}+F_{pr*}$ reach 99 % of F_{ct*} in case of $\Lambda_1 = 0.6$ (#47) and 93 % for $\Lambda_1 = 2.8$ (#61). Both calculated velocity curves and the experimental data reach the dimensionless velocity $\dot{h}_* = 1$ representing $v_c = 48$ mm/s. Like before the two velocity curves merge into one curve with slope $-1/2$ for $t_* > 1$. In contrast to Fig. 13.4 and Fig. 13.5, the excess pressure loss F_{ftS*} reaches its maximum after F_{pr*} since F_{ftS*} depends not only on the flow velocity but also on the increasing flow length which is smaller than the entrance length L_e (see Table 11.1) for this configuration. With increasing tube diameter the inertia force of the liquid mass inside the tube F_{it*} becomes more important and increases up to 25 % of F_{ct*}.

For the lowest Ohnesorge number (Oh $= 5.8 \times 10^{-4}$) of our study (Fig. 13.7) the experiment duration is very small compared to the characteristic time $t_c = 221.3$ s and the experimental data points are shifted further towards the first region with increasing velocity. The scatter of the experimental

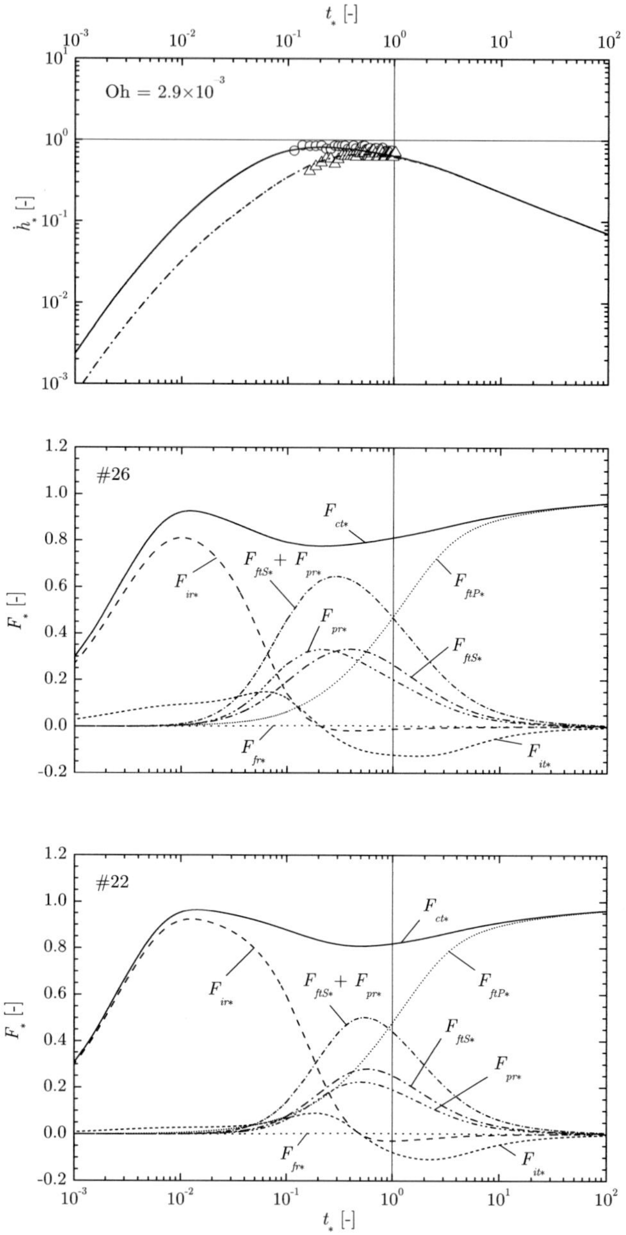

Fig. 13.5. Dimensionless representation of experimental and numerical results for tube diameter d = 11 mm and FC-77 liquid with initial liquid heights h_0 = 10 mm (#26, *open circles*) and h_0 = 50 mm (#22, *open triangles*). *Top*: experimental meniscus velocity $\dot{h}_*$ (*symbols*) versus time t_* compared with the corresponding numerical solutions of (12.23) (*lines*). *Middle* and *bottom*: force progressions F_* versus time t_* for #26 and #22 calculated with (12.23)

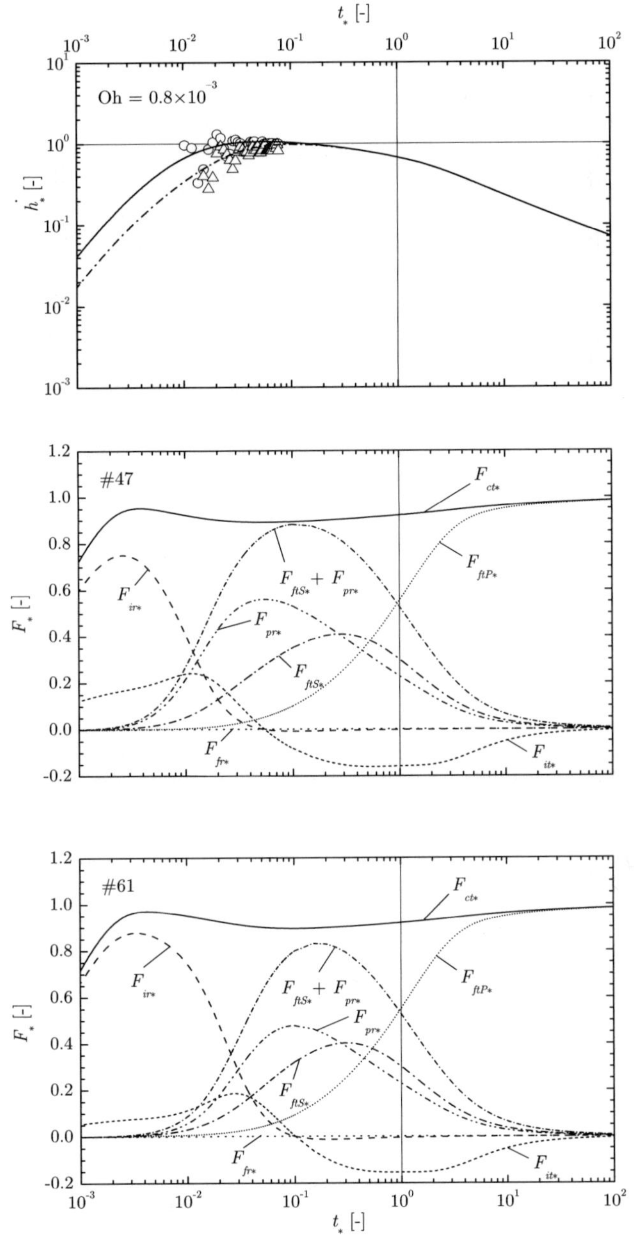

Fig. 13.6. Dimensionless representation of experimental and numerical results for tube diameter $d = 36$ mm and SF 0.65 liquid with initial liquid heights $h_0 = 10$ mm (#47, *open circles*) and $h_0 = 50$ mm (#61, *open triangles*). *Top*: experimental meniscus velocity $\dot{h}_*$ (*symbols*) versus time t_* compared with the corresponding numerical solutions of (12.23) (*lines*). *Middle* and *bottom*: force progressions F_* versus time t_* for #47 and #61 calculated with (12.23)

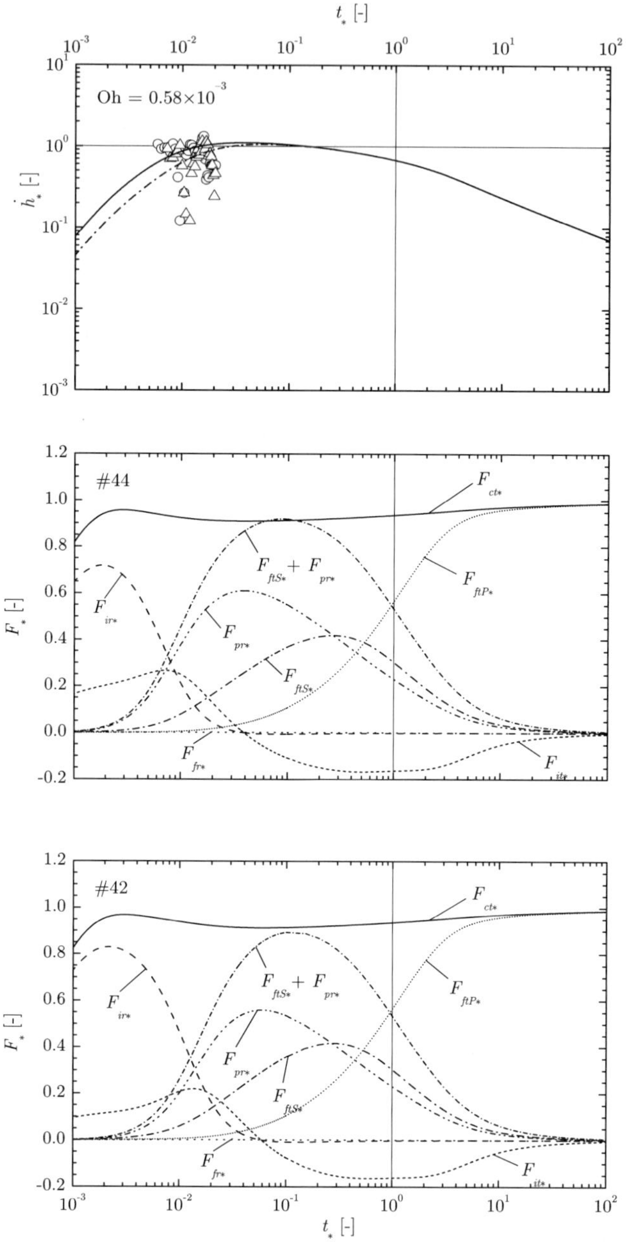

Fig. 13.7. Dimensionless representation of experimental and numerical results for tube diameter $d = 70$ mm and SF 0.65 liquid with initial liquid heights $h_0 = 10$ mm (#44, *open circles*) and $h_0 = 50$ mm (#42, *open triangles*). *Top*: experimental meniscus velocity $\dot{h}_*$ (*symbols*) vs. time t_* compared with the corresponding numerical solutions of (12.23) (*lines*). *Middle* and *bottom*: force progressions F_* vs. time t_* for #44 and #42 calculated with (12.23)

data is higher due to the stronger meniscus oscillation which is also visible in Fig. 13.3 (top, right). Also for this configurations the two velocity curves merge into one curve with slope $-1/2$ for $t_* > 1$.

Due to the large tube diameter of $d = 70$ mm the change from $h_0 = 10$ to 50 mm changes the dimensionless initial liquid height only from $\Lambda_1 = 0.3$ (#44) to $\Lambda_1 = 1.4$ (#42). Consequently the two force plots of Fig. 13.7 show smaller differences than before and the regions with constant meniscus velocity (which again reaches $\dot{h}_* = 1$ representing $v_c = 34$ mm/s) have almost the same extension. For this experiment configuration the inertia fore F_{ir*} and the convective effects F_{pr*} and F_{ftS*} have to be formulated carefully whereas the formulation of the viscous loss F_{ftP*} is less important since the experiments are concluded before F_{ftP*} dominates the flow.

In general Figs. 13.4–13.7 prove that the meniscus rise can be subdivided into three phases where only one force counteracts the driving capillary force F_{ct*}. In the first phase (region I), the inertia force F_{ir*} of the initial liquid height controls the flow and the meniscus velocity increases. The second phase (region II), with constant meniscus velocity, is controlled by the convective losses $F_{pr*} + F_{ftS*}$. Figures 13.4 and 13.5 show that the initial liquid height Λ_1 influences the extension of the first phase of the capillary rise. If the initial liquid height Λ_1 is large enough (Fig. 13.4) this region can be extended even so far that the flow directly enters the final stage without a phase of constant meniscus velocity and without reaching the maximum velocity $\dot{h}_* = 1$ (representing $\dot{h} = v_c$). In the final phase (region III), the capillary flow is controlled by the friction force inside the tube F_{ftP*}. The transition into this region is completed at $t_* = 1$ (representing $t = t_c$). For $t_* > 1$ all velocity curves merge into one curve with a slope of $-1/2$.

The force plots (Figs. 13.4–13.7) explain the physical meaning of the characteristic time $t_c = R^2/(8\nu)$ and the characteristic velocity $v_c = \sqrt{2\sigma/(\rho R)}$ used for the non-dimensionalization of (12.20). The time t_c marks the transition into the friction force controlled flow region III with Lucas-Washburn behavior. The velocity v_c is the theoretical maximum velocity for capillary-driven flows.

13.3 Comparison of the Dominating Forces

Approximate solutions for the meniscus height and velocity for each of the three regions and the transition times between the regions can be derived from the comparison of the two respective dominating forces of each flow phase. For the following comparison some simplifications are made to get analytical solutions. For the driving capillary force $F_{ct*} = s_* \cos\gamma_{d*}$ a completely developed meniscus shape is assumed ($s_* = 1$) and the velocity dependence of the contact angle is neglected ($\cos\gamma_{d*} = 1$). Thus the simplified capillary force reads

$$F_{ct*} = 1 \,. \tag{13.1}$$

The counter force of region I is the inertia force F_{ir*} which is simplified here to

$$F_{ir*} = 8\,(\Lambda_1 + 1)\,\mathrm{Oh}\ddot{h}_* \ . \tag{13.2}$$

According to the force plots (Figs. 13.4–13.7) all other forces can be neglected in region I and the force balance reads $F_{ct*} = F_{ir*}$. Integrating with the initial conditions $h_* = \dot{h}_* = 0$ at $t_* = 0$ yields a relation for the dimensionless meniscus velocity in the inertia dominated region I

$$\dot{h}_{I*} = \frac{t_*}{8\,(\Lambda_1 + 1)\,\mathrm{Oh}} \ , \tag{13.3}$$

and for the dimensionless meniscus height

$$h_{I*} = \frac{t_*^2}{16\,(\Lambda_1 + 1)\,\mathrm{Oh}} \ . \tag{13.4}$$

In region II the capillary force is counteracted mainly by the sum of the convective losses $F_{ftS*} + F_{pr*}$ which are simplified to

$$F_{ftS*} + F_{pr*} = \dot{h}_*^2 \ . \tag{13.5}$$

The balance $F_{ct*} = F_{ftS*} + F_{pr*}$ shows that the meniscus velocity in region II is

$$\dot{h}_{II*} = 1 \ , \tag{13.6}$$

thus the meniscus height in region II reads

$$h_{II*} = t_* \ . \tag{13.7}$$

Region III is controlled by the friction force inside the tube F_{ftP*}. For large times ($t_* \gg 1$) the initial liquid height Λ_1 can be neglected compared to the meniscus height h_* and the friction force can be simplified to

$$F_{ftP*} = h_* \dot{h}_* \ . \tag{13.8}$$

Integrating the balance $F_{ct*} = F_{ftP*}$ using the initial condition $h_{III*}(t_* = 0) = 0$ (representing a flow situation with short regions I and II) yields for the meniscus height in region III

$$h_{III*} = \sqrt{2t_*} \ . \tag{13.9}$$

Differentiation leads to the meniscus velocity in region III

$$\dot{h}_{III*} = \frac{1}{\sqrt{2t_*}} \ . \tag{13.10}$$

The dimensionless times which separate the three flow regions can be calculated from the simplified solutions for the respective meniscus velocities. The inertia controlled region I is terminated at time t_{i*} when the increasing meniscus velocity reaches the constant value of region II. Thus $\dot{h}_{I*} = \dot{h}_{II*}$ yields the dimensionless inertia time

$$t_{i*} = 8\,(\Lambda_1 + 1)\,\mathrm{Oh}\ . \tag{13.11}$$

The values of t_{i*} for the experiment configurations are listed in Table 11.1. The transition into the final viscous controlled region III begins at t_{v*}. It can be calculated from $\dot{h}_{II*} = \dot{h}_{III*}$ and reads

$$t_{v*} = \frac{1}{2}\ . \tag{13.12}$$

These transition times which separate the three regions can also be formulated with dimensional quantities. The inertia time

$$t_i = \frac{h_0 + R}{v_c} = (h_0 + R)\sqrt{\frac{\rho R}{2\sigma}} \tag{13.13}$$

is comparable to t_{pu} (10.1) but additionally describes the influence of the initial liquid height h_0 (t_i is listed in Table 11.1). The viscous time

$$t_v = \frac{1}{2} t_c = \frac{R^2}{16\nu} \tag{13.14}$$

is proportional to t_{vu} (10.2). Since t_{i*} and t_{v*} (or t_i and t_v) are calculated from the simplified forces, they do not predict the exact intersections of F_{ir*} with $F_{ftS*} + F_{pr*}$ or $F_{ftS*} + F_{pr*}$ with F_{ftP*}, respectively, as shown in Figs. 13.4–13.7. Nevertheless (13.11)–(13.14) are useful to calculate the transition times analytically without solving the complete equation for the meniscus rise numerically.

With this results we can re-examine the $d = 4$ mm experiment configurations #40 and #43 with the maximum Ohnesorge number of $\mathrm{Oh} = 4.7 \times 10^{-3}$ shown in Fig. 13.1 (bottom left) and Fig. 13.4. Since $t_v = 0.29$ s the regions I and II are completed very rapidly and the flow directly enters the friction controlled region III. For the higher initial liquid height the inertia time t_i is larger than the viscous time t_v, thus region II with a constant flow velocity does not exist in this case. The flow enters the final state so early that the theoretical maximum velocity v_c cannot be reached. For $\mathrm{Oh} = 2.9 \times 10^{-3}$ (#22, #26, Fig. 13.1 bottom right and Fig. 13.5) t_i is smaller than t_v and region II occurs in this case. All three flow regions are run through during the experiment since $t_v = 2.2$ s is smaller than the experiment duration of 4.7 s. In the case of the minimum Ohnesorge number of $\mathrm{Oh} = 0.58 \times 10^{-3}$ (#42, #44, Fig. 13.3 top right and Fig. 13.7) is far above the experiment duration, thus mainly the meniscus reorientation and the initial stages of the flow become visible during the experiment time.

References

1. B. Fornberg: Math. Comp. **51**, 699 (1988)
2. E.M. Sparrow, S.H. Lin, T.S. Lundgren: Phys. Fluids **7.3**, 338 (1964)

14

Summary

The dynamics of capillary-driven flows of perfectly wetting liquids into circular cylindrical tubes were studied under microgravity condition using a 4.7 s drop tower. Tubes with internal diameters from $d = 4$ to 70 mm were tested with three liquids. The initial liquid height was varied from $h_0 = 10$ to 50 mm. The Ohnesorge number covered a range $5.8 \times 10^{-4} < \mathrm{Oh} < 4.7 \times 10^{-3}$. The dimensionless initial liquid height ranged $0.3 < \Lambda_1 < 25$.

An improved differential equation for the meniscus height, the meniscus velocity, and the progression of the single forces was derived from an integral relation for linear momentum. The reorientation of the free surface at the start of the capillary rise and the dynamic contact angle are considered. Furthermore the equation considers inertia forces, convective forces, and viscous forces inside the tube and the experiment container including the excess pressure loss within the entrance length L_e. The influence of each force on the total meniscus rise is explained in detail using a dimensionless representation. Very good agreement between theoretical and experimental data is found in the complete range of experiment parameters.

Figure 14.1 summarizes the general results of this investigation. During the capillary-driven flow three regions are run through successively. In region I the meniscus height is proportional to the square of time and the flow is dominated by the inertia force of the initially accelerated liquid mass. The dimensionless formulation shows that only this region is influenced by the initial liquid height Λ_1 and the Ohnesorge number. In region II, which is controlled by the convective losses at the tube entrance and within the entrance length of the tube, the meniscus height increases linear with time. The constant meniscus velocity in region II is the theoretical maximum velocity v_c. In the friction controlled region III Lucas-Washburn behavior is found and the meniscus height is proportional to the square root of time. Due to the wide range of characteristic parameters all three regions are verified by experiments.

The transition from region I to region II is described by the inertia time $t_{i*} = 8\,(\Lambda_1 + 1)\,\mathrm{Oh}$. The initial liquid height Λ_1 controls the extension t_{i*} of region I. If Λ_1 is large enough it is possible to suppress region II and the flow

Michael E. Dreyer: *Free Surface Flows under Compensated Gravity Conditions*
STMP **221**, 169–170 (2007)
DOI 10.1007/3-540-44628-1_14

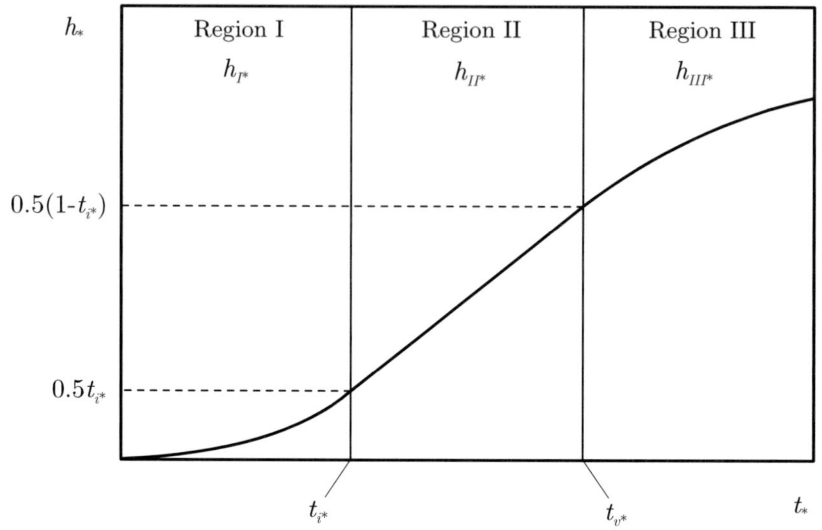

Fig. 14.1. General representation of capillary-driven flow of liquid in circular cylindrical tubes. Dimensionless meniscus height $h_* = h/(t_c v_c)$ plotted versus dimensionless time $t_* = t/t_c$

changes directly from region I into region III. In this case the inertia time equals the viscous time ($t_{i*} = t_{v*}$) and the corresponding initial liquid height reads $\Lambda_1 = 1/(16\mathrm{Oh}) - 1$, which occurred in experiment #37 (Table 11.2).

The transition from region II to region III always begins at the viscous time $t_{v*} = 0.5$. The characteristic time t_* marks the intersection of the convective forces with the friction force in the tube. For $t_* > 1$ all dimensionless velocity progressions collapse into a single curve. The extension of the three regions during a given experiment duration is determined by the Ohnesorge number. In case of high Ohnesorge numbers regions I and II are completed quickly and the Lucas-Washburn behavior of region III predominates. With decreasing Ohnesorge numbers the early phases of the flow are enlarged up to the order of the experiment duration and thus become observable. If the Ohnesorge number is decreased further also the initial meniscus reorientation and the following damped meniscus oscillation becomes visible.

Part IV

Critical Velocities in Open Capillary Channel Flows

In this chapter we consider the flow of a liquid through an open capillary channel. An open capillary channel is a structure that establishes a liquid flow path at low BOND numbers, i.e. when the capillary pressure caused by the surface tension force dominates in comparison to the hydrostatic pressure induced by gravitational or residual accelerations. The cross section of the flow path is partly confined by free surfaces. As an example the capillary channel consisting of two parallel plates is shown in Fig. 15.1. The liquid flows along the x-axis from the inlet to the outlet and forms free surfaces at the sides between the plates. The flow is maintained by external pumps and the free surfaces deform according to the pressure along the flow path.

The aim of the project is to determine the shape of the free surfaces and to find the maximum flow rate which may be achieved in the channel without a collapse of the free surface. Due to convective and viscous momentum transport the pressure along the flow path decreases and forces the free surface to bend inwards. The maximum flow rate is achieved when the free surfaces collapse and gas ingestion occurs at the outlet. This critical flow rate depends on the geometry of the channel and the properties of the liquid.

Similarities exist to compressible gas flows in ducts and open channel flows under terrestrial conditions. Each of these flows is governed by similar equations. The flow rate of these flows is limited due to choking. The theory of choked flow predicts a limiting velocity corresponding to a characteristic signal velocity of the flow. Once this critical velocity is reached the mass flow is maximal and cannot be increased further. In compressible gas duct flows the limiting speed is defined by the speed of sound. The characteristic number is the MACH number (Ma), and the maximal flow passes through the duct when $\mathrm{Ma} = 1$ is reached. In open channel flows the speed of shallow water waves defines the limiting velocity (FROUDE number problem). For the open capillary channel flows investigated in our experiment a limiting speed defined by longitudinal waves is expected (WEBER number problem). This velocity is derived from the capillary pressure and requires a measurement of the contour of the free surface.

Experiments in a drop tower and on a sounding rocket flight have been performed to reach the unique conditions to establish an inertial dominated flow in a capillary channel. The results corroborate the hypothesis that the flow rate limitation is caused by choking. The wave speed in the channel is derived from the general theory of longitudinal wave motion and compared to the local velocity in the constriction of the flow path. The ratio of both velocities is plotted versus the flow rate and approaches unity when gas ingestion occurs. In this case the liquid flow rate cannot be increased by decreasing the downstream pressure. This theory is valid for the inertial dominated regime where viscous effects can be neglected. Analytical solutions are given to predict the maximum flow rate if the viscous forces are dominant. A dimensionless length of the channel has been introduced which clearly enables to distinguish between the inertial and the viscous dominated regime.

15

Introduction and Background

In this part we consider a liquid flow path (in the following referred to as the capillary channel) whose perimeter is partially confined by a free surface. A constant flow rate Q is established by a suction pump at the outlet of the channel (see Fig. 15.1). A perfectly wetting liquid is provided at the inlet by a reservoir and a supply pump. Small, unavoidable fluctuations in the pump rates are compensated by a tube connected to the reservoir with a free surface which in addition determines the pressure in the reservoir. The liquid surfaces of the open channel and the compensation tube are exposed to the same ambient gas pressure, p_a. Both p_a and the temperature of the liquid T are kept constant.

The liquid pressure inside the channel decreases in flow direction due to convective and viscous momentum transport. The mean curvature of the free surfaces adjusts to the pressure inside the channel in the case of a steady flow situation. The maximum flow rate is achieved when the free surfaces collapse and gas ingestion occurs.

The parallel plate channel consists of two plates of breadth b at distance a apart. It was chosen as a representative for capillary vanes used in surface tension tank technology. The mathematical modeling shows that the dimensionless critical flow rate is given by a function

$$Q^*_{crit} = f(\mathrm{Oh}, \Lambda, \tilde{l}) \tag{15.1}$$

with the non-dimensional flow rate Q^*, the aspect ratio Λ and the non-dimensional flow length $\tilde{l}$. The static contact angle γ_s which determines the surface shape in the $x = \text{const}$ plane is assumed to be zero. We consider perfect wetting liquids only. The corresponding numbers for the parallel plate geometry are:

$$\mathrm{Oh} = \sqrt{\frac{\rho\nu^2}{\sigma D_h}}; \; \Lambda = b/a; \; \tilde{l} = \frac{\mathrm{Oh}\, l}{2D_h}; \; Q^* = \frac{Q}{v_0 A_0}; \; v_0 = \sqrt{\frac{4\sigma}{\rho D_h}}; \; A_0 = ab \, . \tag{15.2}$$

Michael E. Dreyer: *Free Surface Flows under Compensated Gravity Conditions*
STMP **221**, 173–175 (2007)
DOI 10.1007/3-540-44628-1_15

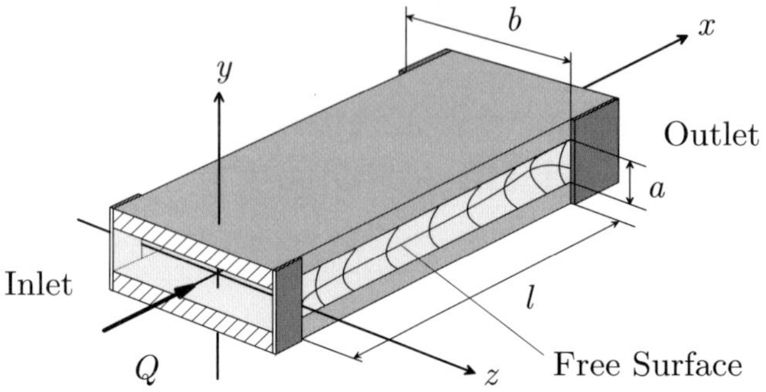

Fig. 15.1. Drawing of the parallel plate channel. The flow is parallel to the $x-$axis, enters the system at the inlet and leaves it at the outlet. The boundaries at the sides are formed by free surfaces which bend inwards due to the decreasing pressure in flow direction

The hydraulic diameter is defined as

$$D_h = \frac{4A_0}{L_w} = 2a \tag{15.3}$$

with the wetted perimeter L_w of the cross section and the initial cross section A_0. This concept is widely used in non-circular duct flow and helps to organize the friction factors for different geometries. The OHNESORGE number Oh in our case is the reciprocal value of the REYNOLDS number with v_0 as the characteristic velocity:

$$\mathrm{Oh} = \sqrt{\frac{\rho\nu^2}{\sigma D_h}} = \frac{2\nu}{D_h v_0} = \frac{2}{\mathrm{Re}_0} \quad \text{with} \quad \mathrm{Re}_0 = \frac{D_h v_0}{\nu} \,. \tag{15.4}$$

The characteristic velocity v_0 can be interpreted as a wave speed of the system and it also scales the maximum velocity of self driven capillary flows [1]. The dimensionless flow length corresponds to the length introduced in entrance flow problems (see for example [2] among others):

$$\tilde{l} = \frac{\mathrm{Oh}\; l}{2D_h} = \frac{l}{D_h \mathrm{Re}_0} \,. \tag{15.5}$$

The solution of (15.1) is achieved by numerical treatment of the governing equations which will be derived in Chap. 16. It gives the limiting flow rate in a numerical sense, i.e. the solution algorithm does not converge for flow rates higher than the critical value. The critical flow rate in the experiment is realized when the free surface becomes unstable and gas ingestion occurs at the outlet. In our theory and in experiments we performed thus far we only considered quasi-steady increases of the flow rate.

15.1 Application for Liquid Management in Space

Open capillary channels are used in space technologies to transport and position liquids. An open capillary channel is a structure of arbitrary geometry which establishes a liquid flow path at low BOND numbers. The free surface of the liquid is always exposed to the ambient pressure and the liquid is fixed inside the channel due to capillary forces. Prominent applications are heat pipes and surface tension tanks for satellites. In the latter case open capillary channels, so called vanes, provide a flow path from the bulk of the propellant to the tank outlet. Simple vanes consist of a thin metal sheet mounted parallel or perpendicularly to the tank shell. Whenever the vane is in contact with the propellant the vane fills itself due to capillary forces. Driven by a gradient of the capillary pressure between the liquid head in the vane and the bulk liquid, the propellant flows to the tank outlet. Thus, vanes are a simple device for direct supply to the tank outlet independent of the location of the bulk liquid in the tank.

However, during operation of the satellite, two undesired effects occur. Firstly, accelerations increase the BOND number to a degree the liquid flow is destabilized inside the vane. Secondly, if the vane flow limit is exceeded the flow collapses and gas will be ingested into the engine. Depending on the requirement, two technical solutions are currently used to avoid gas ingestion. In the case of low volume flow rate and low acceleration the vanes can be used for direct supply. For tank design purposes the maximal flow rate is only crudely estimated requiring high safety factors. For higher flow rates and higher BOND numbers the propellant is buffered in a refillable liquid reservoir which feeds the engine during the maneuver. After the maneuver the reservoir can be refilled by means of the vanes.

References

1. M.E. Dreyer, A. Delgado, H.J. Rath: J. Colloid Interface Sci. **163**, 158 (1994)
2. E.M. Sparrow, S.H. Lin, T.S. Lundgren: Phys. Fluids **7.3**, 338 (1964)

16

Mathematical Model for the Flow Between Parallel Plates

In our theoretical approach we consider the flow through a channel of two parallel plates. The flow enters the channel via the inlet and leaves it at the outlet. The flow is maintained by an external pump at the outlet. The inlet is connected to a reservoir which provides the necessary flow rate. For the analysis we follow the steady state approach introduced by [1] for the flow rate limitation in different capillary vanes. We assume a one-dimensional flow along the channel axis x characterized by the mean velocity v with

$$v = \frac{1}{A} \int_A u_x \mathrm{d}A \tag{16.1}$$

and the liquid pressure p. The symbol A denotes the cross section of the flow path perpendicular to the longitudinal axis x. No gradients of v or p are considered in y and $z-$direction. The origin of the coordinate system is located in the center of the cross section at the channel inlet (see Fig. 15.1). The channel inlet is defined at the location where the open cross section of the test channel starts (seen in flow direction).

The basic equations to be solved are the momentum

$$\mathrm{d}p + \rho v \mathrm{d}v - \mathrm{d}w_f = 0 \tag{16.2}$$

and the mass conservation equation

$$\mathrm{d}(Av) = 0 \ , \tag{16.3}$$

both given in differential form. The term dw_f of (16.2) takes into account irreversible pressure losses of the flow due to molecular momentum transport. The hydrostatic pressure is neglected due to the vanishing body force under microgravity conditions. The criterion for this neglect is that the Bond numbers along the three coordinate directions are smaller than 1%, i.e.

$$Bo_x = \frac{\rho k_x l a}{2\sigma} < 0.01 \ , \tag{16.4}$$

Michael E. Dreyer: *Free Surface Flows under Compensated Gravity Conditions*
STMP **221**, 177–192 (2007)
DOI 10.1007/3-540-44628-1_16

$$Bo_y = \frac{\rho k_y a^2}{2\sigma} < 0.01 \; , \tag{16.5}$$

$$Bo_z = \frac{\rho k_z b a}{2\sigma} < 0.01 \; . \tag{16.6}$$

16.1 Pressure Term and Curvature

For the first term of (16.2) we need a relation between the liquid pressure and the position of the free surface z. The liquid pressure p is related to the curvature of the free surface by the GAUSS-LAPLACE equation

$$\frac{p - p_a}{\sigma} = -2H = -\left(\frac{1}{R_1} + \frac{1}{R_2}\right) \; , \tag{16.7}$$

introduced in Chap. 2. H denotes the mean curvature of the surface, R_1 and R_2 the principal radii of curvature. Since the ambient pressure p_a is constant along the channel the pressure gradient becomes

$$\mathrm{d}p = -2\sigma \mathrm{d}H \; . \tag{16.8}$$

In general the mean curvature H of the surface is calculated by

$$H(x,y) = \frac{\left[1 + \left(\frac{\partial z}{\partial x}\right)^2\right]\frac{\partial^2 z}{\partial y^2} - 2\frac{\partial z}{\partial x}\frac{\partial z}{\partial y}\frac{\partial^2 z}{\partial x \partial y} + \left[1 + \left(\frac{\partial z}{\partial y}\right)^2\right]\frac{\partial^2 z}{\partial x^2}}{2\left[1 + \left(\frac{\partial z}{\partial x}\right)^2 + \left(\frac{\partial z}{\partial y}\right)^2\right]^{3/2}} \tag{16.9}$$

if the surface is given as $z = f(x,y)$ [2]. As shown in Figs. 15.1 and 16.1, z is the position of the surface defined by a function $z = f(x,y)$, with $0 \le x \le l$ and $-a/2 \le y \le a/2$.

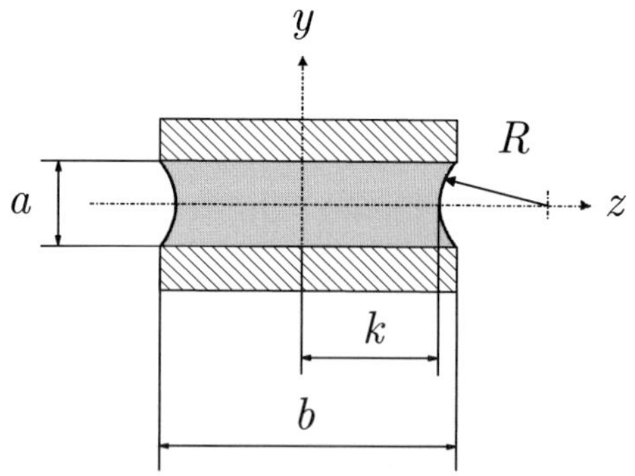

Fig. 16.1. Cross section of the capillary channel with radius of curvature R and contour length k. The optical axis of the camera is aligned parallel to the y-axis

Equation (16.9) can be simplified by applying the basic assumption of one-dimensional flows that the liquid pressure varies negligibly over each cross section. This requires

$$\frac{\partial p}{\partial y} = \frac{\partial p}{\partial z} = 0\,. \tag{16.10}$$

With this assumption the surface curvature in the cross section is constant and defined by the radius of curvature R shown in Fig. 16.1. The radius R can be calculated from the distance $k(x)$ which is defined as the free surface position in the plane $y = 0$

$$k(x) = z(x, y = 0)\,. \tag{16.11}$$

This contour of the free surface can be detected in the experiment by optical means and can be compared with the theoretical solution. The simplification of (16.9) follows because the derivative of z with respect to y vanishes along the contour line

$$\frac{\partial z}{\partial y}(y = 0) = 0\,. \tag{16.12}$$

With (16.11) and (16.12) (16.9) reduces to

$$2H(x, y = 0) = \frac{1}{R}\frac{1}{\left[1 + \left(\frac{\mathrm{d}k}{\mathrm{d}x}\right)^2\right]^{1/2}} + \frac{\frac{\mathrm{d}^2 k}{\mathrm{d}x^2}}{\left[1 + \left(\frac{\mathrm{d}k}{\mathrm{d}x}\right)^2\right]^{3/2}} = \left(\frac{1}{R_1} + \frac{1}{R_2}\right) \tag{16.13}$$

with

$$\left.\frac{\partial^2 z}{\partial y^2}\right|_{y=0} = \frac{1}{R}\,.$$

With the geometrical relation

$$R = \begin{cases} \dfrac{a^2 + (b - 2k)^2}{4(b - 2k)} & |k| \geq \left(\dfrac{b}{2} - \dfrac{a}{2}\right) \\[2ex] \dfrac{a}{2} & |k| < \left(\dfrac{b}{2} - \dfrac{a}{2}\right) \end{cases}\,, \tag{16.14}$$

the radius of curvature R can be calculated. In the first case of (16.14) the surface is assumed to be pinned at the edges of the plates. The radius of curvature may vary from $a/2 < R < \infty$. The second case allows the free surfaces to move between the plates. With (16.14) the cross sectional area can be derived in the following form

$$A = \begin{cases} ab - 2R^2 \arcsin\left(\dfrac{a}{2R}\right) + a\left(R - \dfrac{b}{2} + k\right) & |k| \geq \left(\dfrac{b}{2} - \dfrac{a}{2}\right), \\[2ex] a^2 + 2ka - \dfrac{\pi}{4}a^2 & |k| < \left(\dfrac{b}{2} - \dfrac{a}{2}\right). \end{cases} \tag{16.15}$$

The radius of curvature which appears in the x = const plane between the plates depends on the wetting property of the liquid on the solid. Our model is limited to liquids with a perfect wetting behavior on the solid, thus the static contact angle is zero and the minimal radius of curvature is equal to half the plates distance, $R = a/2$.

Equation (16.13) to (16.15) are valid only for small values of $|\mathrm{d}k/\mathrm{d}x|$, say

$$\cos\phi = \frac{1}{\left[1 + \left(\frac{\mathrm{d}k}{\mathrm{d}x}\right)^2\right]^{\frac{1}{2}}} \geq 0.9 \,, \tag{16.16}$$

thus $\phi \leq 26°$. The angle ϕ is the slope of the curve $k(x)$ or the angle between the tangent on $k(x)$ and the $x-$axis. The errors in our model for the estimation of the first principal radius of curvature (first term in (16.13)) originate from the differences between the plane which contains the principal radius of curvature R_1 (main normal plain) and the plane x = const. These two planes form the angle ϕ. The principal radii of curvature are defined in the two principal normal planes of the surface $z(x, y)$. One principal normal plane is the $y = 0$ plane due to the symmetry of the problem. The other plane is perpendicular to $y = 0$ and to the tangent vector.

16.2 Convective Momentum Transport

With the definition of the flow rate Q from (16.3)

$$Q = Av \,, \tag{16.17}$$

the second term of (16.2) can be rewritten as

$$\rho v \mathrm{d}v = -\rho\frac{Q^2}{A^3}\mathrm{d}A \,. \tag{16.18}$$

16.3 Molecular Momentum Transport

The loss of energy per unit volume dw_f due to viscous forces consists of two parts, the laminar viscous pressure loss and the additional entrance pressure loss which is a result of the change in the velocity profile from the entrance profile to the parabolic velocity distribution (POISEUILLE flow). The additional pressure loss cannot be given in analytical form. Following a paper by [3] we found

$$\frac{\mathrm{d}p}{\mathrm{d}x} = -\frac{1}{D_h \mathrm{Re}_h}\frac{\rho v^2}{2}\left[K_{pl} + 16\frac{dK_{ps}(\check{x}^*)}{d\check{x}^*}\right] \tag{16.19}$$

with $K_{pl} = 96$,

$$K_{ps} = 0.4 + \sum_i \frac{4}{\alpha_i^2} \left[e^{-\alpha_i^2 \breve{x}^*} - 2 \right] e^{-\alpha_i^2 \breve{x}^*} - 2 \int_0^{\breve{x}^*} \varepsilon \left. \frac{\partial U^*}{\partial \breve{y}} \right|_{\breve{y}=1} d\breve{x}^* \qquad (16.20)$$

and $\breve{x}^*$ as a stretched coordinate along the x–axis. The details are given in Appendix A.3.

The laminar pressure loss for two infinite parallel plates, thus the first term of (16.19), can be obtained from an exact solution of the NAVIER-STOKES equation (see for example [4])

$$\frac{\mathrm{d}p_l}{\mathrm{d}x} = -\frac{K_{pl}}{D_h \mathrm{Re}_h} \frac{\rho}{2} v^2 \, . \qquad (16.21)$$

The hydraulic diameter $D_h = 2a$ and the REYNOLDS number formed with the hydraulic diameter

$$\mathrm{Re}_h = \frac{D_h v}{\nu} \, . \qquad (16.22)$$

are used to simplify the general applicability of the data.

The application of the second term of (16.19) requires the knowledge of the velocity distribution at the inlet of the capillary channel. The velocity distribution as well as the inlet pressure can be achieved by numerical calculations for the respective geometry of the channel entrance section (nozzle from the reservoir to the inlet). To show the procedure how this can be performed we use the simplified solution from Sparrow et al. [3] with $\varepsilon = 1$ (details see Appendix A.3). This enables us to write (16.19) in a closed form:

$$\frac{\mathrm{d}p}{\mathrm{d}x} = -\frac{1}{D_h \mathrm{Re}_h} \frac{\rho v^2}{2} \left[K_{pl} + \sum_i \left(192 e^{-16\alpha_i^2 \hat{x}_s} - 128 e^{-32\alpha_i^2 \hat{x}_s} \right) \right] \qquad (16.23)$$

with

$$\hat{x}_s = \frac{x_s}{D_h \mathrm{Re}_h} \, . \qquad (16.24)$$

The coordinate x_s starts at the channel inlet $x = 0$ with a value different from zero to account for the actual velocity profile at the inlet. The work of Sparrow et al. instead considers a constant velocity profile at the inlet which transforms into the parabolic HAGEN-POISEUILLE velocity distribution after the entrance length x_e. The parameter β $(0 \leq \beta \leq 1)$ adapts the ansatz from Sparrow et al. to the actual velocity profile in the experiment set-up. The coordinate x_s reads

$$x_s = x + (1 - \beta) x_e \qquad (16.25)$$

with

$$x_e = 0.0164 D_h \mathrm{Re}_h \, . \qquad (16.26)$$

The actual profile and the corresponding pressure drop depend on the flow before the inlet. These data were computed together with the inlet pressure drop using the commercial CFD package FIDAP. The details are given in the

Appendix A.3. The parameter β is given by an empirical relation similar to the entrance pressure drop (16.35) as

$$\beta = L_1 + \frac{L_2}{\mathrm{Re}_i} \, . \tag{16.27}$$

The inlet REYNOLDS number Re_i is composed with the hydraulic diameter D_i at the inlet

$$\mathrm{Re}_i = \frac{v D_i}{\nu} ; D_i = \frac{2ab}{a+b} \, . \tag{16.28}$$

The values for L_1 and L_2 are given in the appendix in Table A.6. The meaning of (16.27) is that with $\beta = 0$ a laminar parabolic velocity profile at the inlet exists. In this case the additional pressure drop gradient from (16.20) is zero. The probability of this profile increases with decreasing REYNOLDS number. With $\beta = 1$ a block profile at the inlet is assumed, which becomes more likely with increasing REYNOLDS number. The experiment REYNOLDS numbers vary between $163 < \mathrm{Re}_h < 1063$.

Finally, the total irreversible pressure loss dw_f is determined by

$$\mathrm{d}w_f = -\frac{1}{D_h \mathrm{Re}_h} \frac{\rho}{2} \frac{Q^2}{A^2} \left(K_{pl} + 16 K_{ps}^{'} \right) \mathrm{d}x \, . \tag{16.29}$$

with $K_{pl} = 96$ and

$$K_{ps}^{'} = \frac{\mathrm{d}K_{ps}(\breve{x}^*)}{\mathrm{d}\breve{x}^*} \tag{16.30}$$

from (16.20). This expression is differentiated during the numerical solution of the final equation and cannot be given here in closed form. The general assumption for the derivation of (16.29) is a constant cross section of the channel. This assumption is violated by our approach allowing a changing cross section A along the flow path. Nevertheless, (16.29) is used due to the lack of a better modeling of the molecular momentum term.

16.4 General Equation

Substituting (16.2) by (16.8), (16.18) and (16.29) the final differential equation

$$\frac{2\sigma}{\rho} \frac{\mathrm{d}H}{\mathrm{d}x} + \frac{Q^2}{A^3} \frac{\mathrm{d}A}{\mathrm{d}x} - \frac{1}{2} \frac{K_{pf} + 16 K_{ps}^{'}}{D_h \mathrm{Re}_h} \frac{Q^2}{A^2} = 0 \tag{16.31}$$

for the change of pressure along the channel axis is obtained.

16.5 Boundary Conditions

The boundary conditions are defined by the fact that the free surfaces at the inlet and the outlet are pinned to the edges and that the flow enters the test section with a lower pressure than the ambient pressure in the gas:

$$k(x=0) = \pm\frac{b}{2} , \tag{16.32}$$

$$k(x=l) = \pm\frac{b}{2} , \tag{16.33}$$

$$p(x=0) = p_i . \tag{16.34}$$

The pressure drop at the inlet stems from the flow loss in the reservoir, i.e. before the flow enters the test section[1]. This pressure drop must be computed for the respective geometry and can be written in general form

$$p_i = -\left(K_{i1} + \frac{K_{i2}}{\mathrm{Re}_i}\right)\frac{\rho}{2}v^2 + \Delta p_{cb} . \tag{16.35}$$

The values for K_{i1}, K_{i2} and Δp_{cb} are given in the appendix in Table A.4. The capillary pressure

$$\Delta p_{cb} = p_{cb} - p_a = -2H_b\sigma \tag{16.36}$$

(following (2.7)) occurs since the reservoir has a free surface and this surface has the mean curvature H_b. The pressure drop at the inlet leads to an initial curvature of the free surface in the test section

$$-2H_i = \frac{p_i - p_a}{\sigma} . \tag{16.37}$$

16.6 Dimensionless Equations

For the mathematical treatment of the equations all lengths and areas are scaled by half the plate distance $a/2$ and $A_0 = ab$, respectively. The choice of $a/2 = D_h/4$ deviates from (15.5) but it is more suitable for the computation. The values for the flow length will be converted later into

$$\tilde{l} = \frac{\mathrm{Oh}\, l}{2D_h} = \frac{\mathrm{Oh}\, l^*}{8} . \tag{16.38}$$

The liquid velocity is scaled by the characteristic velocity $v_0 = \sqrt{2\sigma/\rho a}$ (derived from [8]). This yields the following dimensionless properties

$$x^* = \frac{2x}{a};\ k^* = \frac{2k}{a};\ l^* = \frac{2l}{a};\ R^* = \frac{2R}{a};\ 2H^* = aH;\ A^* = \frac{A}{A_0}; \tag{16.39}$$

$$v^* = \frac{v}{v_0};\ Q^* = \frac{Q}{v_0 A_0} = \frac{Q}{b\nu}\mathrm{Oh};\ \mathrm{Oh} = \sqrt{\frac{\rho\nu^2}{2\sigma a}};\ \Lambda = \frac{b}{a} . \tag{16.40}$$

The dimensionless form of (16.31) reads

[1] This pressure drop occurs before the liquid enters the test section and is different from the additional pressure loss due to the change of the inlet velocity profile in the channel.

$$\frac{\mathrm{d}H^*}{\mathrm{d}x^*} + \frac{Q^{*2}}{2A^{*3}}\frac{\mathrm{d}A^*}{\mathrm{d}x^*} - \frac{\mathrm{Oh}Q^*}{32A^*}\left(K_{pl} + 16K_{ps}^{'}\right) = 0 \tag{16.41}$$

with the curvature

$$2H^* = \frac{1}{R^*\left[1+\left(\frac{\mathrm{d}k^*}{\mathrm{d}x^*}\right)^2\right]^{1/2}} + \frac{\frac{\mathrm{d}^2k^*}{\mathrm{d}x^{*2}}}{\left[1+\left(\frac{\mathrm{d}k^*}{\mathrm{d}x^*}\right)^2\right]^{3/2}}\ , \tag{16.42}$$

the radius R^* in the cross section plane

$$R^* = \begin{cases} \dfrac{1+(\Lambda-k^*)^2}{2(\Lambda-k^*)} & \text{for} \quad |k^*| \geq \Lambda - 1 \\ 1 & \text{for} \quad |k^*| < \Lambda - 1. \end{cases}\ , \tag{16.43}$$

and the cross section area function

$$A^*(k^*) = \begin{cases} 1 - \dfrac{R^{*2}}{2\Lambda}\arcsin\dfrac{1}{R^*} + \dfrac{1}{2\Lambda}(R^* - \Lambda + k^*) & \text{for} \quad |k^*| \geq \Lambda - 1 \\ \dfrac{1}{\Lambda}(k^* + 1 - \dfrac{\pi}{4}) & \text{for} \quad |k^*| < \Lambda - 1. \end{cases} \tag{16.44}$$

Equations (16.41) and (16.42) have to be integrated numerically with the boundary conditions: (i) position of the surface at the inlet and outlet of the channel, and (ii) surface curvature at the channel inlet

$$k^*(x^* = 0) = k^*(x^* = l^*) = \Lambda \tag{16.45}$$

$$2H^*(x^* = 0) = 2H_i^*\ . \tag{16.46}$$

The non-dimensional inlet curvature reads

$$2H_i^* = -\frac{(p_i - p_a)\,D_h}{4\sigma}\ . \tag{16.47}$$

The numerical solution yields the position of the liquid surface

$$k^*(x^*) = f(\Lambda, \mathrm{Oh}, l^*, Q^*)\ . \tag{16.48}$$

Equation (16.48) shows that the surface position depends on the adjusted flow rate Q^*. For each set of Λ, Oh, and l^* several computations with different Q^* were performed. The critical flow rate was considered to be the maximal flux Q^*_{crit} for which the numerical algorithm converged. The criterion for the approach to the critical flow rate was defined as

$$\frac{|Q^*_{inst} - Q^*_{crit}|}{Q^*_{crit}} < \varepsilon = 10^{-3}\ . \tag{16.49}$$

Finally, the following solution space is obtained

$$Q^*_{crit} = f_1(\Lambda, \mathrm{Oh}, l^*)\ . \tag{16.50}$$

The results are given in Chap. 18.

16.7 Numerical Solution

The mathematical model (16.31) is a third-order ordinary differential equation in k^*. For the numerical solution it is useful to introduce twice the mean curvature as a new variable $h^* = 2H^*$. This splits the differential equation in a coupled system of one second-order equation for the free surface k^* and one first-order equation for the mean curvature h^*:

$$\frac{\mathrm{d}^2k^*}{\mathrm{d}x^{*2}} + \frac{1}{R^*}\left[1+\left(\frac{\mathrm{d}k^*}{\mathrm{d}x^*}\right)^2\right] - h^*\left[1+\left(\frac{\mathrm{d}k^*}{\mathrm{d}x^*}\right)^2\right]^{3/2} = 0 \qquad (16.51)$$

$$\frac{\mathrm{d}h^*}{\mathrm{d}x^*} + \frac{Q^{*2}}{A^{*3}}\frac{\mathrm{d}A^*}{\mathrm{d}x^*} - \frac{\mathrm{Oh}}{32}\frac{Q^*}{A^*}\left(K_{pl} + 16K'_{ps}\right) = 0 \qquad (16.52)$$

with the boundary conditions

$$k^*(x^*=0) = k^*(x^*=l^*) = \Lambda \qquad (16.53)$$

$$h^*(x^*=0) = 2H_i^* \; . \qquad (16.54)$$

With this formulation the system becomes more stable and the boundary conditions are easier to fulfill. For the solution of this nonlinear system we use second-order finite differences and the NEWTON method.

16.8 Theory of Choking

As mentioned before we expect that the limitation of flow rate occurs due to choking. The effect of choking is known from compressible gas and open channel flows under normal gravity. Choking occurs if the flow locally moves at the characteristic wave speed at which physical information propagates along the flow. The theory of choking predicts a limit of mass flux if the ratio of the fluid velocity and the characteristic signal speed becomes unity. In compressible gas duct flows the limiting speed is defined by the speed of sound. The characteristic number is the MACH number (Ma), and the maximal flow passes through the duct when Ma $= 1$ is reached. In open channel flow the speed of shallow water waves defines the limiting velocity (FROUDE number problem). The following section gives a brief introduction to the mechanism of choking. Since this effect is mostly discussed in gas dynamics we summarize the main findings obtained from compressible gas flow. As it turns out later, most relations can be transferred to open channel flow, since there is a close analogy between both flow types based on the basic equations. As a consequence, the characteristic wave speed in compressible gases and open channels, which has a key role in this phenomena, can be derived from a single universal form.

Choking is characterized by the fact that the mass flux $w = \rho v$ of an arbitrary flow has a maximum. In case of a compressible flow the mass flux is related to the MACH number

$$\mathrm{Ma} = \frac{v_g}{c_s} \tag{16.55}$$

and in case of isentropic flow the maximum flow rate per unit area w is reached when the local MACH number becomes unity [6]. In other words, choking occurs if the gas velocity v_g locally reaches the speed of sound c_s.

For a frictionless flow along a streamline this limitation can be derived analytically (see [7]). From the momentum equation

$$\mathrm{d}p = -\rho v \mathrm{d}v \tag{16.56}$$

and the definition of the speed of sound

$$c_s^2 = \left.\frac{\mathrm{d}p}{\mathrm{d}\rho}\right|_{S=\mathrm{const}} \tag{16.57}$$

the relation

$$\frac{\mathrm{d}\rho}{\mathrm{d}v} = -\frac{\rho v}{c_s^2} \tag{16.58}$$

is obtained. ρ is the density, p the pressure and S the entropy, which is kept constant. Applying the differential form of the flow rate per unit area $dw = d(\rho v) = \rho dv + v d\rho$ we find:

$$\frac{\mathrm{d}w}{\mathrm{d}v} = \rho(1 - \mathrm{Ma}^2)\ . \tag{16.59}$$

As long as the flow is subsonic the mass flux per unit area w increases with increasing velocity ($dw/dv > 0$). For supersonic flow the mass flux decreases with increasing flow velocity v ($dw/dv < 0$). At $\mathrm{Ma} = 1$ the flow runs at its maximum flow rate.

Equation (16.59) has fundamental consequences for duct flows. First we consider a frictionless flow through a smooth constriction with changing cross section area A. With (16.56), (16.57) and the equation of continuity in the differential form

$$\frac{\mathrm{d}\rho}{\rho} + \frac{\mathrm{d}v}{v} + \frac{\mathrm{d}A}{A} = 0 \tag{16.60}$$

the equation

$$\frac{\mathrm{d}v}{v} = -\frac{\mathrm{d}A}{A}\frac{1}{1 - \mathrm{Ma}^2} = -\frac{\mathrm{d}p}{\rho v^2} \tag{16.61}$$

is derived, which relates the relative change of the flow properties v and p to the change of cross section $\mathrm{d}A/A$ forced by the geometry of the duct. Obviously the change of properties are of opposite sign for subsonic or supersonic flow. The following four combinations of area change and MACH number exist:

Duct geometry	Subsonic flow	Supersonic flow
$\mathrm{d}A/A > 0$	$\mathrm{d}v/v < 0$ $\mathrm{d}p/\rho v^2 > 0$	$\mathrm{d}v/v > 0$ $\mathrm{d}p/\rho v^2 < 0$,
$\mathrm{d}A/A < 0$	$\mathrm{d}v/v > 0$ $\mathrm{d}p/\rho v^2 < 0$	$\mathrm{d}v/v < 0$ $\mathrm{d}p/\rho v^2 > 0$.

When a subsonic flow passes through a duct with increasing cross section the flow velocity decreases and the pressure increases. But in case of supersonic flow the flow velocity decreases leading to an increase of pressure. The opposite behavior occurs for an area decrease, which speeds up a subsonic flow and slows down a supersonic flow. At the sonic point $\mathrm{Ma} = 1$ a singularity occurs in (16.61) caused by the expression $1 - \mathrm{Ma}^2$ in the denominator. Since infinite acceleration is physically impossible the change in cross section dA/A has to vanish at the sonic point [4]. $dA/A = 0$ is related to the minimal and maximal cross section areas of the duct. But it can be shown that critical flow conditions occur only in minimal cross sections like at the outlet of a nozzle or the throat of a converging/diverging nozzle.

The mechanisms of choking can be discussed for a flow through a nozzle. We consider a nozzle with constant entrance pressure p_0 which discharges into a region where the back pressure p_b is controllable. At a constant pressure ratio $p_b/p_0 < 1$ a steady subsonic flow at a certain exit MACH number appears. If the pressure ratio is increased the flow accelerates associated with an increased ratio $\mathrm{d}v/v$. As (16.61) indicates increased ratios of $\mathrm{d}v/v$ lead to higher values of the MACH number since the local ratio $\mathrm{d}A/A$ does not change. At a certain pressure ratio p_b^*/p_0 the exit MACH number is unity. A further increase of the MACH number is impossible since (16.61) requires a change of sign of dA/A. The pressure drop $\Delta p = p_0 - p_b^*$ along the nozzle is maximal at this state, just as the mass flux per unit area w given by (16.59). Even if the back pressure p_b is reduced strongly below the value p_b^* no additional pressure drop and mass flux are achieved in the nozzle. This effect, that the mass flux becomes maximal if the MACH number is unity, is called choking.

In the case discussed above the flow was assumed to be frictionless and choking occurred due to the change of cross section. But gas flow in a pipe of constant cross section also can be choked. The reason for that is the friction which causes the pressure drop and accelerates the gas. To explain the effect we choose the following three working relations [6] for adiabatic flow

$$\frac{\mathrm{d}p}{p} = -\frac{\kappa \mathrm{Ma}^2[1 + (\kappa - 1)\mathrm{Ma}^2]}{2(1 - \mathrm{Ma}^2)} \frac{4K}{D_h} \mathrm{d}x \,, \tag{16.62}$$

$$\frac{\mathrm{d}\rho}{\rho} = -\frac{\kappa \mathrm{Ma}^2}{2(1 - \mathrm{Ma}^2)} \frac{4K}{D_h} dx = -\frac{\mathrm{d}v}{v} \,, \tag{16.63}$$

$$\frac{\mathrm{dMa}^2}{\mathrm{Ma}^2} = -\frac{\kappa \mathrm{Ma}^2 \left[1 + 0.5(\kappa - 1)\mathrm{Ma}^2\right] \mathrm{Ma}^2)}{1 - \mathrm{Ma}^2} \frac{4K}{D_h} \mathrm{d}x \,, \tag{16.64}$$

where K is the friction factor defined as

$$K = \frac{2\tau_w}{\rho v^2} \,. \tag{16.65}$$

Equation (16.62) to (16.64) depend on wether the MACH number is larger or smaller than unity, since the term $(1 - \mathrm{Ma}^2)$ appears in the denominator of each of these equations. Equation (16.64) shows that with increasing duct length the MACH number tends down toward unity.

Characteristic Velocity

The liquid flow in the capillary channel is treated as a longitudinal motion although transverse velocity components occur due to the change of cross section. This assumption is valid since the liquid acceleration in z-direction is sufficiently smaller than the acceleration in the main direction. For that reason lateral pressure gradients can be neglected and the motion of liquid is essentially caused by longitudinal gradients.

The characteristic wave speed at which physical information in the flow propagates is of the same flow type. The basic assumption in the theory of longitudinal wave motion is that the variation of the driving pressure

$$p^* = p - p_0 \; . \tag{16.66}$$

over the cross section is negligible. p is pressure at position x when the wave passes and p_0 is the undisturbed pressure of the fluid in the channel. If transversal components of the driving pressure p^* are negligible the wave motion is forced by longitudinal gradients of p^*.

For the derivation of the wave speed, the channel is supposed to be uniform and the fluid at rest. With this the flow through cross section area A does not depend explicitly on the x-axis, and is related only to the driving pressure

$$A = A(p^*) \; . \tag{16.67}$$

In order to obtain a universal formulation including waves in gases, compressibility is included. If the entropy S is kept constant the density is related to the driving pressure

$$\rho = \rho(p^*) \; . \tag{16.68}$$

The equations of the one-dimensional motion to be solved are the momentum equation

$$\rho \frac{\partial v}{\partial t} = -\frac{\partial p}{\partial x} - \rho v \frac{\partial v}{\partial x} \tag{16.69}$$

and the equation of continuity,

$$\frac{\partial(\rho A)}{\partial t} = -\frac{\partial(\rho A v)}{\partial x} \; . \tag{16.70}$$

Using the following approach

$$p = p_0 + p^*, \; v = v^*, \; A = A_0 + A^*, \; \rho = \rho_0 + \rho^* \tag{16.71}$$

(16.69) and (16.70) are linearized. The basic idea of (16.71) is that the driving pressure p^* is small compared to the fluid pressure at rest p_0. In response to

p^* all other physical quantities will change within the same order of magnitude. Thus the solution for each quantity can be taken as a superposition of the fundamental solution indicated by index zero and a small disturbance indicated by the asterisk in (16.71). Applying the linearization, (16.69) and (16.70) read

$$\rho_0 \frac{\partial v^*}{\partial t} = -\frac{\partial p^*}{\partial x} , \tag{16.72}$$

$$\frac{\partial}{\partial t} (\rho_0 A^* + \rho^* A_0) = -\rho_0 A_0 \frac{\partial v^*}{\partial x} . \tag{16.73}$$

Differentiating with respect to x and t both equations can be combined

$$\frac{1}{\rho_0 A_0} \frac{\partial^2}{\partial t^2} (\rho_0 A^* + \rho^* A_0) = \frac{1}{\rho_0} \frac{\partial^2 p^*}{\partial x^2} , \tag{16.74}$$

which already has the structure of a wave equation. To introduce the pressure p^* in the left hand side of (16.74) A^* and ρ^* are expanded in Taylor series of first order about $p^* = 0$. Since the partial derivative with respect to t does not act on the Taylor series (16.74) can be written as

$$\frac{1}{A_0} \left(\rho_0 \left. \frac{\partial A^*}{\partial p^*} \right|_{p^*=0} + A_0 \left. \frac{\partial \rho^*}{\partial p^*} \right|_{p^*=0} \right) \frac{\partial^2 p^*}{\partial t^2} = \frac{\partial^2 p^*}{\partial x^2} \tag{16.75}$$

or

$$\frac{\partial^2 p^*}{\partial t^2} = A_0 \left. \frac{\partial p}{\partial (\rho A)} \right|_{p^*=0} \frac{\partial^2 p^*}{\partial x^2} . \tag{16.76}$$

Equation (16.76) has the structure of the well know linear wave equation

$$\frac{\partial^2 p^*}{\partial t^2} = c^2 \frac{\partial^2 p^*}{\partial x^2} \tag{16.77}$$

in which the factor c^2 with

$$c = \sqrt{A_0 \left. \frac{\mathrm{d}p}{\mathrm{d}(\rho A)} \right|_{p^*=0}} \tag{16.78}$$

is the wave speed of longitudinal waves in flows in which the density and the cross section are related to the fluid pressure [5]. Examples are compressible constant-area duct flows, free surface flows in channel with uniform walls or flows in flexible tubes.

For the physical interpretation (16.78) is rewritten as:

$$\frac{1}{\rho_0 c^2} = \left. \frac{1}{\rho} \frac{\mathrm{d}\rho}{\mathrm{d}p} \right|_{p^*=0} + \left. \frac{1}{A} \frac{\mathrm{d}A}{\mathrm{d}p} \right|_{p^*=0} = \kappa + \zeta . \tag{16.79}$$

If the second term on the right hand side is not considered, (16.79) yields the definition of the compressibility κ of the gas. Obviously the relative change of cross section per unit change of pressure, which is the second term, acts like the compressibility. It is called the distensibility ζ and (16.79) can be interpreted as the effective compressibility of compressible fluid flow in a channel with variable cross section. This is one indication for the above mentioned close analogy between compressible gas and open channel flows. Both the motion of longitudinal waves in compressible gases and in open channel are given by the solution of the same wave equation (16.77). As a consequence the wave speeds in both different flow types are calculated from (16.78), and finally (16.79) shows that the variable cross section A in open channel flows acts like the density ρ in compressible duct flows and vice versa. With this correlation many effects known from gas dynamics can be transferred qualitatively to open channel flows just by replacing ρ by A.

In the following, two special cases of (16.78) are derived where either the cross section or the density are kept constant. The first case considers a gas flow at constant cross section area $A = A_0$, and (16.78) reduces to the well known speed of sound

$$c_s = \sqrt{\left.\frac{\mathrm{d}p}{\mathrm{d}\rho}\right|_{p^*=0,\ S=\mathrm{const}}} \ . \tag{16.80}$$

Scaling the velocity of the gas v_g by (16.80) the definition of the MACH number (16.55) is obtained. Note that $S = \mathrm{const}$ was assumed in (16.68).

The second case considers a liquid flow through an open channel at constant density $\rho = \rho_0$. Here the cross section A is variable and the wave speed is obtained from

$$c_{ch} = \sqrt{\left.\frac{A_0}{\rho_0}\frac{\mathrm{d}p}{\mathrm{d}A}\right|_{p^*=0}} \ . \tag{16.81}$$

If a flow under gravity is considered, the driving pressure p is given by the hydrostatic pressure $p = \rho_0 g h$. h is the fill height of a uniform rectangular channel of breadth b. With the hydrostatic pressure and the cross section $A = bh$ (16.81) yields

$$c_{sw} = \sqrt{gh} \tag{16.82}$$

which is the wave speed of water waves in the shallow water approximation. Scaling the liquid flow velocity v_f by (16.82) the definition of the FROUDE number

$$\mathrm{Fr} = \frac{v_f}{c_{sw}} \tag{16.83}$$

is obtained. If capillary forces are dominant, the pressure in (16.81) has to be replaced by the capillary pressure defined by (16.7) which yields

$$v_c = \sqrt{-\frac{2\sigma A}{\rho}\frac{\mathrm{d}H}{\mathrm{d}A}} \ , \tag{16.84}$$

the longitudinal wave speed in an open capillary channel. A problem occurs concerning the application of this equation since the term dH/dA is not determined analytically. For the first approach we solve this problem by neglecting the surface curvature in flow direction x, thus

$$\frac{\mathrm{d}^2 k}{\mathrm{d}x^2} = 0\,, \frac{\mathrm{d}k}{\mathrm{d}x} \ll 1\,. \tag{16.85}$$

Applying this to (16.13) we obtain

$$v_c = \sqrt{\frac{\sigma}{\rho}\frac{A}{R^2}\frac{\mathrm{d}R}{\mathrm{d}A}}. \tag{16.86}$$

Scaling the mean liquid velocity $v = Q/A$ by (16.86) the dimensionless parameter

$$\mathrm{We} = \frac{Q}{Av_c} = \frac{v}{v_c} \tag{16.87}$$

is defined, which can be interpreted as a WEBER number.

In the following, a set of differential equations is derived to show the behavior of the capillary free surface channel flow. Using the definition of the WEBER number (16.31) can be rewritten and the following formulas

$$\frac{\mathrm{d}p}{\rho v^2} = -\frac{1}{2\left(1-\mathrm{We}^2\right)}\frac{K_{pf}+K_{ps}^{'}}{\mathrm{Re}_h}\frac{\mathrm{d}x}{D_h}\,, \tag{16.88}$$

$$\frac{\mathrm{d}A}{A} = -\frac{\mathrm{We}^2}{2(1-\mathrm{We}^2)}\frac{K_{pf}+K_{ps}^{'}}{\mathrm{Re}_h}\frac{\mathrm{d}x}{D_h} = -\frac{\mathrm{d}v}{v}\,, \tag{16.89}$$

$$\frac{\mathrm{dWe}^2}{\mathrm{We}^2} = \frac{3\mathrm{We}^2}{2(1-\mathrm{We}^2)}\frac{K_{pf}+K_{ps}^{'}}{\mathrm{Re}_h}\frac{\mathrm{d}x}{D_h} \tag{16.90}$$

are obtained. (16.88) through (16.90) have the same structure as (16.62) through (16.64) which were derived for a compressible flow in a duct of constant cross section. The direction of change in the stream properties p, A, v and We depend on whether the WEBER number is greater or less than unity. As (16.79) indicates, the density ρ and the cross section A physically act in the same way. For that reason (16.89) is obtained by replacing ρ by A in (16.63). All properties vary with increasing channel length which is equivalent to an increase of flow rate Q if the channel length is constant. Since it is known that the cross section A and pressure p decreases along the channel axis while the liquid accelerates, it follows from (16.88) and (16.89) that the investigated flow must be subcritical. It has to be shown by experiments, if a supercritical flow is possible in open capillary channel flows. In both flow types the WEBER number tends towards unity, as (16.90) shows, if the flow rate is increased. At $\mathrm{We} = 1$ the flow is choked. If the flow rate is increased further by the pump no additional liquid flow rate is obtained, but gas ingestion occurs to satisfy the continuity equation.

References

1. D.E. Jaekle, Jr.: AIAA 91-2172, 1 (1991)
2. I.N. Bronstein, K.A. Semendjajew: *Taschenbuch der Mathematik*, (BSB B.G. Teubner Verlagsgesellschaft 1977)
3. E.M. Sparrow, S.H. Lin, T.S. Lundgren: Phys. Fluids **7.3**, 338 (1964)
4. F.M. White: *Viscous Fluid Flow*, (McGraw-Hill, New York 1974)
5. J. Lighthill: *Waves in Fluids*, (Cambridge University Press, Cambridge 1978)
6. A.H. Shapiro: *The Dynamics and Thermodynamics of Compressible Fluid Flow*, (The Ronald Press Company, New York 1953)
7. L.D. Landau, E.M. Lifschitz: *Lehrbuch der theoretischen Physik, Band VI, Hydrodynamik*, (Akademie Verlag, Berlin 1991)
8. M.E. Dreyer, A. Delgado, H.J. Rath: J. Colloid Interface Sci. **163**, 158 (1994)

17

Experimental Results

Capillary flows in open channels consisting of two parallel plates have been investigated in this research project with a large number of experiments using the drop tower, parabolic aircraft and sounding rocket. The first experiments were carried out in the Bremen drop tower and were based on the experiments of capillary liquid rise between two parallel plates performed by Dreyer [1] and Dreyer et al. [2]. They investigated the capillary rise of various liquids in channels of different geometries and predicted theoretically the time-dependence of the liquid rise.

For the first generation of drop tower experiments, the set-up was modified. A withdrawal device was mounted at the end of the parallel plates which stood upright in a container of liquid. During the experiments the capillary liquid rise was used to fill the channel before the withdrawal device was started to establish a liquid flow between the parallel plates. Several experiments with various liquids and different channel properties were performed[1]. The maximal flow rate determined from the experiments showed good agreement with the theoretical prediction presented in Chap. 16. Nevertheless the flow conditions provided by the set-up were not satisfying. Due to the transition from the self-driven capillary flow to the forced flow, inertia effects occurred which made a controlled approach of the critical value impossible. Furthermore the established flow rate had to be kept constant since the short experimental time did not allow an increase of the flow rate.

To operate under longer microgravity time an experiment module was developed and flown on the sounding rocket TEXUS-37[2]. The preparations for TEXUS led to a new experimental concept. The set-up allows an increase of the flow rate in fine steps up to the critical value. To keep the inertia effects small the flow is established by two synchronized pumps. Three-dimensional

[1] The drop tower experiments were funded by the European Space Agency (ESA).
[2] The sounding rocket experiment was performed with funds of the European Space Agency (ESA) (hardware and the flight opportunity) and the German Aerospace Center (DLR) (research project).

Michael E. Dreyer: *Free Surface Flows under Compensated Gravity Conditions*
STMP **221**, 193–218 (2007)
DOI 10.1007/3-540-44628-1_17

model computations were performed to optimize the flow conditions in the experiment cell and to define the boundary conditions of the flow precisely. At the same time a new experiment module for use in the drop tower and parabolic flights was built. It is based on the TEXUS concept and its operation is analogous to that of the TEXUS experiment. Thus, the concept of the TEXUS experiment module could be tested in advance. The parabolic flights were necessary to get practical experience for the experiment control under the longer microgravity conditions of the TEXUS flight.

The following sections give an overview of the different experiments performed in the drop tower and on board the sounding rocket TEXUS-37. The experiment set-up and the main results are discussed. Since the idea of these experiments was motivated by the application of capillary channels (vanes) in surface tension tanks, the experimental parameters were chosen with respect to propellant data (N_2O_4 and mono-methyl-hydrazine, MMH) and tank geometries. Satellite tanks usually operate in the temperature range $0 \leq T \leq 40°\text{C}$ and typical geometries are: gap distances $1 \text{ mm} \leq a \leq 10 \text{ mm}$, plate breadths $10 \text{ mm} \leq b \leq 30 \text{ mm}$. This leads to the parameter space given by the combination $4.27 \cdot 10^{-4} \leq \text{Oh} \leq 5.881 \cdot 10^{-3}$ and $3.3 \leq \Lambda \leq 10$. Not all desired combinations of Oh and Λ can be obtained in drop tower experiments. The choice of parameters is restricted essentially by two factors, the capillary rise time and the optical resolution. The rise time is required to fill the channel before the actual experiment can be started. It has to be kept small to get maximal observation time of the channel flow. For the evaluation of the free surface a minimum resolution of the video pictures is necessary. To operate with a resolution of 0.1 mm/pixel the breadth and the length of the plates are limited to $b = 30$ mm and $l = 100$ mm, respectively. Based on a defined set of Oh and Λ, the length of the channel has to be chosen with respect to the rise time. Using a low viscosity silicone oil (Silicone Fluid 200 DOW Corning 0.65 cSt) the minimal value of $\text{Oh} = 1.52 \cdot 10^{-3}$ leading to an appropriate rise time was attained.

The parameter space of the experiments is listed in Table 17.1. All fluids show perfect wetting behavior at the capillary walls, thus the static contact angle is zero. The properties of the test liquids are given in Table 17.2. The parameter set TEXUS-37 was chosen for the experiment on board the sounding rocket. Due to technical reasons the channel length was slightly shorter than in the drop tower experiment #35a.

17.1 Experiments in a Drop Tower

The experiment was developed to operate within a drop capsule of the Bremen drop tower. With the 110 m long drop tube evacuated to approximately 10 Pa the drop tower provides a free fall of 4.74 s. The residual acceleration measured in the capsule is less than 10^{-5} g_0. Figure 17.1 shows the schematic drawing of the experimental set up. The capillary channel consists of two parallel

Table 17.1. Experiment parameter for the estimation of critical velocities in open capillary channel

No.	Liquid	a (mm)	b (mm)	l (mm)	t_r (s)	Oh (-)	Λ (-)	$\tilde{l}$ (-)
29	FC-77	2	20	74.9	1.7	$4.72 \cdot 10^{-3}$	10	$4.4 \cdot 10^{-2}$
32	SF 0.65	3	10	94.2	1.3	$1.96 \cdot 10^{-3}$	3.3	$1.5 \cdot 10^{-2}$
33	SF 0.65	3	15	96.3	1.2	$1.96 \cdot 10^{-3}$	5	$1.6 \cdot 10^{-2}$
34	SF 0.65	3	30	94.6	1.2	$1.96 \cdot 10^{-3}$	10	$1.6 \cdot 10^{-2}$
35	SF 0.65	5	25	93.4	1.5	$1.52 \cdot 10^{-3}$	5	$7.0 \cdot 10^{-3}$
35a	SF 0.65	5	25	48.4	0.7	$1.52 \cdot 10^{-3}$	5	$4.0 \cdot 10^{-3}$
TEXUS-37	SF 0.65	5	25	47.0	-	$1.52 \cdot 10^{-3}$	5	$4.0 \cdot 10^{-3}$

Table 17.2. Properties of the test liquids at $T = 20°$C. The static contact angle is given for the test liquids on acrylic glass

Liquid	ρ (kg/m^3)	ν 10^{-6}(m^2/s)	σ 10^{-3}(N/m)	γ_s (°)
FC-77	1789	0.865	15.0	0
SF 0.65	766	0.690	15.8	0

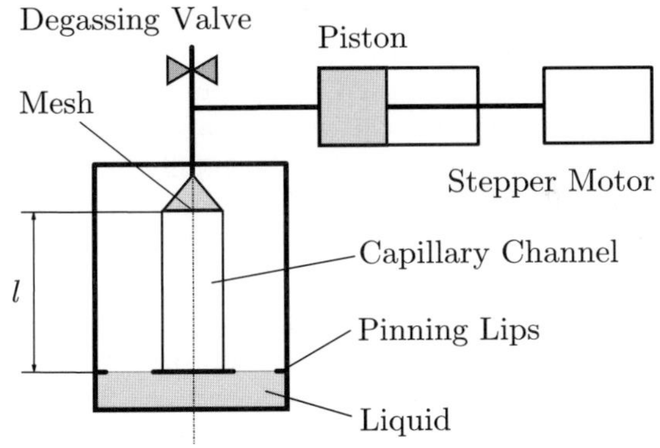

Fig. 17.1. Schematic drawing of the experiment setup for the drop tower. The withdrawal device is filled with liquid up to the mesh before the drop (normal gravity). During the microgravity period the capillary channel fills itself and pumping starts when the meniscus reaches the screen

plates (front view) and is fixed on the bottom of a cylindrical container. For the camera observation the channel and the container are made from acrylic glass (PMMA). On the top of the channel a suction device is mounted. It is connected to a piston driven by a gear system and a stepper motor.

To avoid liquid rise outside the channel during the experiments, circular pinning lips are mounted at a certain height inside the liquid container and outside the channel. The container is filled with the test liquid up to the bottom edges of the lips. Since both plates are dipped into the liquid, the channel is filled up to the wetting lips under 1g-conditions. To operate the experiment under a defined liquid temperature, the temperature of the container is controllable. The channel is observed by two CCD cameras located outside the liquid container. Their optical axes are aligned perpendicular to the front plate of the channel. Both sections of the camera have an overlap in the middle of the channel (only necessary for the large channels). Markings on the outside of the channel define the orientation and the resolution of the picture for the digital image processing. The resolution is 0.1 mm/pixel at a frame rate of 25 frame per seconds. For a homogeneous illumination fluorescent light and a Teflon diffuser were installed behind the back plate.

During the 90 minutes long evacuation of the drop tube, the liquid temperature and the capsule pressure are kept constant. Before the drop the suction side is filled up to the screen located at the inlet of the suction head. Gas bubbles are removed by the degassing valve. With the release of the capsule at time t_0 the liquid starts to rise between the plates. The liquid movement is caused by the capillary pressure gradient between the liquid surface in the container and the liquid surface in the gap of the plates. When the liquid meniscus contacts the suction device at t_r, the piston is started and a defined constant flow rate Q is established. The rise time t_r can be predicted numerically from the meniscus acceleration derived by [2].

The choice of parameters is limited essentially by the rise time t_r and the resolution of the pictures. The lowest Oh numbers can be achieved with the lowest viscosity fluid SF 0.65 and the widest gap a = 5 mm (#35). Larger gap distances lead to longer rise times for the capillary to be filled before the suction starts and thus to shorter observation times of the flow through the capillary itself. To obtain a resolution of 0.1 mm/pixel the channel length was restricted to 100 mm.

A total of 40 experiments were performed with an average of 5 drops for each parameter set. The first drop served to determine the exact rise time t_r of the liquid. This time has to be estimated within ± 0.05 s. If the withdrawal device starts too early, gas is ingested at the beginning of the suction. If the withdrawal device starts too late, the inertia of the rising meniscus tends to form convex menisci at the sides of the capillary. The additional volume contained in this overshoot must be withdrawn and reduces the observation time of the experiment. The following experiments for each parameter set were aimed at the determination of the critical flow rate. Due to the good agreement of the theory with the experiment the critical flow rate could be predicted with good accuracy and the number of drops could be minimized. For each parameter set one stable and one unstable flow rate with gas ingestion could be achieved. The highest stable flow rate was reproduced to minimize experiment errors.

The capillary rise necessary to fill the channel is shown in Fig. 17.2. The pictures show how the liquid meniscus moves into the channel until the

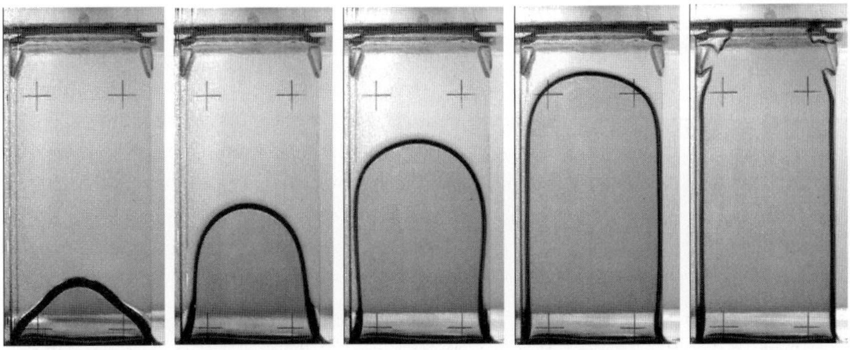

Fig. 17.2. The liquid rises between the parallel plates from the *bottom* to the *top* due to capillary forces (parameter set 35a). No liquid has been withdrawn at the *top* by means of the suction device

meniscus contacts the suction device. Since the camera view is perpendicular to the front plate the free surface appears as a dark contour corresponding to the distance k defined in Fig. 16.1. Note that the driving pressure gradient is determined by the concave radius of curvature in the gap between the plates (visible as a shadow along the free surface).

After the forced flow is established, two typical observations depending on the adjusted Q are made. Below the critical value of flow rate Q_{crit} ($Q < Q_{crit}$) the flow between the plates is steady. The free surface is stable and the corresponding contour of the surface is constant in time, $k = k(x)$. For $Q > Q_{crit}$ the surface collapses and gas ingestion occurs. The free surface becomes time-dependent, $k = k(x, t)$. The withdrawn total flow rate now is a superposition of the maximal liquid flow rate and an additional flow rate of gas. Figure 17.3 shows a stable flow at $Q = 8.75$ ml/s. After the channel was filled by the capillary rise of the liquid the piston was started at $t = 0.78$ s. Surface waves caused by the starting piston travel upstream ($t = 0.84$ s). When the waves are damped out a slight oscillation of the surface can be observed. These oscillations are induced by pressure fluctuations occurring due to the reorientation of the liquid surface in the container under microgravity. The oscillation decays at the end of the experimental time and the surface reaches its final position.

Increasing the flow rate about 3% to $Q = 9.04$ ml/s yields an unstable flow as shown in Fig. 17.4. The surface becomes unstable at the left side of the channel ($t = 1.53$ s), and a gas bubble is ingested into the withdrawal device ($t = 2.25$ s). The surface stabilizes ($t = 2.33$ s) and the process starts again ($t = 3.05$ s). For the determination of the critical flow rate of the channel, several drop experiments are necessary. By adjusting the flow rate alternating above and below the critical value, the upper limit of a stable flow is iterated. All other parameters are kept constant. The critical flow rate is defined as

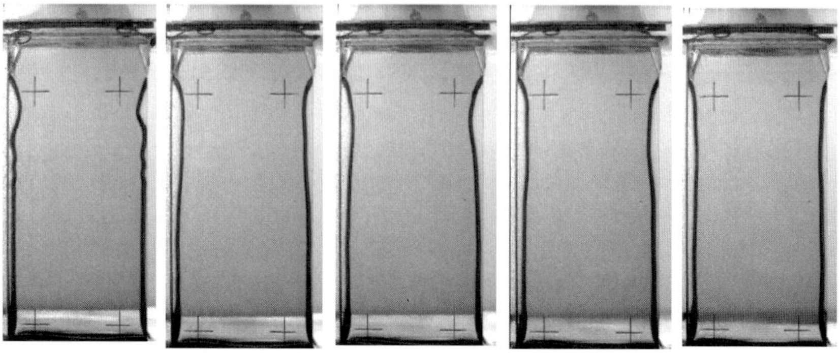

$t = 0.84$ s $t = 1.40$ s $t = 1.68$ s $t = 2.12$ s $t = 3.48$ s

Fig. 17.3. Subcritical flow in a channel of parallel plates (35a) with $Q = 8.75$ ml/s. The flow is from the *bottom* to the *top*. The liquid withdrawal began at $t = 0.78$ s. Surface waves initiated by the suction first travel upstream ($t = 0.84$ s), then the flow stabilizes

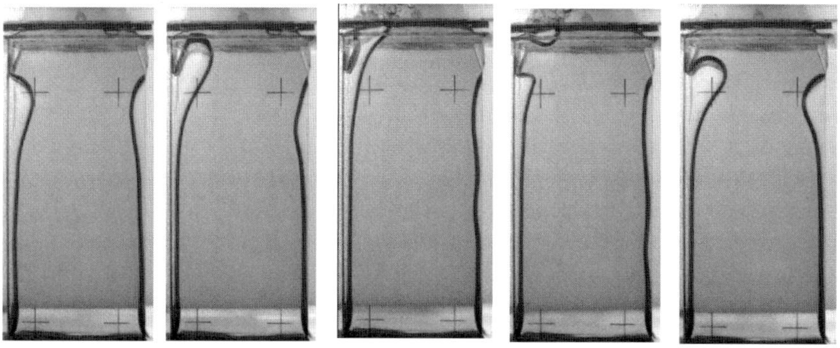

$t = 1.53$ s $t = 2.01$ s $t = 2.25$ s $t = 2.33$ s $t = 3.05$ s

Fig. 17.4. Non-steady liquid flow with gas ingestion (35a) at $Q = 9.04$ ml/s. The free surface becomes unstable ($t = 2.05$ s), a gas bubble is ingested. The surface stabilizes ($t = 2.3$ s) and the cycle starts again ($t = 3.05$ s)

$$Q_{crit}^{\exp} = \frac{1}{2}(Q_{\max}^{stable} + Q_{\min}^{instable})$$

where $Q_{\max}^{stable}$ is the maximum stable flow rate and $Q_{\min}^{instable}$ is the minimum flow rate leading to gas ingestion. The results of the experiments and the theoretical prediction of the critical flow rate are listed in Table 17.3. The temperature for all experiments was set to $T = 20°\text{C}$. The calculations were performed with (16.41). The REYNOLDS number for the critical velocity $\text{Re}_{h,crit} = v_{crit} D_h / \nu$ is well below the transition value for turbulent flow, thus laminar flow conditions are expected. The entrance pressure drop for each parameter set was calculated using the commercial CFD code FIDAP. The

Table 17.3. Results of the drop tower experiments for the determination of critical velocities in open capillary channel flows

No.	$Q_{\max}^{stable}$ (ml/s)	$Q_{\min}^{instable}$ (ml/s)	$Q_{crit}^{\exp}$ (ml/s)	Q_{crit}^{theo} (ml/s)	$Q_{\exp}^{*}$ (-)	Q_{theo}^{*} (-)	$\lvert\Delta Q^{*}\rvert / Q_{theo}^{*}$ (-)	$\mathrm{Re}_{h,crit}$ (-)
29	1.47	1.49	$1.48 \pm 0.60\%$	1.375	0.404	0.375	0.07	163
32	1.75	1.86	$1.81 \pm 3.00\%$	2.03	0.515	0.577	0.11	606
33	2.78	2.93	$2.86 \pm 2.63\%$	3.11	0.542	0.589	0.08	618
34	6.10	6.23	$6.17 \pm 1.00\%$	6.31	0.585	0.598	0.02	626
35	7.44	7.97	$7.71 \pm 3.40\%$	8.09	0.679	0.713	0.05	959
35a	8.75	9.04	$8.90 \pm 1.60\%$	8.83	0.784	0.778	0.01	1063

results have been normalized using (16.35) and the results are given in Table A.4. The capillary pressure drop due to the curvature of the reservoir has been neglected.

The dimensionless flow rate is equal to the dimensionless velocity at the channel inlet and outlet, thus we can plot

$$v_{theo}^{*} = \frac{Q_{crit}^{theo}}{A_0 v_0} = Q_{theo}^{*} \tag{17.1}$$

against

$$v_{\exp}^{*} = \frac{Q_{crit}^{\exp}}{A_0 v_0} = Q_{\exp}^{*} \ . \tag{17.2}$$

The results of the computation compared with the experimental data are depicted in Fig. 17.5. The maximum dimensionless velocity evaluated from the experiment $v_{\exp}^{*}$ is used as the abscissa, the calculated velocity v_{theo}^{*} is plotted on the ordinate. The dotted lines mark the $\pm 10\%$ deviation from the identity, all data points are located within this interval. The labels at the data point refer to the data set number, the dimensionless length of the capillary and the aspect ratio. It can be seen that with decreasing length the velocity increases, and that with increasing gap ratio the velocity increases. The coincidence between experiment and theory is better than $\pm 10\%$. It is shown in Chap. 18 that for large channel lengths $\left(\tilde{l} \geq 10^{-1}\right)$ and $A^{*} = 1$ (or $\Lambda \to \infty$), the flow is dominated by molecular momentum transport and the critical flow velocity is given by

$$v_{crit}^{*} = \frac{2}{K_{pl}\,\tilde{l}}$$

(18.7). If we divide the critical flow rate (or velocity) of the experiment $Q_{\exp}^{*} = v_{\exp}^{*}$ by (18.7) and plot the results versus the channel length, i.e.

$$\frac{v_{\exp}^{*} K_{pl} \tilde{l}}{2} = v_{vis}^{*} = f\left(\tilde{l}\right)$$

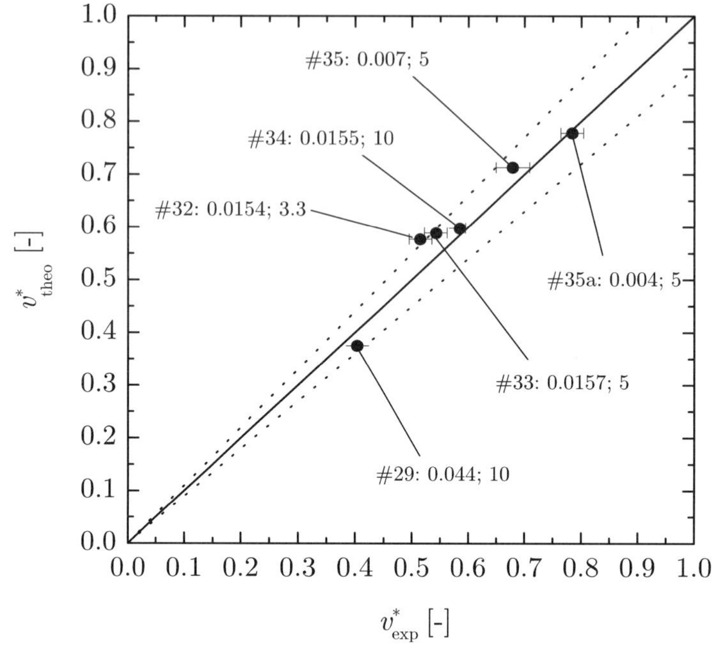

Fig. 17.5. Comparison of experimental and theoretical data for the maximum dimensionless velocity between parallel plates. The labels refer to the data set number, the dimensionless length of the capillary channel $\tilde{l}$ and the gap ratio Λ, respectively. The *dotted lines* mark the ±10% deviation from the identity

we find a power law dependence as shown in Fig. 17.6. The dependence reads

$$v^*_{vis} = 8.4\,\tilde{l}^{\,0.73}$$

with a regression coefficient of 0.995 and a standard deviation of 0.029. If we set the exponent to 3/4, an empirical relation for the critical velocity can be given:

$$v^*_{\text{exp}} = 0.175\,\tilde{l}^{\,0.25} . \tag{17.3}$$

Differences for other configurations may occur from different inlet flow conditions, leading to different H_i and β than those used for the calculation of these results.

17.2 Experiment on a Sounding Rocket

An experiment module was developed for the sounding rocket TEXUS[3] with the parameter set 35a (Table 17.1). TEXUS provides a microgravity

[3] TEXUS = Technologische Experimente unter Schwerelosigkeit, funded by ESA and in cooperation with DLR

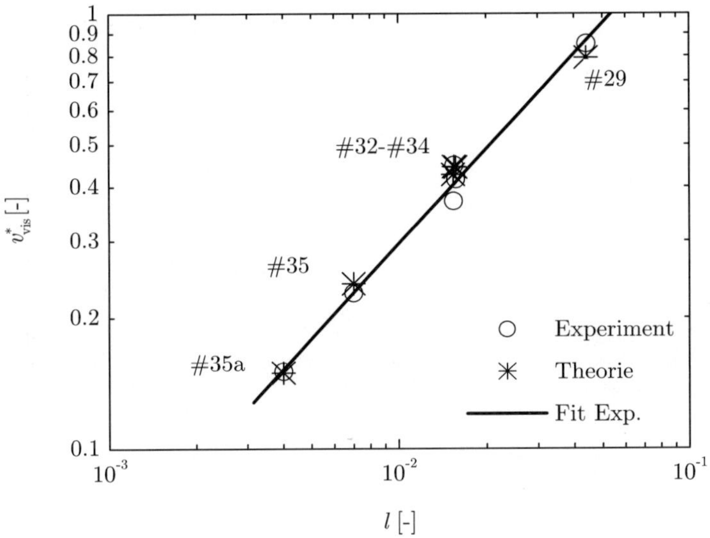

Fig. 17.6. Viscous velocity v^*_{vis} versus channel length $\tilde{l}$

environment of 10^{-4} g_0 in all axes for approximately 6 minutes. A picture of the module is shown in Fig. 17.7. The lower part (the upper part in launch direction) contains the electronic package, the upper part the experiment components. A schematic drawing of the test cell with the compensation tube and the liquid tanks (lower part) is shown in Fig. 17.8. Figure 17.9 shows the main components of the core element, the test section with the reservoir and the inlet nozzle, and the compensation tube. The flow through the test section is established by two gear pumps. One pump supplies the reservoir through a circular gap on the bottom of the reservoir (flow rate Q_1). The flow enters through a screen controlled rectifier to ensure the desired velocity distribution. The liquid flows through the inlet nozzle into the test section, the capillary channel. At the channel outlet the flow rate Q_2 is withdrawn by the suction pump. The difference of the flow rates caused by fluctuations of the rotation speed and varying liquid slip inside the pumps is absorbed by the compensation tube. Furthermore the liquid meniscus in the tube sets the pressure in the reservoir and defines the boundary condition needed to solve (16.51) and (16.52). The same ambient pressure p_a was maintained over the compensation tube and aside the capillary channel. To prevent the meniscus from disappearing into the reservoir the supply pump was adjusted to deliver a 3% higher flow rate than the suction pump. With this adjustment the liquid was supposed to rise continuously in the compensation tube. Each pump is connected to a separate tank for the supply with test liquid and the storage of withdrawn liquid.

The test section is the core element of the experiment set-up. It consists of 2 quartz plates with thickness a_p = 5 mm, breadth b = 25 mm, and

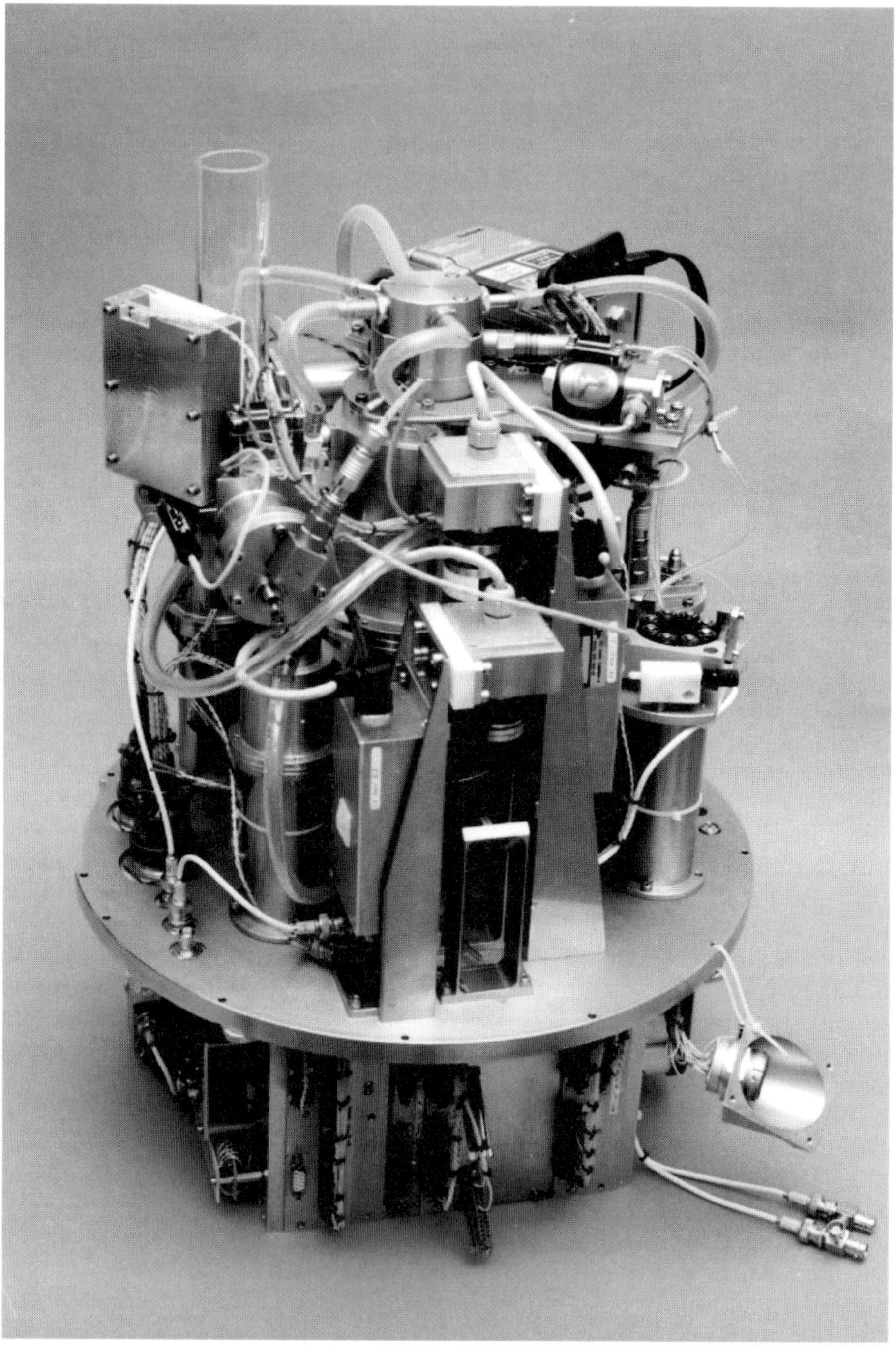

Fig. 17.7. Picture of the TEXUS module TEM06-24. The module is shown in opposite launch direction, thus the gravity vector points upwards

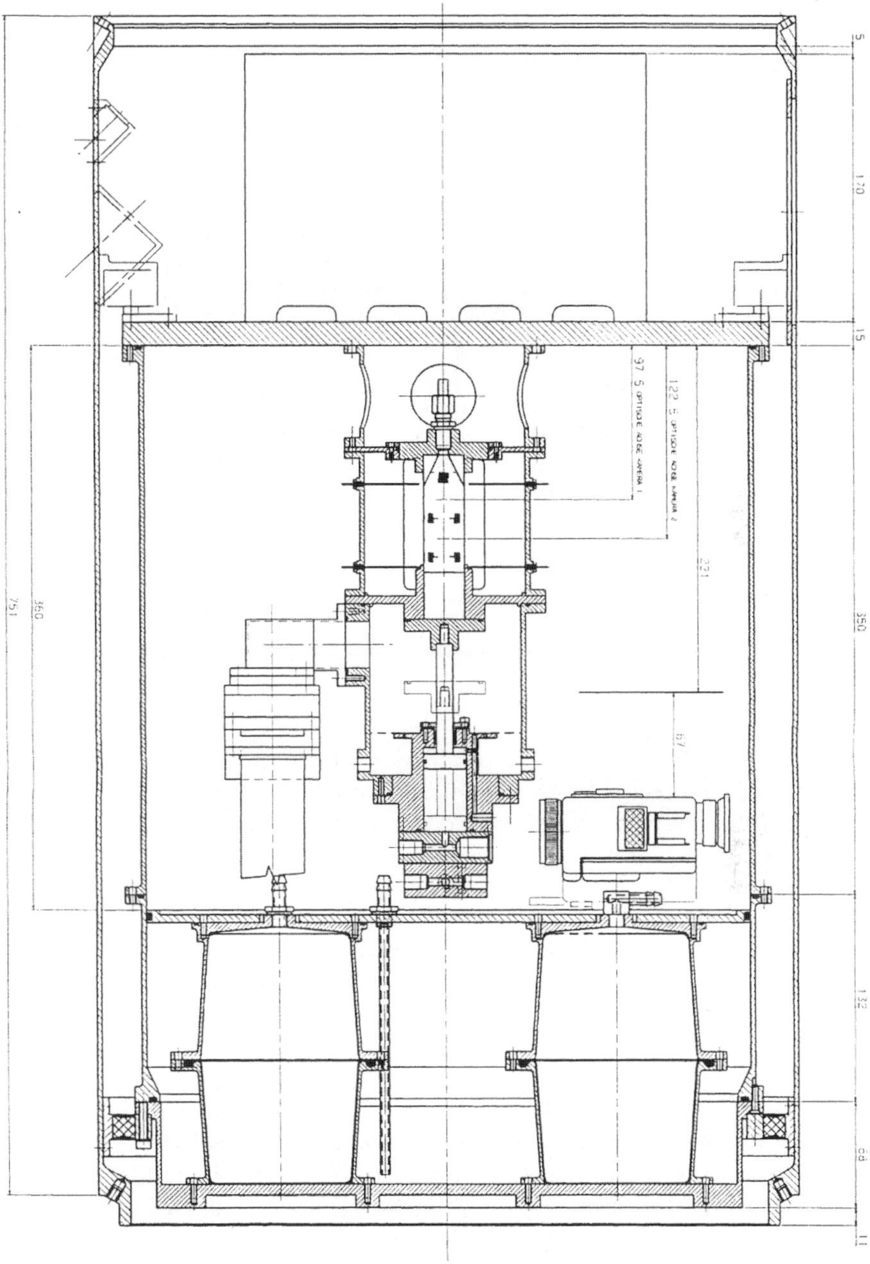

Fig. 17.8. Drawing of the TEM0624. The *lower part* contains the liquid reservoir, the *middle part* the experiment set-up and the *upper part* the electronics. The module is shown here in start position of the rocket, the gravity vector points downwards

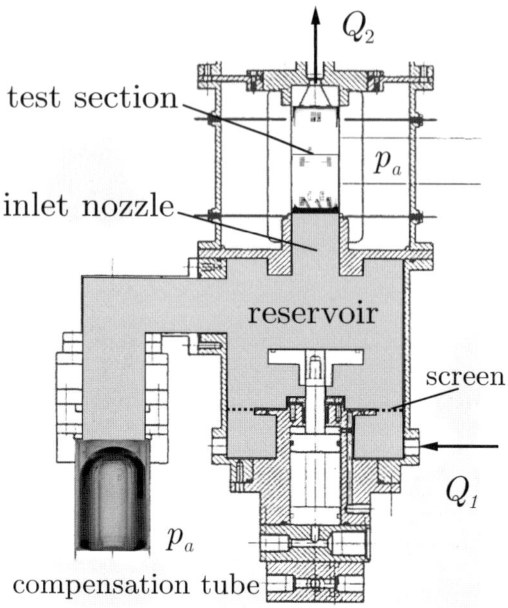

Fig. 17.9. Core element of TEM-0624 with the test section (the capillary channel), the inlet nozzle, the reservoir, the compensation tube and the connections to the pumps Q_1 and Q_2

separation $a = 5$ mm. Figure 17.10a shows a view perpendicular to the plates. The plates have precise measures for calibration and evaluation of the video pictures. Some surfaces of the outer test section were coated with FC-725 (diluted with Butylacetat). The coating film was applied by painting. The surface energy of the film is supposed to be 13–14 mN/m and thus changes the contact angle of silicon fluid 0.65 cSt to $\gamma_s \gg 0$. The aim of the coating is to prevent the spreading of the test liquid away from the capillary channel and in the optical path of the observation. The front side (seen from the camera) contains precise measures etched into the glass surface. The measures are used to calibrate the digital image processing and to control the resolution of the optical arrangement.

The aim of the compensation tube is to yield a defined pressure as a boundary condition for the test section inlet of the capillary channel and to compensate for small differences of the flow rates of pump P1 (supply) and pump P2 (withdrawal). The free surface in the compensation tube was observed with a Sony digital video recorder. The images have been recorded onboard on DV mini cassette and transmitted to the ground station during flight. The upper part of the compensation tube was coated by dipping in FC-725. The compensation tube is shown in Fig. 17.10b.

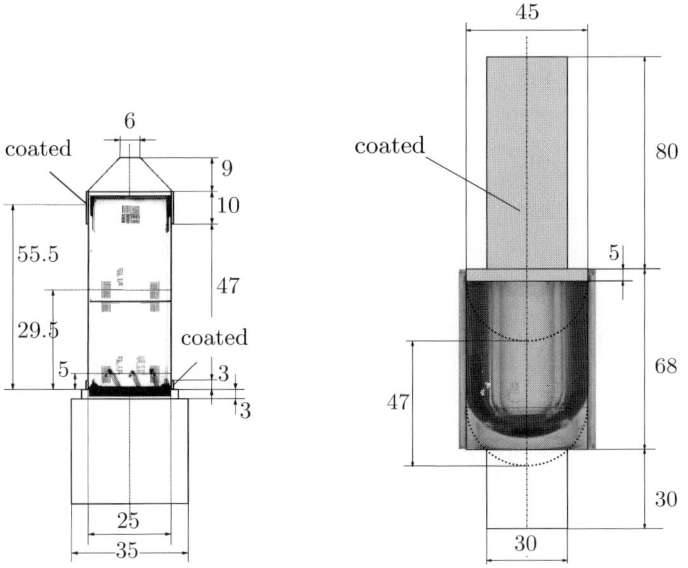

(**a**) Drawing of the test section (**b**) Drawing of the compensation tube

Fig. 17.10. Schematic drawings of the test section and the compensation tube. All measures are given in millimeters

The experiment was observed by three cameras. The video signals from the two CCD cameras and the camcorder were transmitted via the TC module of the rocket to the ground station and recorded on VCR (S-VHS quality). Additionally, the camera view of the compensation tube was recorded on board by a DV camcorder. Two cameras were used for the observation of the capillary channel yielding a resolution of approximately 0.05 mm/pixel. Both pictures had an overlap of 10 mm.

The video cameras were purchased from Aqua-tv (SWM 039, advanced optoelectronic technology, Kempten, Germany). The imaging area (ccd) has the dimensions 7.95 mm (h) × 6.45 mm (v), active picture elements 752 (h) × 582 (v) and a unit cell size of 8.6 μm (h) × 8.3 μm (v). The horizontal resolution (h) is 550 TV lines. The cameras were equipped with Schneider Kreuznach lenses XNP 1.9/35. The optical arrangement is shown in Fig. 17.11. The background illumination is reflected on the back side of the test section by a mirror, the pictures are reflected into the two cameras by two mirrors.

The camera 1 picture is transmitted via the TV 5 channel, the image appears as in Fig. 17.12a, the camera 2 picture and the camcorder picture are transmitted via TV 7 channel and appear as in Fig. 17.12b and c. TV channel 7 could be switched between camera 2 and the camcorder image.

The experiment was controlled manually during the flight of the sounding rocket. All hand-controlled procedures necessary for the TEXUS experiment

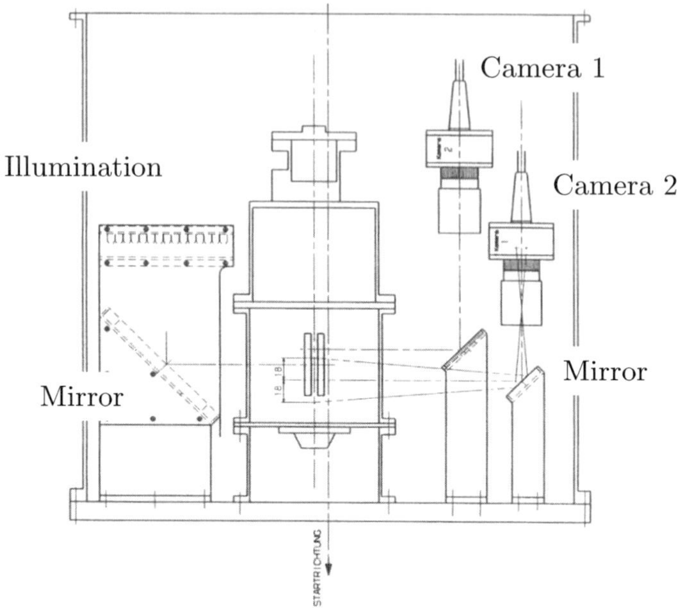

Fig. 17.11. Optical arrangement of the background illumination, the test cell and the two cameras

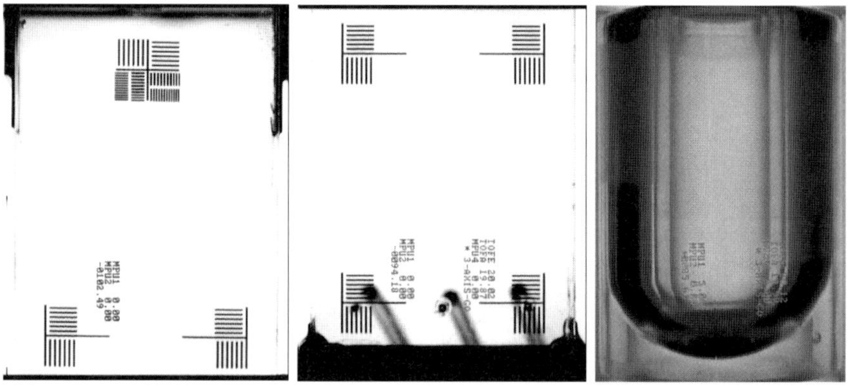

(**a**) Camera 1 video image (**b**) Camera 2 video image (**c**) Camcorder image

Fig. 17.12. Video image of the three cameras. Camera 1 records the *upper part* of the test section, camera 2 records the *lower part* of the test section (seen in flow direction). The camcorder image displays the compensation tube

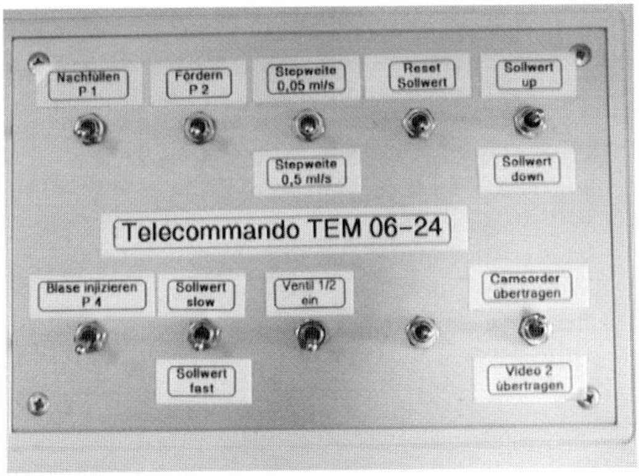

Fig. 17.13. User panel for TEM0624. During the flight the experiment was operated manually via telecommand with this panel

control have been practiced during two parabolic flight campaigns. In the initial configuration before lift off, the compensation tube and the nozzle were each closed with a gate valve.

The experiment procedure was planned with the following parts: the filling of the compensation tube, the filling of the test section, the increase of the flow rate with step width 0.5 ml/s up to the critical volume, the increase with step width 0.05 ml/s up to the critical volume and the repetition of the last step. The experiment was operated from the ground station with a user panel shown in Fig. 17.13.

17.2.1 Filling of the Compensation Tube

The compensation tube is connected to the reservoir (not shown in Fig. 17.9) via the valve V3. The valve opened automatically when the sounding rocket had reached the desired microgravity quality (all axes better than 10^{-4} g_0). Up to the upper limit of the compensation tube a total volume of 97.5 ml had to be filled. The time required for the filling of this volume with $q_{pu1} = 5$ ml/s was estimated to 19.5 s. The expected velocity of the meniscus is 3 mm/s and the corresponding dynamic contact angle of the meniscus is 9°. This procedure was finished when the meniscus had reached the upper limit of the compensation tube. The timeline of this procedure is shown in Table 17.4 .

17.2.2 Filling of the Test Section

To establish a flow through the test section the valves V1 and V2 must be opened. This moved the stemple away from the inlet nozzle with a velocity of 5 mm/s. The stemple fits into the contour of the inlet nozzle and allows the liquid to spread into the test section without bubble enclosure. The following was a capillary driven filling process through the test section into the tube

Table 17.4. Filling procedure for the compensation tube

time (s)	event	comment
	set P1 $q_{pu1} = 5$ ml/s	
	Camcorder transmit	view on compensation tube
0	Lift off	
67	Camcorder start	automatic
76	Valve V3 opens	automatic
78	start P1	filling of compensation tube
98	stop P1	compensation tube filled

Table 17.5. Filling procedure for the test section

time (s)	event
	set $q_{pu2} = 5$ ml/s
	open valve V1 and V2
102.38	start P2
104.39	meniscus in compensation tube moves
107.77	meniscus visible in test section
113.38	start P1

from the outlet of the test section to the pump P2. During this filling process the pump P2 runs with $q_{pu2} = 5$ ml/s to evacuate partly the tube and to suck the meniscus into the pump. The volume to be filled was 28.5 ml/s and the estimated filling time was 5-6 s. After the successful filling of the test section a stable subcritical meniscus surface was visible. The time line for this procedure is shown in Table 17.5.

17.2.3 Coarse Approach to the Critical Velocity

After having established a subcritical velocity at $q_{pu} = 5$ ml/s the critical velocity was approached with step width 0.5 ml/s. At each step bubbles were injected with two different velocities and the movement of the meniscus in the compensation tube was recorded. As the critical velocity was reached, the flow rates of the pumps were reduced with one fine (0.05 ml/s) and one coarse step (0.5 ml/s) to stop the gas ingestion at the outlet. The flow remained stable up to a flow rate of 8.5 ml/s.

After each cycle of the experiment (as prescribed in Table 17.6) the video image of the camcorder was switched to TV 7 and the image was transmitted to the ground. Depending on the position of the meniscus in the compensation tube the decision was taken to stop P1 or not.

Table 17.6. Procedure for the first approach of the critical velocity

time (s)	event
121	set $q_{pu2} = 6$ ml/s
+1	observe meniscus in test section
	set P4 slow
+4	start P4
+6	set P4 fast
+8	stop P4
	set P4 slow
+9	camcorder transmit
	decision to stop P1
+10	video 2 transmit
	increase q_{pu2}

17.2.4 Fine Approach to the Critical Velocity

The aim of this procedure was to approach the critical velocity with a fine step width to approximate quasi-steady flow. The flow rate of the pump was increased from the subcritical value with steps of 0.05 ml/s. Due to the unexpected high flow rate difference between P1 and P2 the position of the meniscus in the compensation tube was corrected several times. The stop and start of P1 led to supercritical flow conditions. This procedure and the interpretation of the video images during flight consumed more time than anticipated. Thus the supercritical value of 9.1 ml/s was reproduced only once. The time line of the flight data is given in Table 17.7 and 17.8. After the end of the guaranteed microgravity period the pump P2 was stopped two times to test the capability of the device to withstand disturbances caused by the inertia of the liquid in the test section. Even with this maneuver it was not possible to destabilize the liquid in the capillary channel or even to push it out of the channel.

17.2.5 Data Evaluation Compensation Tube

The motion of the meniscus in the compensation tube was recorded on camera 3 permanently onboard and frequently on TV 7 channel 2 by switching between camera 2 and camera 3. The pictures from camera 2 contain the flight time, camera 3 has no time display. With some remarkable events on both video tapes at $t = 105.16; 114.72; 176.16; 216.92; 265.08; 323.76; 363.16$ s, the time was synchronized between the two cameras. These events were set at the lower most point of the meniscus, when the pump P1 was switch on, except for the first one. The pictures from camera 3 were calibrated with the inner diameter of the compensation tube (45 mm). An optical correction for the refraction was applied. Figure 17.14 shows the image from the camcorder

Table 17.7. Procedure for the first approach of the critical velocity

time (s) (s)	**SPU1 SPU2** (ml/s)	**MPU1** (-)	**MPU2** (-)
0	launch	off	off
77	begin μg	off	off
78	5.0	on	off
97.85	*5.0*	*off*	*off*
102.38	*5.0*	*off*	*on*
113.83	*5.0*	*on*	*on*
120.53	6.0	on	on
137.59	7.0	on	on
150.34	7.5	on	on
151.63	8.0	on	on
168.26	*8.0*	*off*	*on*
175.18	*8.0*	*on*	*on*
182.52	8.5	on	on
194.62	9.0	on	on
203.26	8.95	on	on
209.74	*8.95*	*off*	*on*
216.0	*8.95*	*on*	*on*
220.54	8.45	on	on
231.34	8.5	on	on
245.38	9.0	on	on

(camera 3) with some image enhancements. The optical axis is marked by the two lines, the reference system for evaluating the meniscus position in time.

The optical correction was performed as follows: the distance between the middle axis of the compensation tube and the lens ($g = 128$ mm) must be corrected for the virtual location of the maximum diameter $g_1 = 119.77$ mm (refractive index of PMMA 1.491, plan parallel plate distortion). The virtual location of the center point of the meniscus is $g_2 = 121$ mm (refractive index of Silicon fluid 1.378, the beam passes two plan parallel plates: the PMMA and the Silicon fluid). Because the virtual point lies $g_2 - g_1$ beyond the virtual point of the calibration mark, the correction factor reads: $f = 1 + (g_2 - g_1)/g_1 = 1.01$. The meniscus location versus flight time is shown in Fig. 17.15. It is clearly visible how the adverse motion of the meniscus corresponds to the telecommands for the pumps. The first motion in the positive direction is due to the filling of the compensation tube. After that the test section was opened and filled itself by capillarity, leading to a motion of the meniscus in the negative direction. In the following period of time both pumps operated nominally with a difference in the flow rates which leads to a rising meniscus. When the meniscus reached the upper limit the pump P1 was stopped by telecommand and thus the meniscus moves downwards until pump P1 was

Table 17.8. Procedure for the second approach of the critical velocity

time (s)	**SPU1 SPU2** (ml/s)	**MPU1** (-)	**MPU2** (-)
258.77	*9.0*	*off*	on
264.17	*9.0*	*on*	on
276.26	8.5	on	on
281.66	8.55	on	on
296.14	8.6	on	on
301.97	8.65	on	on
302.83	8.7	on	on
314.71	*8.7*	*off*	*on*
322.92	*8.7*	*on*	*on*
328.32	8.75	on	on
330.05	8.8	on	on
346.46	8.85	on	on
357.48	*8.85*	*off*	*on*
362.23	*8.85*	*on*	*on*
366.55	8.9	on	on
367.2	8.95	on	on
368.06	8.9	on	on
373.46	8.95	on	on
380.81	9.0	on	on
387.07	9.05	on	on
392.04	9.1	on	on
395.71	*9.1*	*off*	*on*
399.82	*9.15*	*off*	*on*
402.62	*9.2*	*off*	*on*
417.96	**9.2**	**off**	**off**
420.98	**9.2**	**off**	**on**
427.03	**9.2**	**off**	**off**
429.62	**9.2**	**off**	**on**
433.3	*9.2*	*on*	*on*

switched on again. This procedure was repeated several times. The meniscus motion in the compensation tube was evaluated to yield the flow rates in the different time intervals. The velocity of the meniscus center was computed with a linear regression from the meniscus location versus time. The results for the time intervals where both pumps operated nominally are shown in Table 17.9. Except for no. 8.3 the average relative difference between the two pumps is 13.7%. In the time interval 8.3 the contact line of the meniscus touches the coated area and the free surface shape changes drastically. To obtain the pressure drop across the free surface in the compensation tube the dynamic contact angle can be evaluated from the meniscus velocity. The dynamic contact angle can be estimated following a relation from [3] to

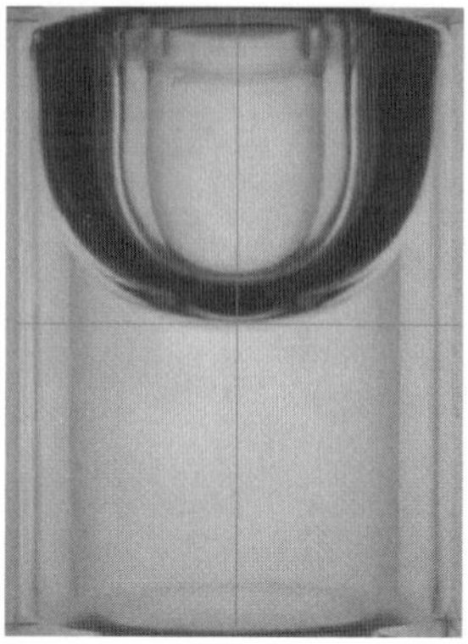

Fig. 17.14. Video picture of the meniscus in the compensation tube. The lines show the coordinate system for the evaluation of the meniscus velocities. The meniscus position is counted *positive left* from the *vertical line* and *negative right* from the *vertical line*

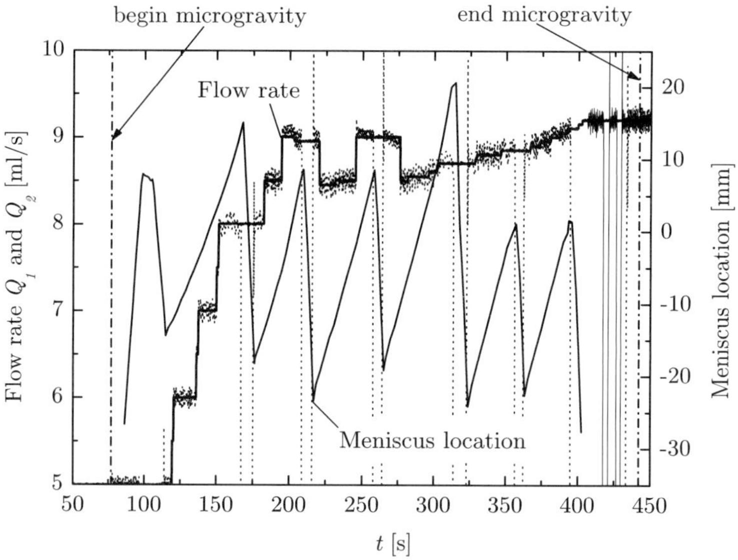

Fig. 17.15. Comparison between the flow rates of the pump and the motion of the meniscus in the compensation tube. The *dotted line* shows the monitor values of pump P1 which drop to zero when the pump was switch of. The set values for P1 and P2 were identical per definition

Table 17.9. Evaluation of the meniscus velocity during the operation of both pumps

No.	t (s)	q_{1s}, q_{2s} (ml/s)	v_m (mm/s)	q_m (ml/s)	Δq (%)	γ_d (°)	Δp_{cb} (Pa)
2.1	122.6-134.6	6	0.48	0.76	12.7	4.9	-1.399
2.2	137.6-149.6	7	0.55	0.87	12.5	5.1	-1.399
2.3	152.6-167.9	8	0.70	1.11	13.8	5.6	-1.398
4.1	182.6-191.6	8.5	0.69	1.10	12.9	5.6	-1.398
4.2	194.6-209.4	9	0.99	1.58	17.5	6.2	-1.396
6	227.5-245.6	8.5	0.73	1.16	13.1	5.6	-1.398
8.1	266.6-275.6	9	0.73	1.16	12.9	5.6	-1.398
8.2	281.6-296.6	8.55	0.76	1.2	14.1	5.7	-1.398
8.3	302.6-311.0	8.7	1.12	1.78	20.5	6.5	-1.395
10	332.6-344.6	8.8	0.78	1.24	14.1	5.8	-1.398

$$\tan \gamma_d = 3.4 \ \mathrm{Ca}^{1/3} \tag{17.4}$$

and the Capillary number $\mathrm{Ca} = u_m \mu / \sigma$. For the meniscus velocities given in Table 17.9 the dynamic contact angle varies between 4.9° and 6.5°. The corresponding capillary pressure reads

$$\Delta p_{cb} = p_{cb} - p_a = -\frac{2\sigma \cos \gamma_d}{R} \tag{17.5}$$

and is given in Table 17.9. The reference pressure for a contact angle $\gamma_s = 0$ is $\Delta p_{cb}(\gamma_s = 0) = -1.404$ Pa.

The results for the time intervals where only one pump was operating are shown in Table 17.10. The values for time interval #1 show a difference between the set value of pump P1 and the measured value of 4.4%. The values for the remaining time intervals show large differences between the set value of pump P2 and the measured values. This cannot be interpreted as a malfunction of pump P2 since the pump P1 was only switched of during this time, but the inlet into the reservoir was not closed by a valve. The differences may be caused by a flow from the inlet side even without pumping. The pumps are tooth wheel pumps which allow a leak flow rate through the pump.

17.2.6 Data Evaluation Test Section

The typical observation at a flow rate below the critical value is shown in Fig. 17.16. The interpretation of the video observation is the same as in the drop tower experiments (Sect. 17.1). The camera view is perpendicular to the front plate. Since the plates are transparent the free surface appears dark corresponding to the distance k introduced in Fig. 16.1. It is obvious that the surface curvature increases with increasing flow rate and the liquid flow path becomes smaller in cross section. If the limit is exceeded the free surface becomes unstable and gas ingestion occurs as shown in Fig. 17.17. In this

Table 17.10. Evaluation of the meniscus velocity during the operation of one pump only

No.	t (s)	Remark	q_{1s} (ml/s)	q_{1m} (ml/s)	Δq (%)	q_{2s} (ml/s)	q_{2m} (ml/s)	Δq (%)
1	86.6-98.6	filling ct	5.0	4.78	4.4			
3	170.6-176.2	PU1 off				8.0	6.62	17.3
5	209.6-216.9	PU1 off				8.95	6.97	22.2
7	258.4-265.1	PU1 off				9.0	6.7	25.6
9	314.6-323.8	PU1 off				8.7	7.51	13.7
11	359.6-363.2	PU1 off				8.85	7.33	17.2

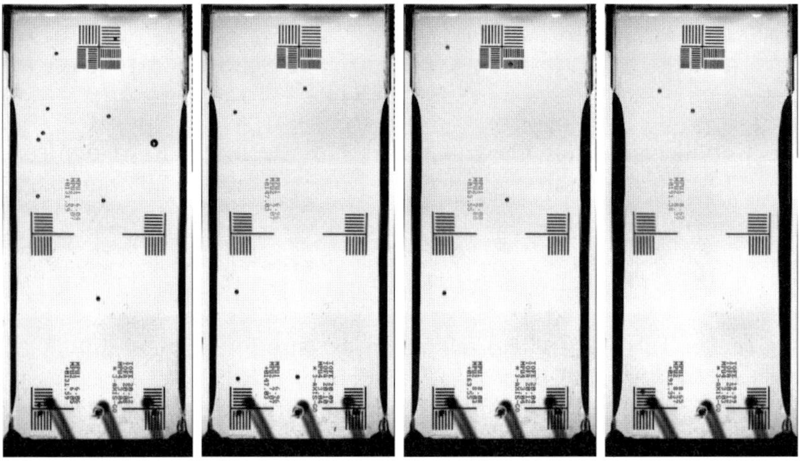

$Q = 6$ ml/s $\quad Q = 7$ ml/s $\quad Q = 8$ ml/s $\quad Q = 9$ ml/s

Fig. 17.16. Stable flow rate. The flow is from the *bottom* to the *top*

case, the liquid pressure at the channel exit is below the maximal value the free surface can withstand. For the comparison with the theoretical prediction of the surface position k the video pictures are processed digitally and the innermost surface lines have been extracted. A comparison of some surface contours with the numerical solution of (16.51) to (16.54) are shown in Fig. 17.18 and 17.19. The inlet surface curvature H_i is required as a boundary condition (16.54). The data could be solved for by a numerical simulation of the flow from the bottom of the reservoir (screen rectifier) through the inlet nozzle to the inlet (for details see appendix A). The values for (16.35) are $K_{i1} = 1.4$ and $K_{i2} = 312$. The pressure drop due to the curvature of the free surface in the compensation tube was taken from Table 17.9 to solve (16.36). Finally the dimensionless inlet curvature $2H_i^*$ can be given as a function of the dimensionless flow rate Q^*

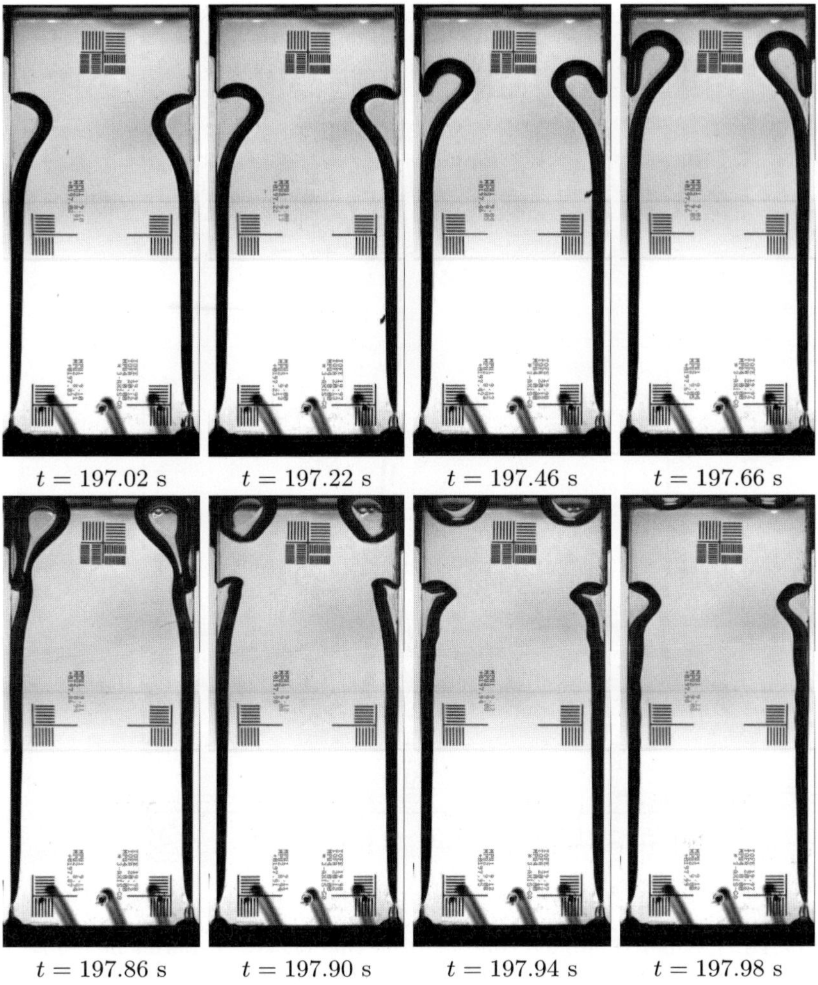

Fig. 17.17. Unstable flow rate. The adjusted flow rate is above the critical one, gas ingestion occurs at the outlet

$$2H_i^* = 0.222 + \frac{K_{i2}}{4}\left(1 + \frac{1}{\Lambda}\right) \mathrm{Oh}\ Q^* + \frac{1}{2}K_{i1}Q^{*2}\ . \qquad (17.6)$$

The comparison of the experimental and theoretical surface contours are shown in Fig. 17.18 for the lowest and in Fig. 17.19 for the highest flow rate. The evaluated surface position is represented by the solid line, the dotted lines give the error of the surface contour due the limited resolution and oscillations of the free surface. The dashed line is the theoretical prediction of k. Obviously the model computations predict the critical flow rate and the surface position in good agreement with the experimental data.

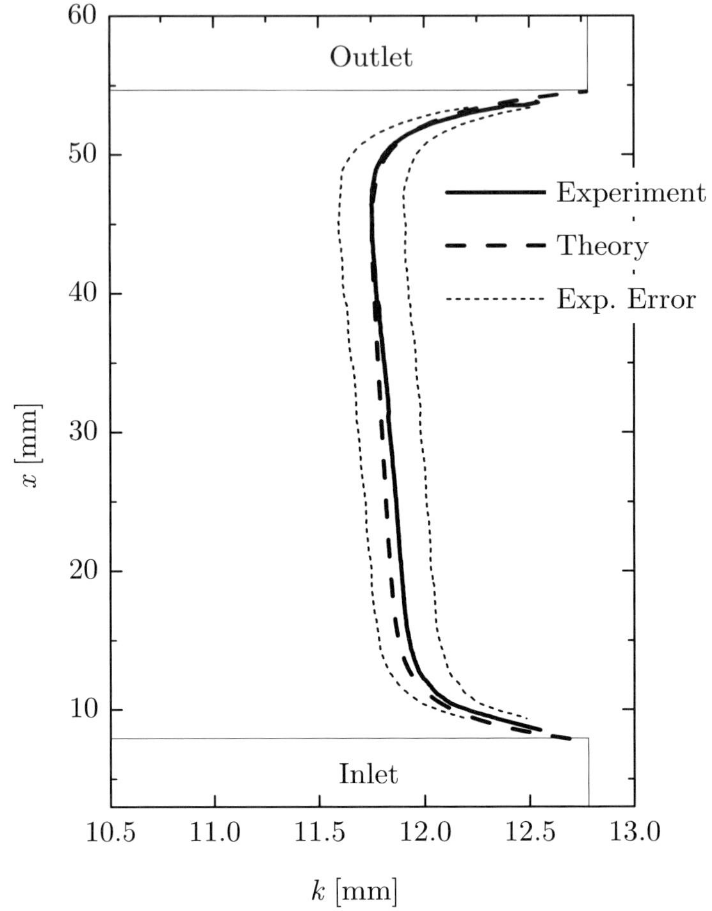

Fig. 17.18. Evaluated surface contour from the TEXUS experiment (*solid lines*) with $Q = 6$ ml/s, error of the contour value k (*dotted lines*) and theoretical prediction (*dashed lines*)

As mentioned in Sect. 16.8 we expect that the flow rate limitation occurs due to choking at the smallest cross section of the flow. The appropriate characteristic number to investigate whether choking is present or not is the WEBER number defined by (16.87). To form this number the characteristic wave speed v_c and the mean liquid velocity v were determined at the minimal cross section area. The results are shown in Fig. 17.20. The critical flow rate was determined from (16.50) with the parameter set TEXUS-37 from Table 17.1 to $Q^*_{crit} = (9.0 \pm 0.1)$ ml/s. The characteristic wave speed v_c can be computed from (16.86) with the differential $\mathrm{d}R/\mathrm{d}A$ known from (16.14) and (16.15). The criterion for the approach of the critical value was set to $\varepsilon = 10^{-14}$ in this case. This implies the simplification of the curvature to one

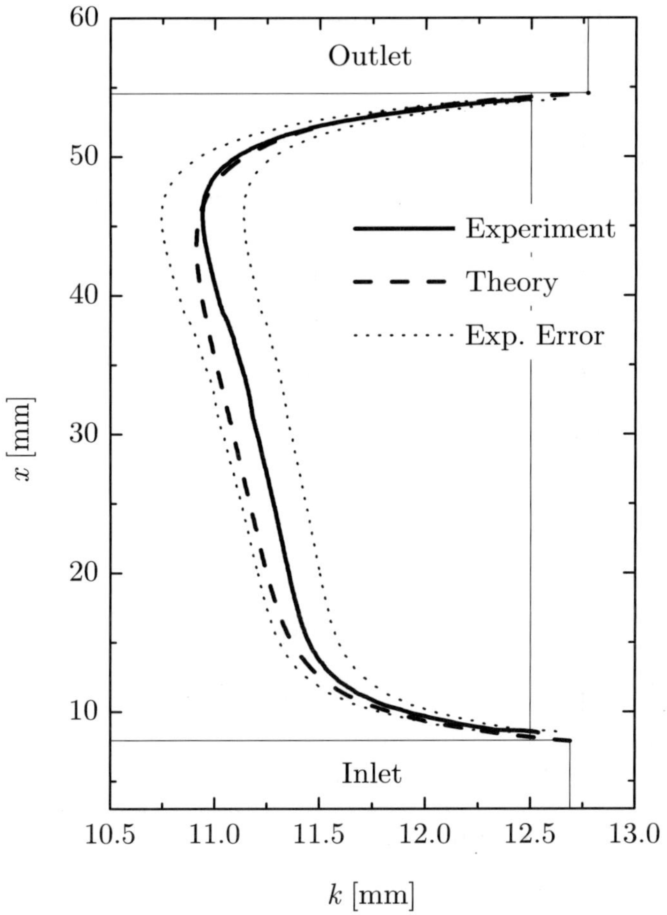

Fig. 17.19. Evaluated surface contour from the TEXUS experiment (*solid lines*) with $Q = 9$ ml/s, error of the contour value k (*dotted lines*) and theoretical prediction (*dashed lines*)

radius R_1. The curve is given as a dashed line in Fig. 17.20. The calculation of the characteristic wave speed v_c from (16.84) requires the evaluation of the differential dH/dA from the numerical solution of (16.48). The solution is more accurate and shown as a solid line in Fig. 17.20. The results confirm the choking effect. The WEBER number increases with increasing flow rate. As the theory predicts, the flow velocity and the wave speed approach the same value at the smallest cross section. At $Q^{\mathrm{exp}}_{crit} = 9.1\,\mathrm{ml/s}$ the flow runs near the critical condition and the WEBER number yields $\mathrm{We} = 0.53$. Figure 17.20 states that the capillary flow in the channel is very close to the critical value since the WEBER number varies strongly with small changes of Q.

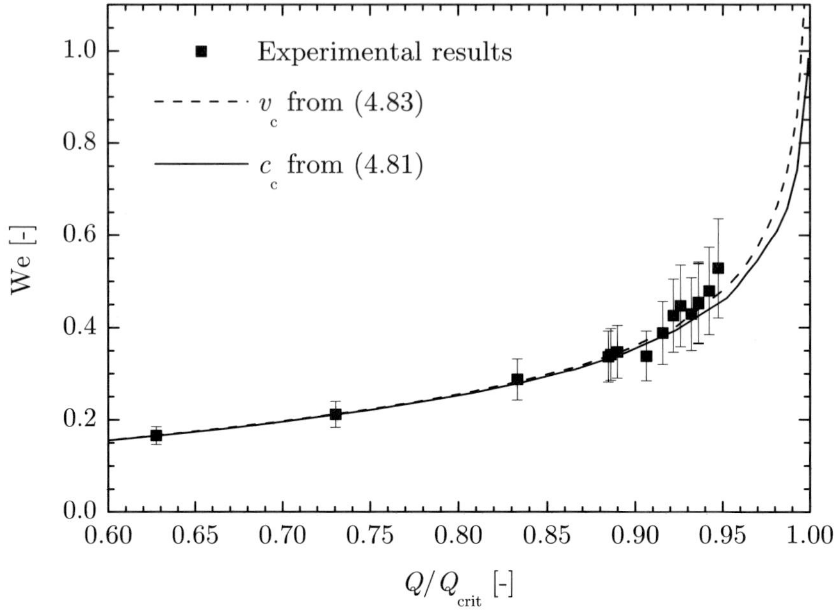

Fig. 17.20. Plot of the WEBER number as a function of the critical flow rate. The dashed line was computed with the characteristic velocity v_c from (16.86), thus using only one radius of curvature R_1. The solid line was computed with (16.84), the differential $\mathrm{d}H/\mathrm{d}A$ was evaluated numerically

References

1. M.E. Dreyer: *Kapillarer Flüssigkeitsanstieg zwischen parallelen Platten unter kompensierter Gravitation*, (VDI-Verlag Düsseldorf 1994)
2. M.E. Dreyer, A. Delgado, H.J. Rath: J. Colloid Interface Sci **163**, 158 (1994)
3. G. Friz: Z. angew. Phys. **19.4**, 374 (1965)

18

Results and Discussion

Figure 18.1 shows results of (16.41) for the free surface shape k^*, the mean velocity v^* and the mean curvature h^* as a function of the flow length $\tilde{l}$. The computed values are close to the critical flow rate. The values in Fig. 18.1a have been computed with a low Oh number $\mathrm{Oh} = 10^{-5}$ and a short channel length $\tilde{l} = 10^{-5}$. The curvature difference between the inlet at $x^* = 0$ and the outlet $x^* = \tilde{l}$ is nearly zero — the flow is dominated by convective momentum transport. The smallest cross section appears in the middle of the channel.

Figure 18.1b shows the data for a large Oh number $\mathrm{Oh} = 10^{-2}$ and a large channel length $\tilde{l} = 10$. The mean curvature increases nearly linear from the inlet to the outlet. The smallest cross section of the channel appears shortly before the outlet. The velocity remains constant and is very small compared to the mean curvature. The flow is dominated by viscous momentum transport.

Figure 18.2 shows the results for the critical flow rate from (16.50) in the parameter space:

$$\begin{aligned} \Lambda &= 1; 2; 5; 50 \\ \mathrm{Oh} &= 10^{-5}; 10^{-4}; 10^{-3}; 10^{-2} \\ \tilde{l} &= 10^{-5}; 10^{-4}; 10^{-3}; 10^{-2}; 10^{-1}; 10^{0}; 10\ . \end{aligned}$$

The dimensionless critical flow rate Q^*_{crit} (in the numerical sense) is plotted as a function of the OHNESORGE number and the dimensionless length $\tilde{l}$. The maps indicate that the critical flow rate Q^*_{crit} is independent from Oh for $\tilde{l} > 0.1$. Please note this statement is true only for the plot $Q^*_{crit} = f(\mathrm{Oh}, \Lambda, \tilde{l})$. The OHNESORGE number still plays a role in the dimensionless length since $\tilde{l} = \mathrm{Oh}\ l/(2D_h)$. For smaller channel lengths the critical flow rate increases with increasing Oh number. This behavior depends slightly on the aspect ratio.

Figure 18.3 shows a plot of the curvature difference $\Delta h^* = h^*_{\mathrm{out}} - h^*_{\mathrm{in}}$ between inlet and outlet of the channel as a function of the dimensionless flow length $\tilde{l}$. The curve was plotted for $\mathrm{Oh} = 10^{-4}$ and $\Lambda = 50$, but it can be considered to show the general behavior of the system. The data points of the

Michael E. Dreyer: *Free Surface Flows under Compensated Gravity Conditions*
STMP **221**, 219–223 (2007)
DOI 10.1007/3-540-44628-1_18

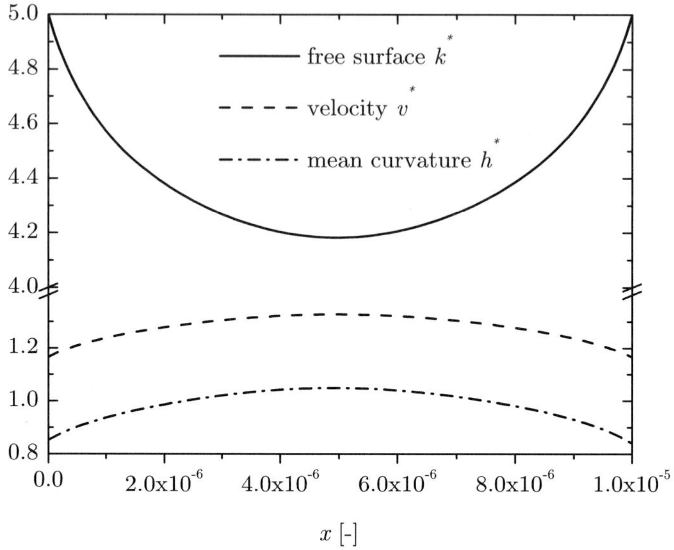

(**a**) Oh $= 10^{-5}$, $\tilde{l} = 10^{-5}$

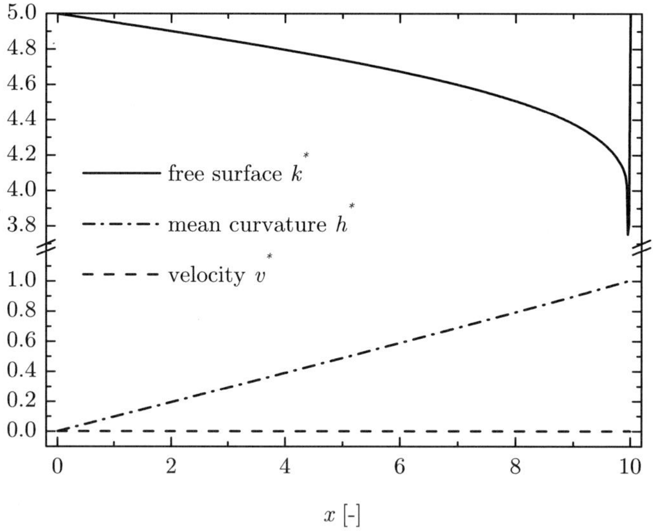

(**b**) Oh $= 10^{-2}$, $\tilde{l} = 10$

Fig. 18.1. The figures show the contour of the free surface k^* (*solid line*), the velocity v^* (*dashed line*) and the mean curvature h^* (*dashed-pointed line*) for $\Lambda = 5$. (**a**) Oh $= 10^{-5}$, $\tilde{l} = 10^{-5}$, $Q^\star = Q^\star_{crit} = 1.165$, $2H^\star_i = 0.852$ (**b**) Oh $= 10^{-2}$, $\tilde{l} = 10$, $Q^\star = Q^\star_{crit} = 0.00198$, $2H^\star_i = 0.00167$. The scaling of the ordinate refers to all plotted properties

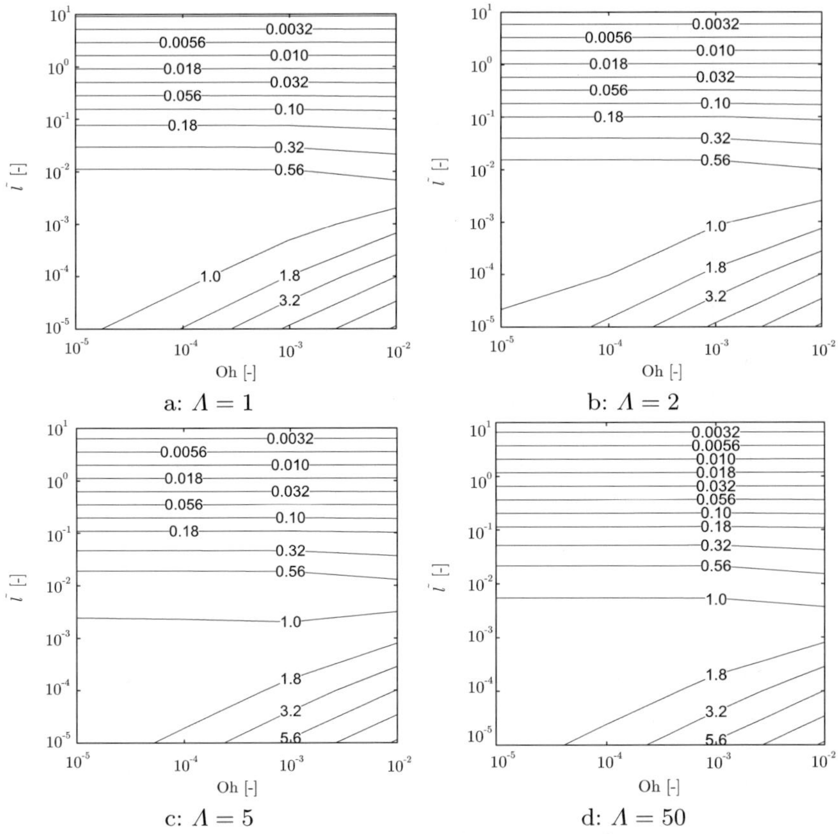

Fig. 18.2. Contour maps of the critical flow rate $Q^\star = Q^\star_{\mathrm{crit}}$ for the parallel plate channel plotted in logarithmic scale as function of the OHNESORGE number Oh and the dimensionless length $\tilde{l}$ for different Λ

experiments are depicted to show the investigated range of $\tilde{l}$. They cannot be used as exact values due to the differences in Oh and Λ.

With the assumption of a constant $A = A_0$ the integration of (16.2) yields

$$\Delta h^* = h^*(x^* = \tilde{l}) - h^*(x = 0) = -\frac{a}{2\sigma}\Delta p = -\frac{a}{2\sigma} w_f(x = l) \; . \tag{18.1}$$

This assumption leads to a smaller contribution of the viscous pressure loss, since the actual flow path is narrower and therefore the local flow velocity higher. The assumption becomes true for $\Lambda \to \infty$. Equation (18.1) indicates that the curvature difference between inlet and outlet is a measure for the viscous pressure loss in the channel. If the flow length is small $\tilde{l} \leq 10^{-3}$, the pressure loss between inlet and outlet is small, thus the flow is dominated by convective momentum transport. In the border case of a very short length the

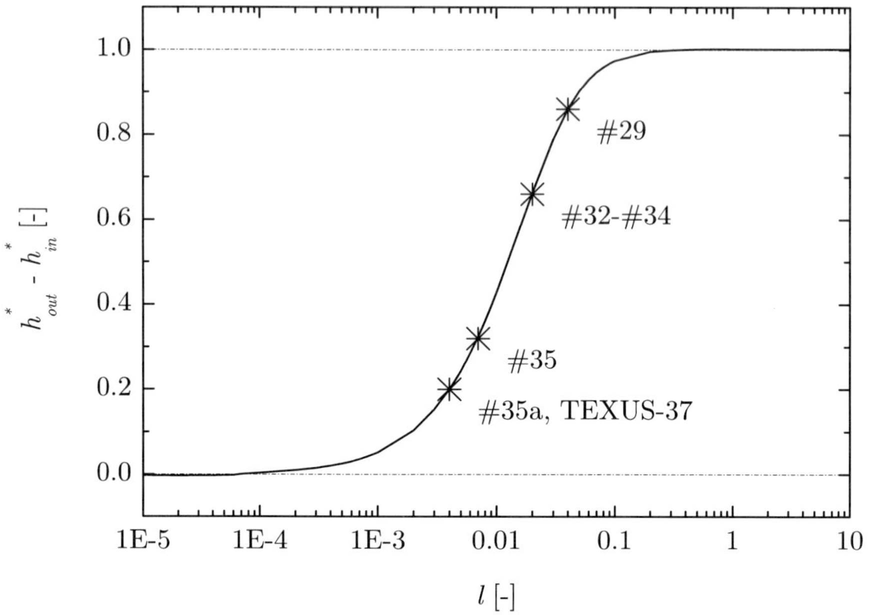

Fig. 18.3. Difference of the dimensionless curvature between inlet and outlet of the channel as a function of the dimensionless flow length $\tilde{l}$ computed for Oh $= 10^{-4}$ and $\Lambda = 50$. The data points indicate the location of the experiments

curvature difference Δh^* tends to zero (see Figs. 18.1a and 18.3). In the other extreme of a very long channel $\tilde{l} \geq 10^{-1}$, the flow is dominated by viscous momentum transport and the curvature difference tends to unity. The limit

$$\lim_{\tilde{l} \to \infty} \Delta h^* = 1 \tag{18.2}$$

can be taken to predict the maximal flow rate for a pure viscous dominated flow. From the integration of (16.29) follows

$$w_f(x = l) = -\frac{K_{pl}}{D_h \mathrm{Re}_h} \frac{\rho}{2} \frac{Q^2}{A^2} l \, . \tag{18.3}$$

The additional pressure loss K'_{ps} was neglected for this purpose. Substituting w_f in (18.1) by (18.3) yields

$$\Delta h^* = \frac{a}{2\sigma} \frac{K_{pl}}{D_h \mathrm{Re}_h} \frac{\rho}{2} \frac{Q^2}{A^2} l \, . \tag{18.4}$$

At $x = l$ the area is $A = A_0 = ab$ and we may write with $v = Q/(ab)$

$$\Delta h^* = 2 K_{pl} v^* \tilde{l} \, , \tag{18.5}$$

or in the limit of (18.2)

$$v^*_{crit} = \frac{2}{K_{pl}\,\tilde{l}} \,. \tag{18.6}$$

Since $v^* = Q^*$ at the channel outlet, the corresponding critical flow rate is

$$Q^*_{crit} = \frac{2}{K_{pl}\,\tilde{l}} \,. \tag{18.7}$$

An empirical fit on the data in Fig. 18.2 for $l > 10^{-2}$ shows the dependence on Λ which was neglected to derive (18.7).

$$Q^*_{crit} = \left(1 - \frac{0.2}{\Lambda}\right) \frac{2}{K_{pl}\,\tilde{l}} \,. \tag{18.8}$$

This equation evolves into (18.7) for $\Lambda \rightarrow \infty$. Since $\Lambda < 1$ are not interesting due to technical unimportance, the bandwidth of the critical flow rate is about -20% from (18.7). It should be repeated here that this discussion is valid only for the case that the inlet flow profile parameter H_i and β have the same order of magnitude as discussed. Higher inflow resistances may drastically reduce the critical flow rate, assuming that the ambient pressures above the free reservoir surface (or the compensation tube) and the free surface in the capillary channel are equal.

Finally, an interesting similarity between the free (driven by the capillary pressure at the meniscus) and the forced (as discussed in this chapter) flow in a capillary channel should be mentioned. From the theory of the capillary rise between parallel plates it is known that for a flow length

$$l > \frac{a}{12\mathrm{Oh}} \quad \text{or} \quad \tilde{l} > \frac{2}{K_{pl}} \gtrsim 2 \cdot 10^{-2}$$

the flow is viscous dominated and the meniscus velocity is given by

$$v^* = \left(1 - \frac{1}{\Lambda}\right) \frac{2}{K_{pl}\,\tilde{l}} \,. \tag{18.9}$$

From this we conclude that for large Λ and large flow lengths $\tilde{l} > 10^{-1}$ the self-driven capillary flow velocity is the maximal velocity that cannot be exceeded by other means. As shown by (18.9) an increasing aspect ratio leads to an increasing velocity. For $\Lambda \rightarrow \infty$ the self driven velocity becomes similar to the critical velocity. In contrast to that, $\Lambda = 1$ causes two equal radii of curvature with different signs and thus no capillary pressure. For $\Lambda = 1$ no capillary rise occurs. This behavior may not be transferred to the forced flow through the parallel plates since the critical flow rate for $\Lambda = 1$ is larger than zero as shown in Fig. 16.50a. Please note that for this consideration the additional pressure loss due to the velocity profile change in the channel has been omitted for the sake of simplicity.

Part V

Perspectives

19

Conclusion

This book covers some propellant management related problems, such as free surface oscillations, capillary rise, and capillary channel flows. Some of the results are directly applicable, i.e. the characteristics of the surface oscillation (rise velocity, frequency, and damping) and the capillary channel flow which occurs in liquid acquisition devices (LAD's). Others, like the capillary rise, address a fundamental phenomena which always plays a role in the changing acceleration environment of spacecrafts. In addition, the theory for the tube is widely used in porous media to describe wicking, the imbibition of a wetting liquid into a porous structure. Open capillary channels also encounter a capillary rise whereas the surface curvature is more complex than in a right circular tube. One or more aspect ratios come into play and may inhibit the capillary rise even if a perfectly wetting liquid is used. The data presented in this book could be used in these cases to benchmark computational fluid mechanics (CFD) codes and to apply them to more complex shapes.

Modern design of propellant tanks relies on the usage of approved and validated codes, which could be either commercial codes as the ones mentioned and used in this book, or in-house codes written by industry or academia. These codes must be benchmarked by experiments performed within the relevant space of dimensionless numbers (if possible) and in a relevant environment. The author participated to some of these benchmark tests in a French-German project called COMPERE[1]. One experience is that not only the code itself has been benchmarked but also the correct use of the many numerical parameters necessary to tune the code to the specific problem under investigation. In addition, two test cases extracted from this book have been published in a special issue dedicated to validation of two-phase flow codes [1, 2].

[1] Comportement des ergols dans les reservoir. The support by the French space agency CNES and the funding by the German space agency DLR is gratefully acknowledged

Michael E. Dreyer: *Free Surface Flows under Compensated Gravity Conditions*
STMP **221**, 227–228 (2007)
DOI 10.1007/3-540-44628-1_19

Even if the motivation for the work described in this book stems from propellant management in space, the results are important for the understanding of the fluid mechanics of capillary dominated flows, in particular in the inertial dominated regime. To work in compensated gravity is not only important from a technology readiness level (TRL) point of view where testing is required in a relevant environment to demonstrate the functionality of a breadboard, a subsystem or system. It also enlarges the time and length scales discussed in this book and enables a look in the early development of transient processes before viscous forces become dominant and the governing equation become linear. Capillary driven flows on earth are always limited to small length scales defined by the capillary length $\sqrt{\sigma/\rho g}$ (2.15). If the time scale for viscous momentum diffusion L^2/ν (2.23) is formed with this length, the time $(\sigma/\mu g)$ scales the beginning of the dominance of viscous momentum transport and thus linear behavior. It goes without saying that the coefficient for that relation must be determined in each case. This has been demonstrated in the case of the capillary rise where the start of the so-called Washburn behavior was identified precisely. When the capillary rise theory is used for wicking into a porous media under normal gravity conditions, these findings can be used to identify the different flow regimes.

20

Future Projects

The discovery of the choking effect in capillary channels could be made thanks to a propellant management paper by Jaekle [3] and experiments in the compensated gravity environment of a drop tower and a sounding rocket. After a thorough analysis of the governing equations taking into account inertial effects and the sophisticated image processing of the data, the results could be finally published by Rosendahl et al. [4]. But we are far away from the fully unsteady modeling and experimental corroboration of this flow phenomenon. For that reason, the American (NASA) and German (DLR) space agencies have decided to operate the *Capillary Channel Flow* (CCF) experiment onboard the space station in the Microgravity Science Glovebox. The CCF team (M. Weislogel, Portland State University, Oregon, USA and the author) will investigate steady, transient, and oscillatory flows in different capillary channels. Two experiment units are foreseen: one parallel plate channel (with fixed distance a and breadth b) where the lengths l of the open surfaces can be independently changed by moveable sliders[1], and a channel consisting of two tilted plates of fixed breadth b and length l where the plate distance a, wedge angle α, and taper angle β can be changed independently. The second unit will have the possibility to inject bubbles in the test section to study the passive liquid-gas separation capabilities of open capillary channels. An important variety of inertial-capillary dominated flows will be studied that are of basic interest in fluid physics and at the same time key to certain space applications, but cannot studied on the ground. Applications of the results are directed to designers challenged by the containment, storage, and handling of large liquid inventories (fuels, cryogens, water) aboard spacecraft. The results are expected to be useful for designing and testing, and instrumentation for verification and validation of liquid management systems of current orbiting, design stage, and advanced spacecraft. The results will also be useful to improve life support system design, phase separation, and enhancing current system reliability by designing into the system passive redundancies.

[1] One slider may be closed to convert the parallel plate channel into a groove

Michael E. Dreyer: *Free Surface Flows under Compensated Gravity Conditions*
STMP **221**, 229–233 (2007)
DOI 10.1007/3-540-44628-1_20

In the frame of an ESA-MAP project[2] on convective boiling and condensation, another sounding rocket experiment is in preparation to investigate two-phase flows in open capillary channels. The occurrence of two-phase flow is in many cases unavoidable due to mechanical disturbances (sloshing) as well as due to temperature and pressure changes. As in the CCF experiment, the flow will be forced through a grooved capillary channel to study the maximum flow rate which can be withdrawn if gas bubbles are present. Furthermore, the distribution of the gas bubbles in the channel, the coalescence with each other and with the free surface, and the corresponding pressure drop are of major interest. Besides this, the techniques to create and handle the two-phase flow in a microgravity environment will be verified by this experiment. The experiences from the two sounding rocket flights TEXUS 41 and TEXUS EML-1 (TEXUS 42) will be very helpful to performed the planned operations on TEXUS 45.

Apart from two-phase flows, non-isothermal conditions and cryogenic environments have not been discussed in this book on purpose. Isothermal phenomena are by now well predictable by commercial codes. The remaining task for the PMD designer is to encounter the mission requirements for a storable propellant tank with good concepts and to validate them by analysis and similarity. If cryogenic propellants have to be used, the situation is different. Heat and mass transfer has to be considered as well and appropriate tools are still to be validated. Several projects are ongoing to contribute to this topic.

Static free surface shapes can be calculated from the GAUSS-LAPLACE equation (2.7) either by in-house codes or the *Surface Evolver.* The boundary condition at the wall is the contact angle which in isothermal cases differs around the static contact angle by the advancing and the receding contact angle. If a non-condensable gas is present and a temperature gradient between wall and liquid is imposed, for instance by the heat flow towards a cryogenic tank from the environment, a surface tension driven flow occurs (frequently called MARANGONI convection) which is independent of gravity. The flow field in the vicinity of the contact line deforms the free surface and causes a different contact angle on the macroscopic scale. Different approaches have been tested to provide quantitative data to be used as a boundary condition for static solvers. Based on the experiments described in Part II a set-up was developed to expose the contact line to a different temperature than the bulk liquid by dividing the right circular cylinder in two parts from which the upper part was heated and put on top of the lower part short before the release of the drop capsule. Upon step reduction of the BOND number the surface oscillation extends over the warmer upper part and causes a different behavior than in the isothermal case.

The set-up has been modified for better visualization of the contact line to an annular gap geometry where a cartridge heater in the middle of the test cell leads to a capillary ascent and the outer contact line is pinned at a

[2] Microgravity application programm AO-2004-111

sharp edge. A constant heater power provides a linear temperature increase of the surface and the contact line dynamics can be correlated with the increasing temperature difference. The data is now being used in the COMPERE project to benchmark codes to be used for cryogenic tanks. Spreading of a cold liquid over a heated wall occurs during end of thrust since the wall in contact with the ullage gas is warmer than the wall in contact with the liquid. Physical similarity was aimed by adjusting the relevant dimensionless numbers REYNOLDS-MARANGONI and WEBER-MARANGONI with the liquid properties and temperature gradients towards the original situation[3].

Refillable reservoirs are being used for storable liquid in the aerospace industry with a high reliability based on experiments and analysis. The functional design relies on the driving capillary force where the contact angle is a boundary condition. If a cryogenic environment is considered the bulk liquid may be supposed to flow into a warmer structure. Temperature gradients may arise and a flow driven by surface tension gradients may occur depending on the amount of non-condensable gas in the system. It has been shown in experiments that a liquid rising between two heated plates encounters a different contact angle which can be correlated the same way as describes above by the REYNOLD-MARANGONI and WEBER-MARANGONI number. The upper limit of capillary transport is reached when the wall temperature reaches the saturation temperature of the liquid. The modeling of this complex phenomenon has to take into account heat and mass transport of the liquid as well as heat conduction in the solid. The interaction of liquid flow, evaporation, heat flow in the solid, flow of the vapor, and eventually condensation does not only occur on the larger length scales of liquid acquisition devices for cryogenic environment.

PMD components like woven metallic screens, which are used to separate liquid and gas, rely on the bubble point. This is the pressure needed to create a bubble at the largest pore of the screen and push it away from it. The capillary force is responsible for this pressure and the contact angle is again the boundary condition. If liquid evaporates from a wet screen it must be replaced by wicking from the bulk liquid. Detailed knowledge of the wicking velocity as a function of temperature gradient and evaporation rate is mandatory to design the device properly. Liquid flow occurring in the screen, either due to wicking or local temperature gradients in the pores, degrades the bubble

[3] The WEBER-MARANGONI number reads

$$\mathrm{We}_M = \frac{\sigma}{\sigma_T \theta} , \qquad (20.1)$$

and the REYNOLDS-MARANGONI number is

$$\mathrm{Re}_M = \frac{\sigma_T \theta D}{\rho_0 \nu^2} \qquad (20.2)$$

with the temperature coefficient of the surface tension σ_T, the temperature difference θ and a characteristic length D.

point and calls for larger safety margins. Investigations on dynamics aspects of the bubble point and wicking of porous screens within a cryogenic environment are planned to provide more information to design LAD's for cryogenic propellants.

Since drop tower experiments suffer from the short experiment time which often constrains smaller length scales to be able to reach certain equilibrium situation, a sounding rocket experiment SOURCE[4] is in planning which will investigate the surface deformation in a cylindrical tank. A preheated test cylinder will be filled with a cold liquid during the microgravity phase of the sounding rocket, the MARANGONI convection will be observed and the contact angle as a function of the temperature difference will be evaluated from the free surface shape. The results shall confirm the hypothesis drawn from the short term experiments that a large free surface will be deformed by temperature gradients across its contact line.

In other cases drop tower experiments are the easiest and cheapest way to develop techniques for other microgravity platforms. This is the case for the study of the free surface oscillation of cryogenic liquids. As shown in Part I and discussed in Part II the use of liquid nitrogen decreases the OHNESORGE number one order of magnitude and enables the improved study of particular phenomena such as the layer formation in the beginning of the surface reorientation and the first decent of the center point. This justifies the high investment of time and money to build a dewar for drop tower operations. This task has been successfully completed and the set-up delivers good optical data with small temperature gradients in the order of 1 K in a 70 K environment. First results have been presented by Stief et al. [5] and differences between prediction (numerical calculation and extrapolation of experimental correlations) and experiment are higher than expected. Further tests with isothermal conditions are underway to create the database for the comparison with non-isothermal experiment. Tests with other cryogens like argon, neon, and methane, are envisioned to prepare the handling of liquid hydrogen. Substantial progress in the technology readiness level of cryogenic components cannot be made without using a cryogenic environment. The feasibility of doing these test in a compensated gravity environment has been demonstrated successfully.

The design of vehicles, facilities and habitats for future space travel and exploration needs a consistent database which considers the acceleration as a variable between the low values in coasting flight and stationary orbits (10^{-6} g_0 to 10^{-3} g_0), the intermediate levels on Moon ($\approx 1/6$ g_0) and Mars ($\approx 1/3$ g_0), and the high values of several g_0 during launch (g_0 is the gravity acceleration on earth). Furthermore, cryogenic environments for the storage of propellants and consumables have to be considered. The underlying physical mechanisms must be understood accurately to developed tools that do not need large safety factors to account for uncertainties. Safety factors increase

[4] Sounding rocket COMPERE experiment

mass which is one of the most critical budget factor in space technology. Low safety factors call for accurate and reliable prediction tools which have to be benchmarked against a large experimental database achieved in a relevant environment within the relevant space of dimensionless numbers. It is trivial to say that the reproducibility of the experimental data has to be proven to quantify the probability of the prediction. Many microgravity experiments suffer from this restriction since flight opportunities are rare and expensive, sometimes short, and reflights are often impossible. To achieve the same technology level as for earth bound technologies one day, a certain part of the budget for space activities has to be used to set up the required database.

References

1. M. Michaelis, M.E. Dreyer: Test-case No 31: Reorientation of a Free Liquid Interface in a Partly Filled Right Circular Cylinder upon Gravity Step Reduction (PE). In: *Validation of Advanced Computational Methods for Multiphase Flow*, ed by H. Lemmonier, D. Jamet, O. Lebaigue (Bergell House Inc., Redding 2005) pp. 235-253
2. A. Ohlhoff, U. Rosendahl, M.E. Dreyer: Test-Case No 35: Flow Rate Limitation in Open Capillary Channels (PE). In: *Validation of Advanced Computational Methods for Multiphase Flow*, ed by H. Lemmonier, D. Jamet, O. Lebaigue (Bergell House Inc., Redding 2005) pp. 277-290
3. D.E. Jaekle, Jr.: AIAA 91-2172, 1 (1991)
4. U. Rosendahl, A. Ohlhoff, M.E. Dreyer: Choked flows in open capillary channels: theory, experiment and computations, J. Fluid Mech. **518**, 187 (2004)
5. M. Stief, J. Gerstmann, M.E. Dreyer: Proc. Appl. Math. Mech. **5**, 553 (2005)

Part VI

Appendix

A

Entrance Flow Problems

The flow problems dealt within this work often encounter an inlet flow from an infinite reservoir (in the mathematical sense) into the test section and an entrance flow in the test section itself. The inlet flow takes place before the closed cross section of the test channel and is characterized by strong convective accelerations and weak molecular momentum transport. In the entrance region of the test section a boundary layer with strong molecular momentum transport develops from the wall towards the center and the increasing displacement thickness leads to a weak convective acceleration. Depending on the REYNOLDS number of the flow problem and the length of the test section, the laminar flow profile may not be developed and the boundary layer may characterize the velocity distribution. The two different sections of the flow, the inlet and the entrance flow in the above defined sense, are modeled in the following by analytical and numerical means with the emphasize to have analytical or empirical relations which can be implemented in the complex flow models.

A.1 Analytical Calculation of the Inlet Pressure Drop

In this section a one-dimensional momentum balance is applied on a control volume around the inlet of a circular cylindrical tube. The pressure is averaged over the respective cross sections. The inlet area of the tube with radius R is shown in Fig. A.1. A cylindrical coordinate system (r, z, φ) is attached to the tube axis at the inlet and counts positive into the tube. A spherical coordinate system $(\tilde{r}, \theta, \tilde{\varphi})$ originates at the same point and counts positive into the control volume CV. In principle, we follow the approach given by [1], the differences are discussed along with the respective equations.

A1 a semispherical control volume (CV) around the inlet with radius R, inlet area $\mathrm{CS}_1 = A_1 = 2\pi R^2$, outlet area $\mathrm{CS}_2 = A_2 = \pi R^2$. The flow rate is

$$q = \bar{v}_z A_2 = -v_r(\tilde{r} = R) A_1$$

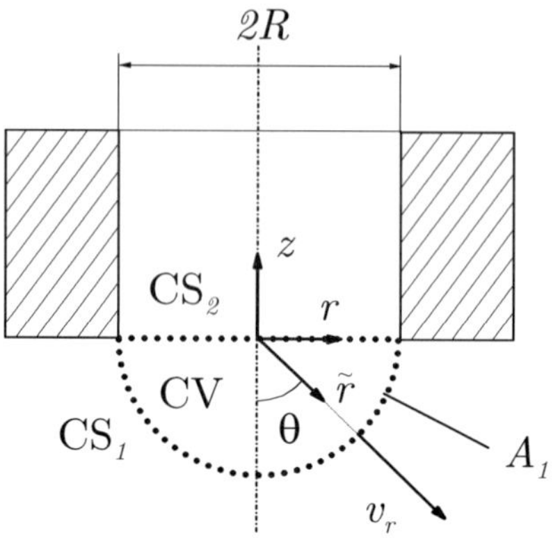

Fig. A.1. Schematic drawing for the inlet section of a cylindrical tube. CV: control volume, CS: control surface

with

$$\bar{v}_z = \frac{1}{A_2}\int\limits_0^{2\pi}\int\limits_0^{R} v_z(r)\big|_{z=0}\, r\mathrm{d}r\mathrm{d}\varphi\ ,$$

A2 a sink flow towards the origin outside the control volume with the velocity field

$$v_r = -\frac{q}{2\pi\tilde{r}^2}\ ;\tilde{r} \geq R\ , \tag{A.1}$$

A3 a velocity field at the tube entrance $z = 0$ assuming POISEUILLE flow (which is not reasonable except for very low REYNOLDS numbers, but will be used partly to compare the results with other authors)

$$v_z(z=0) = 2\bar{v}_z\left(1 - \frac{r^2}{R^2}\right)\ , \tag{A.2}$$

A4 a velocity field at the tube entrance $z = 0$ used for numerical calculations

$$v_z(z=0) = \bar{v}_z\ . \tag{A.3}$$

The linear momentum conservation in $z-$direction reads for the CV with control surfaces CS_1 and CS_2

$$\sum F_z = \frac{\mathrm{d}}{\mathrm{d}t}\left\{\iiint\limits_{CV} \rho v_z \mathrm{d}V\right\} + \iint\limits_{CS} \rho v_z\left(\boldsymbol{v}_{rel}\cdot\boldsymbol{n}\right)\mathrm{d}A \tag{A.4}$$

taking the outward normal of the control surface (see [2]). The velocity $\boldsymbol{v}_{rel}$ is the difference between the velocity of the control surface and the flow velocity. Since the CV is fixed in this case, the relative velocity $\boldsymbol{v}_{rel}$ is the flow velocity $\boldsymbol{v}$ on the CS. The force at the inlet of the control volume reads:

$$F_{z1} = \int_0^{2\pi}\int_0^{\pi/2} \left(-\tau_{rr}\cos\theta\,\tilde{r}^2\right)\big|_{\tilde{r}=R}\sin\theta\mathrm{d}\theta\mathrm{d}\tilde{\varphi} \tag{A.5}$$

with

$$\tau_{rr} = -p + 2\mu\frac{\partial v_r}{\partial\tilde{r}} \tag{A.6}$$

for $\tilde{r} \geq R$ and $v_\varphi = v_\theta = 0$.

The pressure at $\tilde{r} = R$ can be calculated from the Navier- Stokes equation in spherical coordinates (see for example [3])

$$\frac{\partial v_r}{\partial t} + v_r\frac{\partial v_r}{\partial\tilde{r}} = -\frac{1}{\rho}\frac{\partial p}{\partial\tilde{r}} + \nu\left\{\frac{1}{\tilde{r}^2}\frac{\partial}{\partial\tilde{r}}\left[\tilde{r}^2\frac{\partial v_r}{\partial\tilde{r}}\right] - \frac{2v_r}{\tilde{r}^2}\right\}. \tag{A.7}$$

Levine et al. [1] neglect the non-linear term in their derivation of the pressure outside the CV and thus end up without the $\sim q^2$ term in (A.9) and (A.10). The molecular momentum flux term in (A.7) cancels out with the velocity distribution given in (A.1) and thus the simplified equation of Levine et al. reads

$$\frac{\partial v_r}{\partial t} = -\frac{1}{\rho}\frac{\partial p}{\partial\tilde{r}}. \tag{A.8}$$

The velocity field outside the control volume (see (A.1)) has only one component v_r in radial direction, the other components are equal to zero. Therefore the pressure depends only on the $\tilde{r}-$ direction and the gradient reads

$$\frac{\mathrm{d}p}{\mathrm{d}\tilde{r}} = \rho\left\{\frac{\dot{q}}{2\pi\tilde{r}^2} + \frac{q^2}{2\pi^2\tilde{r}^5}\right\}. \tag{A.9}$$

Integration from $p(\tilde{r} = \infty)$ to $p(\tilde{r} = R)$ yields

$$p(R) = p_\infty - \rho\left\{\frac{\dot{q}}{2\pi R} + \frac{q^2}{8\pi^2 R^4}\right\}. \tag{A.10}$$

The pressure far from the control volume consists of the ambient pressure above the liquid surface and an additional pressure, e.g. a capillary pressure due to a curved free surface

$$p_\infty = p_a + \Delta p_c\,. \tag{A.11}$$

The capillary pressure has to be specified for the respective geometry of the experiment container. The meaning and the sign of Δp_c has been defined in Chap. 2. The force F_{z1} follows from (A.10) to

$$F_{z1} = \pi R^2\left\{p_\infty - \rho\left[\frac{\dot{q}}{2\pi R} + \frac{q^2}{8\pi^2 R^4} + \frac{2\nu q}{\pi R^3}\right]\right\}. \tag{A.12}$$

The force at the outlet of the control volume and thus at the inlet of the tube (containing the desired pressure drop) is calculated with

$$F_{z2} = -\int_0^{2\pi}\int_0^{R} p(z=0) r \mathrm{d}r \mathrm{d}\varphi = -\pi R^2 p(z=0) \; . \tag{A.13}$$

In the following the local acceleration in the control volume and the momentum fluxes across the boundaries are computed. The right hand side of (A.4) reads in a more specific form

$$\begin{aligned} \sum F_z = & \frac{\mathrm{d}}{\mathrm{d}t}\left\{\int_0^{2\pi}\int_0^{\pi/2}\int_0^{R} \rho\, v_z \tilde{r}^2 \sin\theta \,\mathrm{d}\tilde{r}\mathrm{d}\theta\mathrm{d}\tilde{\varphi}\right\} \\ & + \int_0^{2\pi}\int_0^{\pi/2} \rho \left[v_z(\boldsymbol{v}_{rel}\cdot\boldsymbol{n})\tilde{r}^2\right]\Big|_{\tilde{r}=R} \sin\theta \,\mathrm{d}\theta\mathrm{d}\tilde{\varphi} \\ & + \int_0^{2\pi}\int_0^{R} \rho\, v_z^2 \, r\mathrm{d}r\mathrm{d}\varphi \; . \end{aligned} \tag{A.14}$$

The computation of the change of momentum in the CV requires the knowledge of a velocity distribution within the CV, which is not available. The change of momentum will be therefore computed from the averaged values at the boundaries of the CV. Since the CV is constant in this case the differential operator can be applied on the velocity within the integral signs

$$\frac{\mathrm{d}}{\mathrm{d}t}\left\{\int_0^{2\pi}\int_0^{\pi/2}\int_0^{R} \rho\, v_z \tilde{r}^2 \sin\theta \,\mathrm{d}\tilde{r}\mathrm{d}\theta\mathrm{d}\tilde{\varphi}\right\} = \rho \int_0^{2\pi}\int_0^{\pi/2}\int_0^{R} \frac{\mathrm{d}v_z}{\mathrm{d}t}\tilde{r}^2 \sin\theta \,\mathrm{d}\tilde{r}\mathrm{d}\theta\mathrm{d}\tilde{\varphi} \; . \tag{A.15}$$

The acceleration is taken from the mean acceleration calculated with the velocity distributions at the control surfaces

$$\frac{\mathrm{d}v_z}{\mathrm{d}t} = \dot{v}_z = \frac{1}{2}(\dot{v}_{z1} + \dot{v}_{z2}) \; . \tag{A.16}$$

The acceleration at the CS_1 is estimated as

$$\dot{v}_{z1} = \frac{1}{A_1}\int_0^{2\pi}\int_0^{\pi/2} \frac{\mathrm{D}v_z}{\mathrm{D}t}\tilde{r}^2\Big|_{\tilde{r}=R} \sin\theta \;\mathrm{d}\theta\mathrm{d}\tilde{\varphi} \; . \tag{A.17}$$

The total acceleration in the tube coordinate system has an opposite sign compared to the acceleration in the reservoir ($v_z = -v_r\cos\theta$). The total differential reads

$$\frac{\mathrm{D}v_r}{\mathrm{D}t} = \frac{\partial v_r}{\partial t} + v_r\frac{\partial v_r}{\partial \tilde{r}} = -\frac{\dot{q}}{2\pi\tilde{r}^2} - \frac{q^2}{2\pi^2\tilde{r}^5} \tag{A.18}$$

and the acceleration becomes

$$\dot{v}_{z1} = \frac{\dot{q}}{4\pi R^2} + \frac{q^2}{4\pi^2 R^5} \,. \tag{A.19}$$

The calculation of the acceleration at the CS_2 leads to

$$\dot{v}_{z2} = \frac{1}{A_2} \int_0^{2\pi} \int_0^{R} \frac{\mathrm{D}v_z}{\mathrm{D}t} r \bigg|_{z=0} \mathrm{d}r\mathrm{d}\varphi = \frac{\dot{q}}{\pi R^2} \,. \tag{A.20}$$

For the position $z = 0$, the total differential reduces to

$$\frac{\mathrm{D}v_z}{\mathrm{D}t} = \frac{\partial v_z}{\partial t} \,.$$

The mean acceleration is from (A.16)

$$\dot{v}_z = \frac{5}{8} \frac{\dot{q}}{\pi R^2} + \frac{1}{8} \frac{q^2}{\pi^2 R^5} \,. \tag{A.21}$$

Levine et al. made a different approach to computed the mean acceleration in the CV. They assumed that the acceleration at the CS is convected into the CV with the velocity at the CS. This leads to a slightly different term for the mean acceleration

$$\dot{v}_{z,Levine} = \frac{19}{24} \frac{\dot{q}}{\pi R^2} + \frac{1}{8} \frac{q^2}{\pi^2 R^5} \,. \tag{A.22}$$

The difference occurs in the non-stationary term only and is $|5/8 - 19/24| = 0.167$. Equation (A.21) leads to the total change of momentum in the CV

$$\rho \int_0^{2\pi} \int_0^{\pi/2} \int_0^{R} \frac{\mathrm{d}v_z}{\mathrm{d}t} \tilde{r}^2 \sin\theta \, \mathrm{d}\tilde{r}\mathrm{d}\theta \mathrm{d}\tilde{\varphi} = \frac{2}{3}\rho\pi R^3 \left\{ \frac{5}{8} \frac{\dot{q}}{\pi R^2} + \frac{1}{8} \frac{q^2}{\pi^2 R^5} \right\} \,. \tag{A.23}$$

The inlet momentum flux is computed with

$$\begin{aligned} &\int_0^{2\pi} \int_0^{\pi/2} \rho \left[v_z (\boldsymbol{v}_{rel} \cdot \boldsymbol{n}) \tilde{r}^2 \right] \big|_{\tilde{r}=R} \sin\theta \, \mathrm{d}\theta \mathrm{d}\tilde{\varphi} \\ &= \int_0^{2\pi} \int_0^{\pi/2} \rho \left[(-v_r \cos\theta) \, v_r \tilde{r}^2 \right] \big|_{\tilde{r}=R} \sin\theta \, \mathrm{d}\theta \mathrm{d}\tilde{\varphi} \end{aligned} \tag{A.24}$$

and leads to

$$\int_0^{2\pi} \int_0^{\pi/2} \rho \left[(-v_r \cos\theta) \, v_r \tilde{r}^2 \right] \big|_{\tilde{r}=R} \sin\theta \, \mathrm{d}\theta \mathrm{d}\tilde{\varphi} = -\frac{1}{4} \frac{\rho q^2}{\pi R^2} \,. \tag{A.25}$$

The outlet momentum flux becomes

$$\int_0^{2\pi}\int_0^R \rho\, v_z^2 \tilde{r}\, \mathrm{d}\tilde{r}\mathrm{d}\tilde{\varphi} = K_{ec}\frac{\rho q^2}{\pi R^2} , \tag{A.26}$$

where K_{ec} is the kinetic-energy correction factor ($K_{ec} = 4/3$ for assumption A3, $K_{ec} = 1$ for assumption A4). Inserting all terms in (A.4) and reorganizing for the unknown pressure at the outlet of CV gives

$$p(z=0) = p_\infty - \rho\left\{\frac{11}{12}\frac{\dot{q}}{\pi R} + \frac{q^2}{\pi^2 R^4}\left[K_{ec} - \frac{1}{24}\right] + \frac{2\nu}{R}\frac{q}{\pi R^2}\right\} . \tag{A.27}$$

Levine et al. end up with a slightly different assumption for the pressure at the tube inlet:

$$p(z=0)_{Levine} = p_\infty - \rho\left\{\frac{37}{36}\frac{\dot{q}}{\pi R} + \frac{7}{6}\frac{q^2}{\pi^2 R^4} + \frac{2\nu}{R}\frac{q}{\pi R^2}\right\} . \tag{A.28}$$

The difference in the inertia term $|11/12 - 37/36| = 0.116$ stems from the different assumptions for the derivation of the change of momentum in the CV, the difference $|31/24 - 7/6| = 0.124$ is due to the neglect of the convective terms in the reservoir by [1]. Since they assumed the parabolic velocity distribution at the entrance, $K_{ec} = 4/3$ was taken for the comparison with our results.

Introducing a normalized pressure, the REYNOLDS number with the hydraulic diameter of the inlet and a dimensionless number for the acceleration

$$\hat{p} = \frac{2p}{\rho \bar{v}_z^2}; \mathrm{Re}_i = \frac{\bar{v}_z D_i}{\nu} = \frac{2q}{\pi R \nu}; D_i = 2R; \Pi_1 = \frac{\dot{q}R}{\pi \nu^2} \tag{A.29}$$

leads to the dimensionless pressure

$$\hat{p}(z=0) = \hat{p}_\infty - \left\{\frac{22}{3}\frac{\Pi_1}{\mathrm{Re}_i^2} + 2\left[K_{ec} - \frac{1}{24}\right] + \frac{8}{\mathrm{Re}_i}\right\} . \tag{A.30}$$

The steady case reduces to $\Pi_1 = 0$. The equation given by Levine et al. reads in dimensionless form

$$\hat{p}(z=0) = \hat{p}_\infty - \left\{\frac{74}{9}\frac{\Pi_1}{\mathrm{Re}_i^2} + \frac{7}{3} + \frac{8}{\mathrm{Re}_i}\right\} . \tag{A.31}$$

The differences are $|22/3 - 74/9| = 0.88$ in the non-steady term and $|62/24 - 7/3| = 0.25$ in the convective term. The viscous contribution remains the same. To implement the equation in the mathematical model for the capillary rise it is written here in terms of the meniscus velocity $\dot{h}$ and the meniscus acceleration $\ddot{h}$ $\left(\text{with } q = \pi R^2 \dot{h} \text{ and } \dot{q} = \pi R^2 \ddot{h}\right)$:

$$p(z=0) = p_\infty - \rho\left\{\frac{11}{12}R\ddot{h} + \frac{23}{24}\dot{h}^2 + \frac{2\nu}{R}\dot{h}\right\} . \tag{A.32}$$

The kinetic-energy correction factor K_{ec} was set to $K_{ec} = 1$.

A.2 Numerical Calculation of the Inlet Pressure Drop

A.2.1 Computation of the Inlet Pressure Loss into a Cylindrical Tube from an Infinite Reservoir

Calculations of the inlet pressure drop into a cylindrical tube were performed with FIDAP Version 8.55 to compare the data with the analytical model presented above. The geometry is shown in Fig. A.2 . The radius of the hemisphere R_s was chosen such that $v_s \approx 0.01\bar{v}_z$. The velocity v_s is the inlet boundary condition of CS_1, $\bar{v}_z$ is the cross section averaged velocity at the inlet of the tube (CS_2). Since the radius of the tube ($R = 0.25$ cm) was not changed in the computations, the radius R_s is $R_s = 2$ cm. The properties of the liquid are: $\rho = 1$ g/cm^3, $\mu = 10^{-2}$ g/(cm s). The length l of the tube is always larger than the entrance length $z_e = 0.0575\ \mathrm{Re}_i D_i$. The parameters of the calculation are given in Table A.1. The constant flow rate through the tube is $q = \pi R^2 \bar{v}_z = 2\pi R_s^2 v_s$.

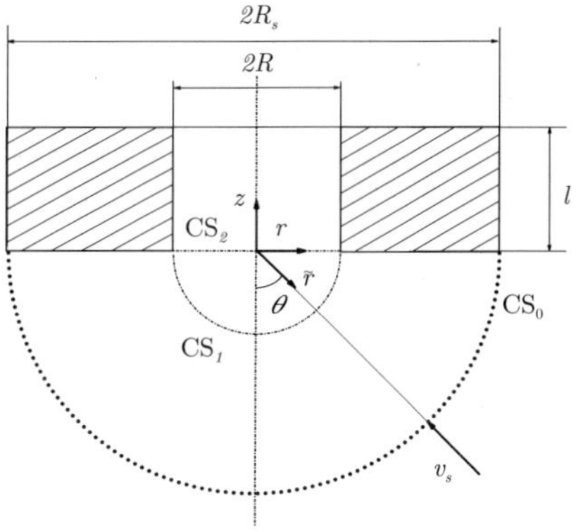

Fig. A.2. Computational domain for the calculation of the inlet pressure drop of a cylindrical tube (not to scale, in particular for l)

Table A.1. Parameter for the FIDAP calculations

Re_i (-)	l (cm)	q (cm^3/s)	$\bar{v}_z$ (cm/s)	v_s (cm/s)
63	2	0.247	1.3	0.0098
160	5	0.628	3.2	0.025
400	12	1.571	8.0	0.063
1000	30	3.927	20.0	0.156
2500	75	9.818	50.0	0.391

The results of the numerical calculation are nondimensionalized in the following way:

$$\hat{p} = \frac{2p}{\rho \bar{v}_z^2}; \hat{v}_z = \frac{v_z}{\bar{v}_z}; \hat{z} = \frac{z}{D_i \mathrm{Re}_i}; \hat{r} = \frac{r}{D_i}; D_i = 2R \, .$$

For simplification the superscripts of the non-dimensional quantities are dropped for the rest of this section. Figure A.3 shows the pressure plot at the center line versus the z-coordinate. Due to a pressure distribution in radial direction (see Fig. A.4) in the entrance section of the tube this pressure is not the mean pressure of the cross section and can not be compared with the analytical considerations above. The pressure across the inlet section of the tube decreases towards the tube wall due to the higher accelerations at the sharp corner. Furthermore, the displacement of liquid from the wall due to the boundary layer leads to velocities in radial direction and thus to a pressure gradient. The cross section averaged pressure is somewhat lower than the center line pressure and is discussed in detail later. The differences in the radial pressure distribution decrease to less than 1% after $z \geq 1/\mathrm{Re}_i$ (one diameter after the entrance) and the center line pressure can be used as the cross section averaged pressure.

The center line pressure was set to zero at $p(z = z_e) = 0$ with $z_e = 0.0575$ (taken from [4]) where the pressure gradient achieves its linear decrease due to the fully developed laminar profile (HAGEN-POISEUILLE). The pressure decreases from infinity to the inlet approximately about unity. This corresponds

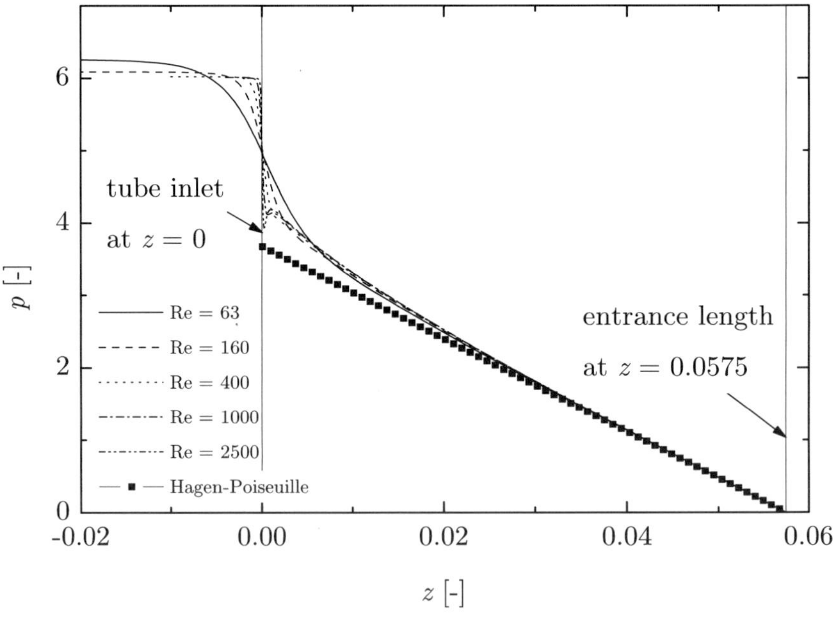

Fig. A.3. Plot of the pressure $p(r = 0)$ versus the z-coordinate

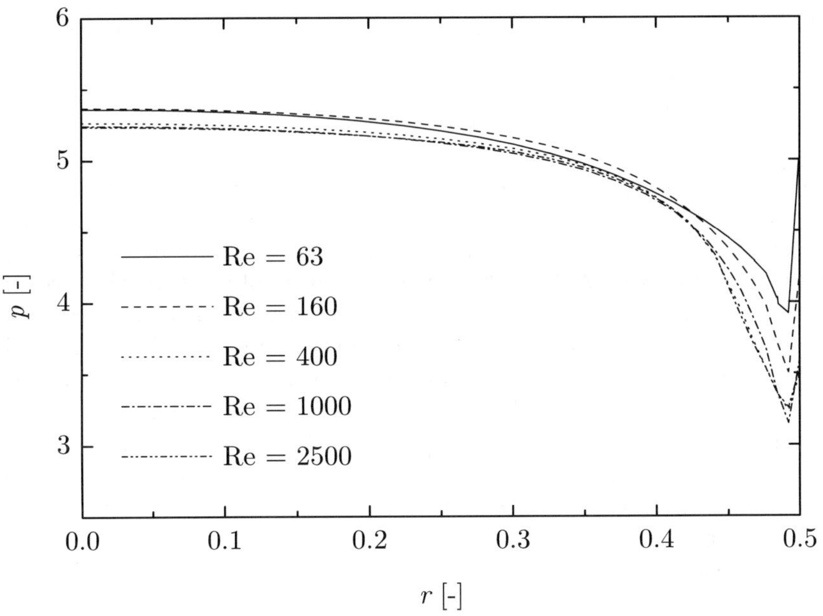

Fig. A.4. Pressure $p(z = 0)$ at the tube inlet versus the radial coordinate r

to the convective acceleration from the reservoir velocity $v_k = 0.01\bar{v}_z$ to the inlet velocity $\bar{v}_z(z = 0) = 1$. Depending on the REYNOLDS number the pressure decreases further and reaches the constant gradient of the laminar flow distribution (see (12.14)). For $\mathrm{Re}_i \geq 1000$ a sharp decrease is followed by an increase of the pressure in flow direction, which can be explained by a vena contracta due to the sharp inlet. A dead water region narrows the effective tube diameter and leads to an additional acceleration at the throat, followed by a deceleration and a pressure increase. This phenomena is well known from flows through obstructions ([2]). The resulting pressure gradient is depicted in Fig. A.5.

Figure A.6 shows the v_z-velocity along the center line. The velocity increases towards the entrance of the tube approximately about unity and achieves its maximum after the entrance length. The entrance length is usually defined with

$$v_z(z = z_e) = 0.99\ v_{z,\max}\ .$$

This criterion for the estimation of the entrance length z_e is fulfilled within 0.5%. Figure A.7 shows the velocity profile at the inlet of the tube, which can be assumed to have constant values within 20% in the tube center and decrease sharply towards the tube wall due to the boundary layers.

To compare the pressure values with the analytical data the pressure from the numerical calculations must be cross section averaged. For the entrance into the control volume this value reads in dimensional form

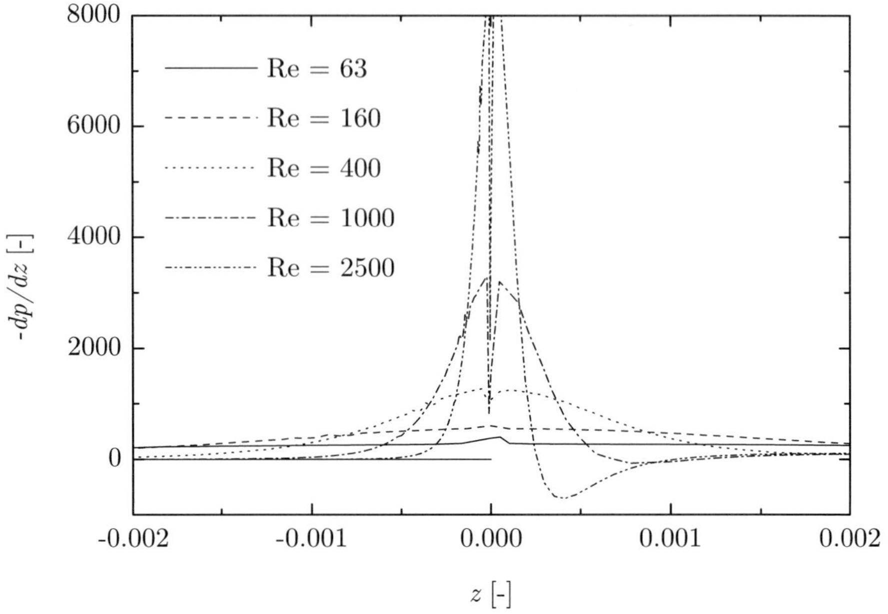

Fig. A.5. Plot of the negative pressure gradient $-dp/dz$ versus the z-coordinate. The strong changes of the plots for Re = 1000 and Re = 2500 at $z = 0$ are numerical artefacts due to mesh incompatibilities between the inlet region and the tube entrance

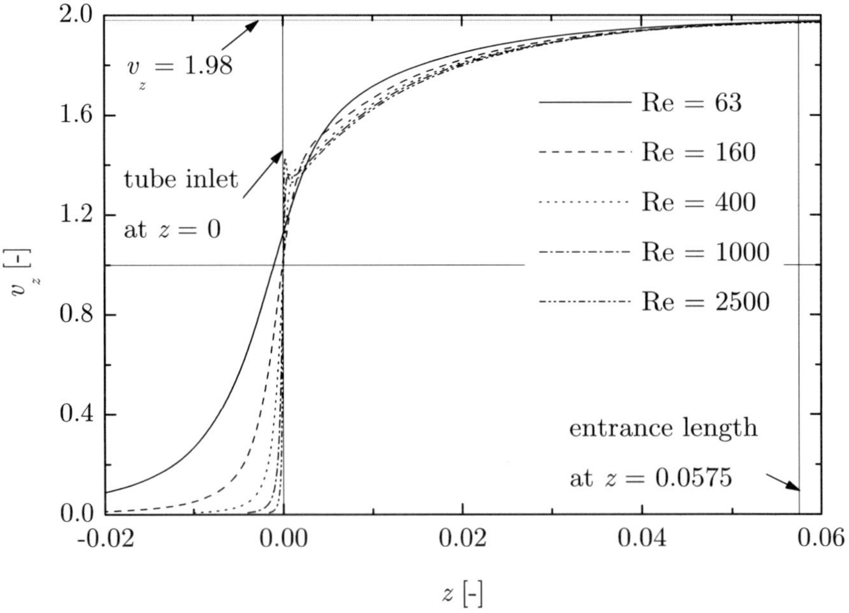

Fig. A.6. Velocity distribution along the center line of the tube: $v_z(r = 0)$

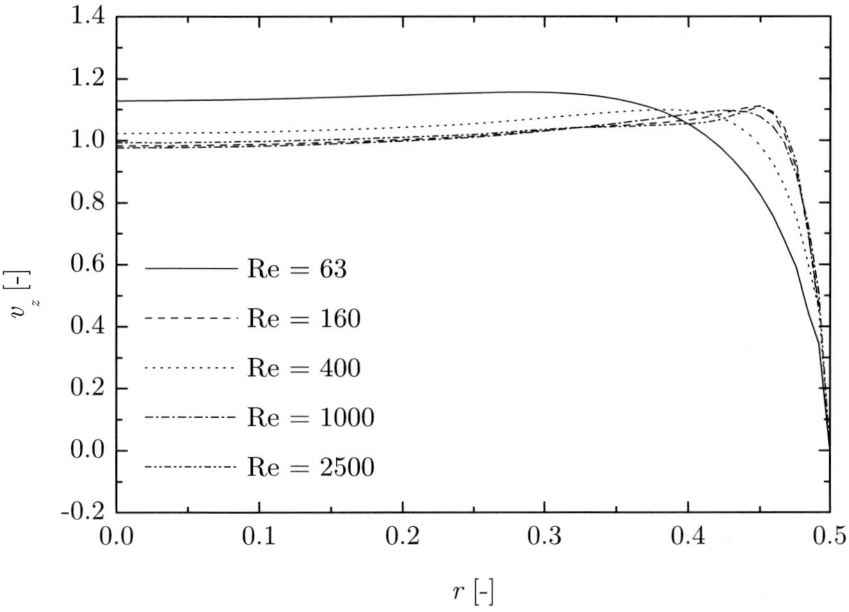

Fig. A.7. Velocity distribution at the entrance of the tube: $v_z(z = 0)$

$$p(z=-R) = \frac{1}{\mathrm{CS}_1} \int_0^{2\pi} \int_0^{\pi/2} p(\theta)\tilde{r}^2\big|_{\tilde{r}=R} \sin\theta \, \mathrm{d}\theta \mathrm{d}\varphi \; , \tag{A.33}$$

and the entrance pressure into the tube is calculated by

$$p(z=0) = \frac{1}{\mathrm{CS}_2} \int_0^{2\pi} \int_0^{R} p(r)\big|_{z=0} \, r \, \mathrm{d}r \mathrm{d}\varphi \; . \tag{A.34}$$

From these values the pressure differences against p_∞ are calculated and compared with the analytical predictions in Table A.2. The equations from the analytical calculation for the pressure drop at the entrance into the CV read for the steady case from (A.12)

$$p_\infty - p(\tilde{r} = R) = \Delta p_{CS1}\left(z = -\frac{R}{D_i \mathrm{Re}_i}\right) = \frac{1}{4} + \frac{8}{\mathrm{Re}_i} \; , \tag{A.35}$$

and the pressure at the tube inlet from (A.30) with $K_{ec} = 1$

$$p_\infty - p(z=0) = \Delta p_{CS2}(z=0) = \frac{46}{24} + \frac{8}{\mathrm{Re}_i} \; . \tag{A.36}$$

The differences between the analytical and numerical pressure drops at the entrance into the CV can be explained by the fact that the sink flow is

Table A.2. Comparison of the pressure drops

	Δp_{CS1} (inlet CV)		Δp_{CS2} (inlet tube)	
Re_i	num.	theo.	num.	theo.
63	0.411	0.377	1.771	2.044
160	0.379	0.300	1.572	1.967
400	0.357	0.270	1.489	1.937
1000	0.339	0.258	1.478	1.925
2500	0.330	0.253	1.481	1.920

disturbed by the wall perpendicular to the tube inlet. This leads to additional accelerations outside the boundary layer and the displacement thickness of this wall, and thus to a higher pressure decrease than predicted by the analytical calculation, which assumes an undisturbed sink flow outside the control volume. The pressure drop at the tube inlet is smaller in the numerical calculation than in the theoretical assumption. The differences may stem from the averaging in the CV (analytical model) and the flow field of a vena contracta observed in the numerical solution.

Plotting the pressure drop at the tube entrance versus $1/\mathrm{Re}_i$ (Fig. A.8) yields

$$\Delta p_{CS2}(z=0) = K_{it1} + \frac{K_{it2}}{\mathrm{Re}_i} \, .$$

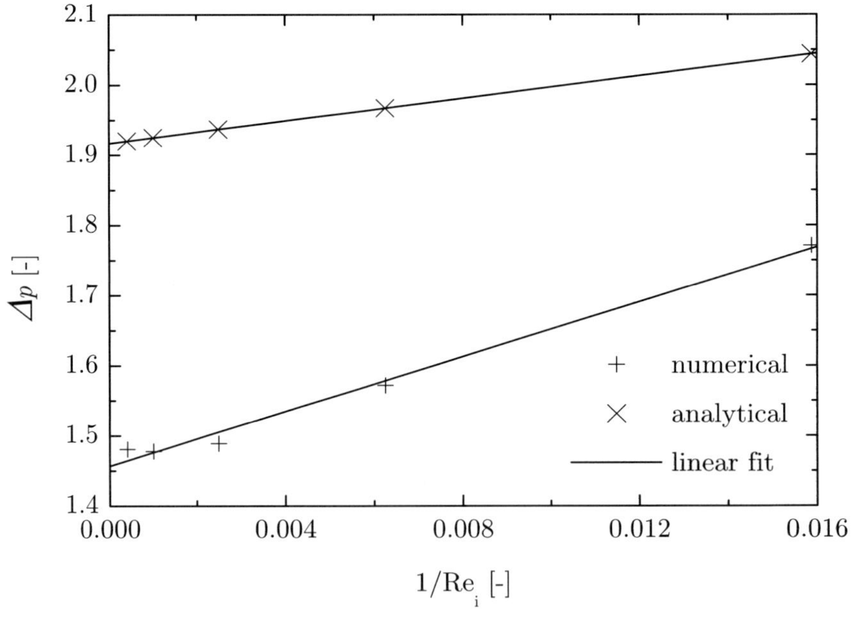

Fig. A.8. Pressure drop at the tube inlet versus the Reynolds number

Table A.3. Pressure drop factor comparison

	numerical	theoretical
K_{it1}	1.46	1.92
K_{it2}	19.5	8.00

The values for the convective pressure loss K_{it1} and the viscous pressure loss K_{it2} are given in Table A.3.

The analytical estimation overpredicts the pressure drop in the control volume, may be due to the averaging of the total change of momentum. It should be emphasized here that the foregoing consideration holds only up to the tube inlet. The evolution of the pressure inside the tube is discussed after the next section.

A.2.2 Computation of the Inlet Pressure Loss into a Rectangular Channel from an Infinite Reservoir through a Nozzle

The pressure drop for the inlet flow into the capillary channel has been calculated with the CFD program FIDAP. The parameters for the calculation correspond to the data given in Table 17.1. The calculations for the drop tower experiments have been performed with a closed reservoir and an outlet (thus the inlet of the test section) with $A = ab$. The withdrawn flow rate was provided by a change of the free surface in the reservoir. The capillary pressure due to this curvature was neglected. The calculations for the TEXUS-37 module have been performed with the original reservoir dimensions as described in Sect. 17.2. The reservoir was designed such that the inflow was provided over a large surface on the bottom of the reservoir. The results of these computations are given in Table A.4 and in Fig. A.9. The pressure was cross section averaged over the inlet and the outlet cross sections (thus the inlet of the test section).

Table A.4. Coefficients for the pressure drop at the inlet of the channel. The calculations have been performed with the flow rates indicated in the table

No	K_{i1} (-)	K_{i2} (-)	Q (ml/s)
FT-29	2.7	152	$0.25 < Q < 1.5$
FT-32	2.2	70	$0.25 < Q < 1.8$
FT-33	2.2	115	$1.0 < Q < 3.0$
FT-34	2.3	195	$2.0 < Q < 6.0$
FT-35/35a	2.0	150	$5.0 < Q < 10$
TE-35a	1.4	312	$5.0 < Q < 10$

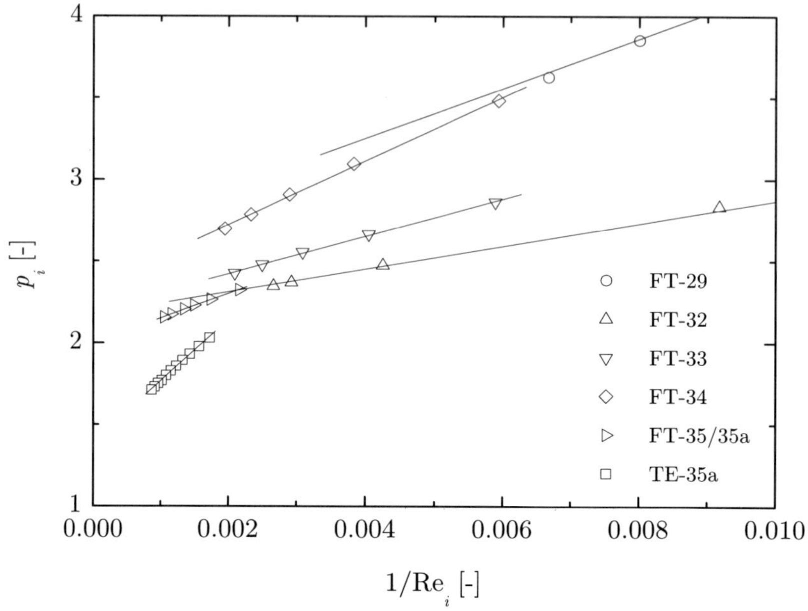

Fig. A.9. Dimensionless inlet pressure drop versus $1/\mathrm{Re}_i$. The symbols show the results of the computations for different flow rates, the lines denote the linear regression to achieve the coefficients K_{i1} and K_{i2}

The values in Table A.4 were evaluated using

$$p_i = -\left(K_{i1} + \frac{K_{i2}}{\mathrm{Re}_i}\right)\frac{\rho}{2}v^2 \tag{A.37}$$

with

$$\mathrm{Re}_i = \frac{vD_i}{\nu} = \frac{2Q}{\nu(a+b)} \tag{A.38}$$

and

$$D_i = \frac{2ab}{a+b} \ . \tag{A.39}$$

The plots in Fig. A.9 show a different behavior between the TEXUS-37 module and the drop tower experiment. The convective contribution to the pressure drop is much smaller for the TEXUS-37 module due to the optimization of the inlet nozzle. The flow in the reservoir is mainly a sink flow towards the inlet of the nozzle and a duct in the nozzle itself with decreasing cross section. The velocity distribution is shown in Fig. A.10. Most of the acceleration takes place in the duct, but no separation occurs. The inflow conditions in the drop tower experiment were different due to the fact that the parallel plate channel starts at the bottom of the reservoir and emerges from the free surface. Thus the liquid flows into the channel from the sides and separation could be observed in the vicinity of the test section inlet. The values presented here are

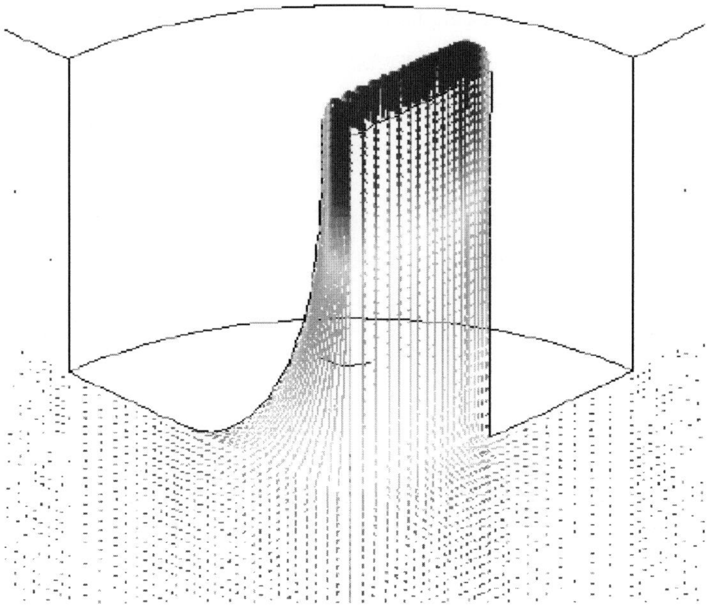

Fig. A.10. Velocity plot of the flow in the reservoir of the TEXUS-37 modul. Only a quarter of the flow domain is shown. The flow is from the bottom to the top. It can be seen that the acceleration takes place in the nozzle. A boundary layer develops along the wall of the nozzle and contributes to the $1/\mathrm{Re}_i$ part of the pressure loss

valid for the given geometry and the calculated flow rates only. The empirical equation (A.37) is needed for the implementation into the mathematical model. A generalized investigation of the entrance pressure drop was not the scope of this work. But due to the complex shapes of the entrance sections, coefficients from literature are not available and could only be achieved by means of three-dimensional calculations of the flow before the test section inlet.

A.3 Determination of the Entrance Pressure Distribution

Since most of the problems we deal with in this work do not have a fully developed laminar flow field, additional information about the entrance pressure loss due to the change of the inlet flow field to the HAGEN-POISEUILLE flow field after the entrance length x_e are needed. One relevant paper [5] for simple geometries, such as the circular cylindrical tube and the parallel plate channel, is reported here. The results were compared with solutions from CFD calculations.

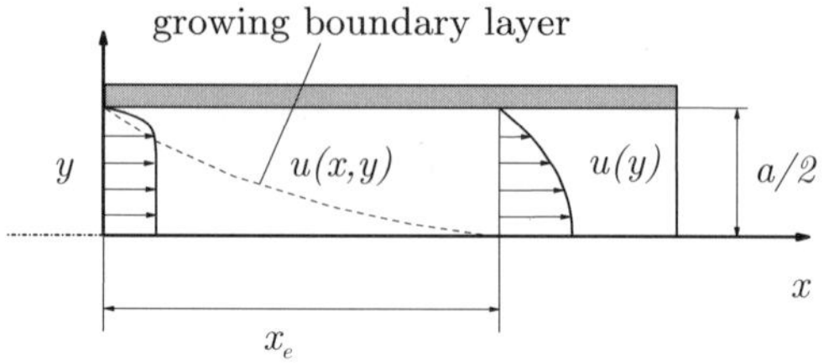

Fig. A.11. Entrance flow profile in a channel of two parallel plates

A.3.1 Parallel Plates

Primarily, we consider the flow between two parallel infinite plates (two dimensional flow). In the ideal case (used for most of the numerical computation in the literature) the flow profile is constant at the inlet $x = 0$ and develops towards the parabolic velocity distribution at x_e (see Fig. A.11). In the general case the flow profile is different from the block profile at $x = 0$. The difference can be expressed by a profile factor

$$\Gamma = \frac{u(y=0)}{\bar{u}} \tag{A.40}$$

where $u(y = 0)$ is the center line velocity and

$$\bar{u} = \frac{1}{a} \int_{-a/2}^{a/2} u(y)\mathrm{d}y \tag{A.41}$$

the two dimensional cross section averaged mean velocity in $x-$ direction. The profile factor is $\Gamma = 1$ for a constant flow field at the entrance ($u(x = 0) = \bar{u}$) and $\Gamma = 1.5$ for the fully developed laminar flow distribution (HAGEN-POISEUILLE). The assumptions for the following considerations are given by [5] as:

A5 The flow is incompressible and laminar.
A6 The cross section is constant.
A7 No pressure gradients exist (boundary layer assumption) perpendicular to the flow direction

$$\frac{\mathrm{d}p}{\mathrm{d}y} = 0\,. \tag{A.42}$$

A8 Shear is negligible in flow direction compared to shear perpendicular to the flow direction

$$\frac{\partial^2 u}{\partial x^2} \ll \frac{\partial^2 u}{\partial y^2}\,. \tag{A.43}$$

Sparrow et al. [5] introduce a stretched coordinate

$$\mathrm{d}x = \varepsilon \mathrm{d}x^* \qquad \text{with} \qquad x = \int_0^{x^*} \varepsilon \mathrm{d}x^* , \tag{A.44}$$

and write the velocity

$$u = u_{fd} + u^* \tag{A.45}$$

in terms of the fully developed flow field u_{fd} and the difference velocity u^*. With the dimensionless variables

$$\breve{x} = \frac{4x\nu}{a^2\bar{u}} = \frac{16x}{D_h \mathrm{Re}_h}; \; \breve{x}^* = \frac{4x^*\nu}{a^2\bar{u}}; \; \breve{y} = \frac{2y}{a}; \; \breve{u} = \frac{u}{\bar{u}}; \; D_h = 2a; \; \mathrm{Re}_h = \frac{\bar{u}D_h}{\nu} \tag{A.46}$$

the solution for the velocity reads

$$\breve{u} = 1.5\left(1 - \breve{y}^2\right) + \sum_i \frac{2}{\alpha_i^2}\left\{\frac{\cos(\alpha_i \breve{y})}{\cos \alpha_i} - 1\right\} e^{-\alpha_i^2 \breve{x}^*} . \tag{A.47}$$

The first term describes the fully developed HAGEN-POISEUILLE profile, the second term the difference velocity u^*. The eigenvalues of α_i are solutions of the equation

$$\tan \alpha_i = \alpha_i , \tag{A.48}$$

the values are given in Table A.5. The plot of $\breve{u}$ versus $\bar{y}$ is shown in Fig. A.12. The flow starts with an inviscid core region at the inlet ($\bar{x}^* = 0$), which is accelerated and reduced by a growing boundary layer from the wall. The corresponding derivatives of the velocity read

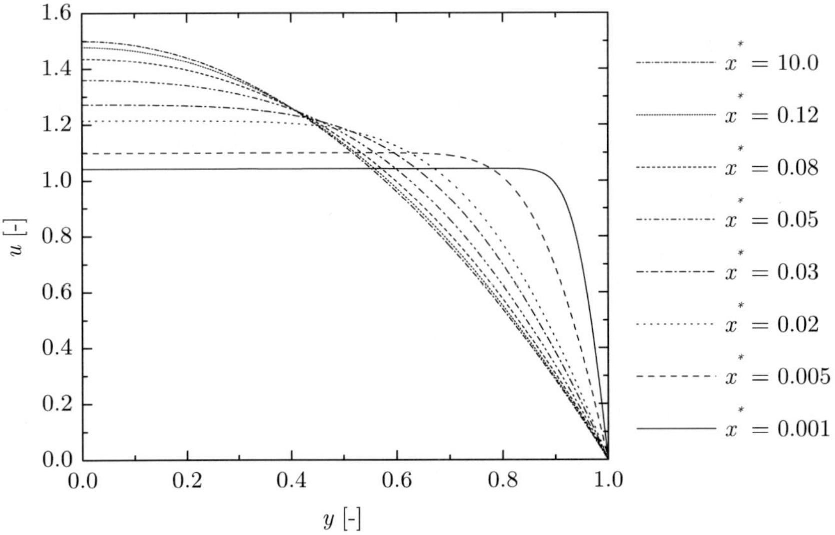

Fig. A.12. Plot of the dimensionless velocity $\breve{u}$ versus $\breve{y}$ as a funtion of x^*

Table A.5. Eigenvalues α_i for the pressure gradient calculation in the entrance region

i	circular tube	parallel-plate channel
1	5.13562	4.49341
2	8.41724	7.72525
3	11.6198	10.9041
4	14.7960	14.0662
5	17.9598	17.2207
6	21.1170	20.3712
7	24.2701	23.5193
8	27.4206	26.6660
9	30.5692	29.8116
10	33.7165	32.9563
11	36.8629	36.1005
12	40.0085	39.2438
13	43.1535	42.3873
14	46.2980	45.5304
15	49.4422	48.6731
16	52.5860	51.8164
17	55.7296	54.9593
18	58.8730	58.1010
19	62.0162	61.2437
20	65.1593	64.3860
21	68.3022	67.5269
22	71.4450	70.6693
23	74.5877	73.8105
24	77.7303	76.9548
25	80.8728	80.0963

$$\frac{\partial \breve{u}}{\partial \breve{x}^*} = -2\sum_i \left\{ \frac{\cos(\alpha_i \breve{y})}{\cos \alpha_i} - 1 \right\} e^{-\alpha_i^2 \breve{x}^*} , \tag{A.49}$$

$$\frac{\partial \breve{u}}{\partial \breve{y}} = -3\breve{y} - \sum_i \frac{2}{\alpha_i} \frac{\sin(\alpha_i \breve{y})}{\cos \alpha_i} e^{-\alpha_i^2 \breve{x}^*} , \tag{A.50}$$

$$\left.\frac{\partial \breve{u}}{\partial \breve{y}}\right|_{\breve{y}=1} = -3 - 2\sum_i e^{-\alpha_i^2 \breve{x}^*} . \tag{A.51}$$

The coefficient ε for the stretching of the $x-$coordinate is defined as

$$\varepsilon = \frac{\int_0^1 \left(2\breve{u} - 1.5\breve{u}^2\right) \frac{\partial \breve{u}}{\partial \breve{x}^*} \mathrm{d}\breve{y}}{\left(\frac{\partial \breve{u}}{\partial \breve{y}}\right)_{\breve{y}=1} + \int_0^1 \left(\frac{\partial \breve{u}}{\partial \breve{y}}\right)^2 \mathrm{d}\breve{y}} . \tag{A.52}$$

Once the velocity profile is know, the pressure loss can be given in dimensionless form as

$$\frac{2\,(p_i - p)}{\rho \bar{u}^2} = 6\breve{x} + K_{ps}(\breve{x}) \tag{A.53}$$

with

$$K^*_{ps1} = 0.4 + \sum_i \frac{4}{\alpha_i^2}\left\{e^{-\alpha_i^2 \breve{x}^*} - 2\right\} e^{-\alpha_i^2 \breve{x}^*} - 2\int_0^{\breve{x}^*} \varepsilon \left.\frac{\partial \breve{u}^*}{\partial \breve{y}}\right|_{\breve{y}=1} d\breve{x}^* \;, \tag{A.54}$$

if we consider the stretched coordinate, and

$$K_{ps2} = 0.8 + \sum_i \frac{4}{\alpha_i^2}\left\{e^{-\alpha_i^2 \breve{x}} - 3\right\} e^{-\alpha_i^2 \breve{x}} \;, \tag{A.55}$$

if we set $\varepsilon = 1$. The pressure at $x = 0$ is denoted by p_i. The first term of (A.53) describes the linear pressure loss of a fully developed flow field, the additional pressure loss K^*_{ps1} or K_{ps2} vanishes after the entrance length has reached. The derivative of the second term of (A.47), thus $\breve{u}^* = u^*/\bar{u}$, at $\breve{y} = 1$ reads

$$\left.\frac{\partial \breve{u}^*}{\partial \breve{y}}\right|_{\breve{y}=1} = -2\sum_i e^{-\alpha_i^2 \breve{x}} \;. \tag{A.56}$$

The comparison of the different friction factors is shown in Fig. A.13. Since we are seeking for the pressure gradient, the derivatives of the additional friction factors are shown in Fig. A.14. The picture shows that the simplified solution K_{ps1} with $\varepsilon = 1$ underpredicts the gradient. For this reason, the exact solution with K_{ps2} has been used for the computation of the critical velocity in capillary channels. For the prediction of the capillary rise in tubes the simplified solution is sufficient. Now we can write the profile factor Γ^*

$$\Gamma^*(\breve{x}^*) = 1.5 + \sum_i \frac{2}{\alpha_i^2}\left\{\frac{1}{\cos\alpha_i} - 1\right\} e^{-\alpha_i^2 \breve{x}^*} \;. \tag{A.57}$$

The plot in Fig. A.15 shows the profile factor Γ^* as a function of the stretched coordinate $\breve{x}^*$. The value of Γ^* corresponds to the ordinate in Fig. A.12, which shows the velocity on the center line of the channel. The profile factor calculated with $\varepsilon = 1$ underpredicts the velocity increase in the entrance section, but this has no influence on the definition of the entrance length x_e. Both profile factors approach $1.5 - 0.015$ at x_e (which is the definition of the entrance length). Due to momentum transport before the inlet (in the nozzle and/or the reservoir) the flow enters the channel inlet with a profile factor $1 < \Gamma < 1.5$. From FIDAP calculations with the respective inlet geometry of the experiments (TEXUS-37 and drop tower) we evaluated the velocity $u(0,0,0)$, i.e. at the origin of the coordinate system for the parallel plate channel and derived a profile factor

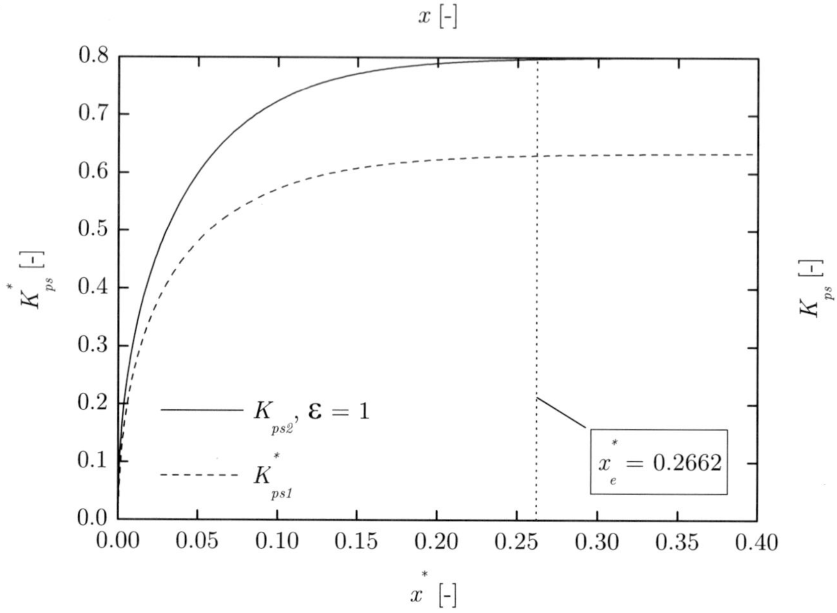

Fig. A.13. Comparison of the different friction factors K^*_{ps1} and K_{ps2} versus the flow length $\breve{x}^*$ and $\breve{x}$, respectively

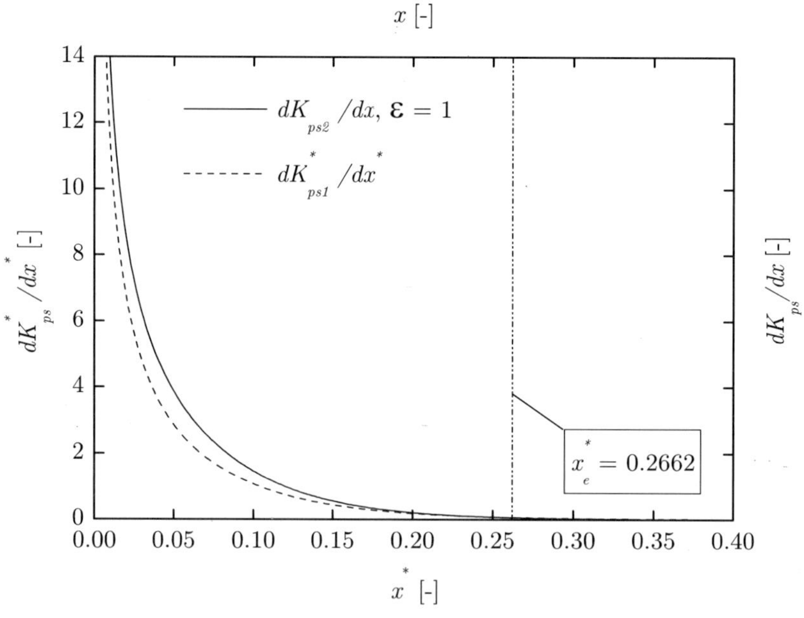

Fig. A.14. Plot of the derivative of the additional friction factors K^*_{ps1} and K_{ps1} versus the dimensionless flow length $\breve{x}^*$ and $\breve{x}$, respectively

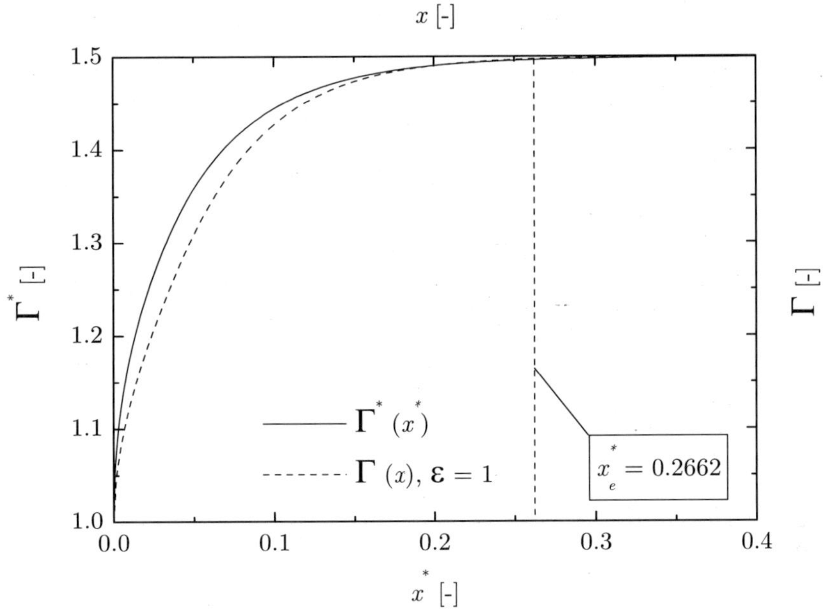

Fig. A.15. Plot of the profile factor Γ^* as function of the stretched coordinate $\breve{x}^*$ with $\varepsilon \neq 1$, and plot of $\Gamma(\breve{x})$ for $\varepsilon = 1$, both in dimensionless form

$$\Gamma_i = \frac{u(0,0,0)}{\bar{u}} . \tag{A.58}$$

Comparison with (A.57) yields $\breve{x}^*$ and the corresponding β

$$\beta = \frac{\breve{x}_e^* - \breve{x}_s^*}{\breve{x}_e^*} \tag{A.59}$$

with $x_e = 0.01639 D_i \mathrm{Re}_i$ and $\breve{x}_e^* = 0.2622$. A fully developed flow has $\beta = 0$ and $\Gamma_i = 1.5$, a constant velocity profile at the inlet $u = \bar{u}$ gives $\beta = 1$ and $\Gamma_i = 1$. The actual coordinate to start the profile development is than

$$x_s = x + (1 - \beta) x_e . \tag{A.60}$$

The results given in Table A.6 show that the β factor can be plotted versus the inlet REYNOLDS number. This gives a relation

$$\beta = L_1 + \frac{L_2}{\mathrm{Re}_i} . \tag{A.61}$$

similar to the entrance pressure drop (see (16.35)). The inlet REYNOLDS number Re_i is composed with the hydraulic diameter D_i and the cross section averaged flow velocity at the inlet

$$\mathrm{Re}_i = \frac{v D_i}{\nu}; D_i = \frac{2ab}{a+b} . \tag{A.62}$$

Table A.6. Coefficients for the entrance pressure drop at the inlet of the channel. The calculations have been performed with the flow rates indicated in the table

No	L_1 (-)	L_2 (-)	Q (ml/s)
FT-29	1	-7	$0.25 < Q < 1.5$
FT-32	1	-12	$0.25 < Q < 1.8$
FT-33	1	-21	$1.0 < Q < 3.0$
FT-34	1	-34	$2.0 < Q < 6.0$
FT-35/35a	1	-18	$5.0 < Q < 10$
TE-35a	1	-75	$5.0 < Q < 10$

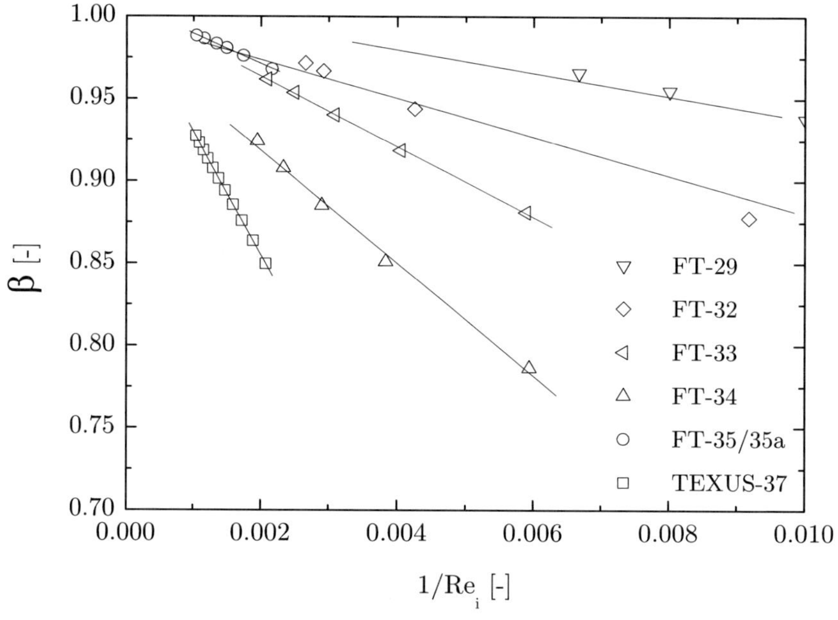

Fig. A.16. Plot of β as a function of $1/\mathrm{Re}_i$. The symbols show the results of the computations for different flow rates, the lines denote the linear regression to achieve the coefficients L_1 and L_2

The final equation to be implemented into the differential equation for the determination of the critical flow velocity (see Chap. 16) is the derivative of (A.53) in dimensionless form

$$\frac{\mathrm{d}p}{\mathrm{d}x} = -\frac{\rho\bar{u}^2}{2}\left\{6\frac{\mathrm{d}\breve{x}}{\mathrm{d}x} + \frac{\mathrm{d}K_{ps}(\breve{x}^*)}{\mathrm{d}x}\right\} \qquad \text{with} \qquad \breve{x}^* = \frac{16x^*}{D_i\mathrm{Re}_i} \tag{A.63}$$

or with dimensions

$$\frac{\mathrm{d}p}{\mathrm{d}x} = -\frac{1}{D_h\mathrm{Re}_h}\frac{\rho\bar{u}^2}{2}\left\{K_{pl} + 16\frac{\mathrm{d}K_{ps}(\breve{x}^*)}{\mathrm{d}\breve{x}^*}\right\}, \tag{A.64}$$

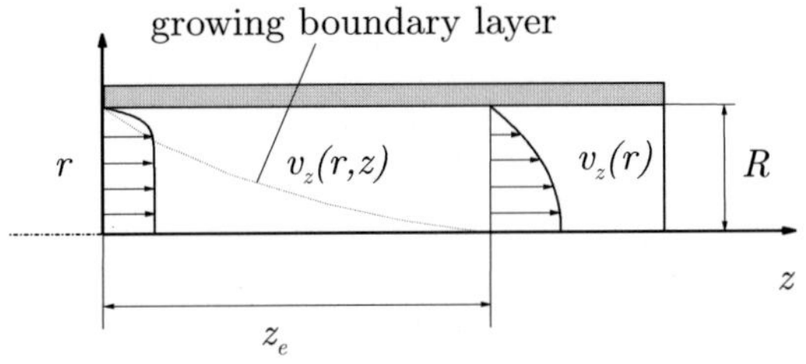

Fig. A.17. Entrance flow profile in a circular cylindrical tube (not to scale)

where $K_{pl} = 96$, and K_{ps} is given by (A.54) or (A.55). In the simplified case with $\varepsilon = 1$, the pressure gradient reads

$$\frac{\mathrm{d}p}{\mathrm{d}x} = -\frac{1}{D_h \mathrm{Re}_h} \frac{\rho \bar{u}^2}{2} \left\{ K_{pl} + \sum_i \left[192 e^{-16\alpha_i^2 \hat{x}} - 128 e^{-32\alpha_i^2 \hat{x}} \right] \right\} \tag{A.65}$$

with $\hat{x} = x/(D_h \mathrm{Re}_h) = \check{x}/16$.

A.3.2 Circular Cylindrical Tube

In the case of a circular cylindrical tube (see Fig. A.17) the scaling of the equation was performed by [5] the following way:

$$\hat{r} = \frac{r}{R}; \; \check{z} = \frac{z\nu}{R^2 \bar{v}_z} = \frac{4z}{D_h \mathrm{Re}_h}; \; \check{z}^* = \frac{z^* \nu}{R^2 \bar{v}_z}; \; \hat{v}_z = \frac{v_z}{\bar{v}_z}; \; D_h = 2R; \; \mathrm{Re}_h = \frac{\bar{v}_z D_h}{\nu} \tag{A.66}$$

The definition of the stretched coordinate z^* is similar to (A.44). The velocity profile reads

$$\hat{v}_z = 2\left(1 - \hat{r}^2\right) + \sum_i \frac{4}{\alpha_i^2} \left\{ \frac{J_0(\alpha_i \hat{r})}{J_0(\alpha_i)} - 1 \right\} e^{-\alpha_i^2 \check{z}^*} \; . \tag{A.67}$$

The first term describes the fully developed HAGEN-POISEUILLE profile, the second term the difference velocity. The eigenvalues of α_i are solutions of the equation

$$J_1(\alpha_i)/J_0(\alpha_i) = 0.5\alpha_i \; , \tag{A.68}$$

the values are given in Table A.5. The coefficient ε for the stretching of the $z-$ coordinate is defined as

$$\varepsilon = \frac{\int_0^1 \left(2\hat{v}_z - 1.5\hat{v}_z^2\right) \frac{\partial \hat{v}_z}{\partial \check{z}^*} \hat{r} \mathrm{d}\hat{r}}{\left(\frac{\partial \hat{v}_z}{\partial \hat{r}}\right)_{\hat{r}=1} + \int_0^1 \left(\frac{\partial \hat{v}_z}{\partial \hat{r}}\right)^2 \hat{r} \mathrm{d}\hat{r}} \; . \tag{A.69}$$

Once the velocity profile is known, the pressure loss can be given in dimensionless form as

$$\frac{2\,(p_i - p)}{\rho \bar{v}_z^2} = 16\check{z} + K_{ts}(\check{z}) \tag{A.70}$$

with

$$K_{ts1}^{*} = \frac{2}{3} + \sum_i \frac{8}{\alpha_i^2}\left\{e^{-\alpha_i^2 \check{z}^*} - 2\right\} e^{-\alpha_i^2 \check{z}^*} - 4 \int_0^{\check{z}^*} \varepsilon \left.\frac{\partial \hat{v}_z}{\partial \hat{r}}\right|_{\hat{r}=1} \mathrm{d}\check{z}^* \,, \tag{A.71}$$

if we consider the stretched coordinate, and

$$K_{ts2} = \frac{4}{3} + \sum_i \frac{8}{\alpha_i^2}\left\{e^{-\alpha_i^2 \check{z}} - 3\right\} e^{-\alpha_i^2 \check{z}} \,, \tag{A.72}$$

if we set $\varepsilon = 1$. The pressure at $z = 0$ is denoted by p_i. The first term of (A.70) describes the linear pressure loss of a fully developed flow field, the additional pressure loss K_{ts1}^{*} or K_{ts2} vanishes after the entrance length has reached.

The final equation implemented in the model of the capillary rise reads now

$$\frac{2(p_i - p)}{\rho \bar{u}^2} = K_{tl}\hat{z} + \frac{4}{3} + \sum_{i=1}^{25} \frac{8}{\alpha_i^2}\left\{e^{-4\alpha_i^2 \hat{z}} - 3\right\} e^{-4\alpha_i^2 \hat{z}} \tag{A.73}$$

with $K_{tl} = 64$ and

$$\hat{z} = \frac{z}{D_h \mathrm{Re}_h} = \frac{\check{z}}{4} \,. \tag{A.74}$$

For the comparison of this solution with our numerical calculations we need the derivative

$$\frac{\mathrm{d}p}{\mathrm{d}z} = -\frac{1}{D_h \mathrm{Re}_h}\frac{\rho \bar{v}_z^2}{2}\left\{K_{tl} + 64 \sum_{i=1}^{25}\left[\frac{3}{2} e^{-4\alpha_i^2 \hat{z}} - e^{-8\alpha_i^2 \hat{z}}\right]\right\} \tag{A.75}$$

or in dimensionless form

$$\frac{\mathrm{d}\hat{p}}{\mathrm{d}\hat{z}} = -K_{tl}\left\{1 + \sum_{i=1}^{25}\left[\frac{3}{2} e^{-4\alpha_i^2 \hat{z}} - e^{-8\alpha_i^2 \hat{z}}\right]\right\} \tag{A.76}$$

with the pressure $\hat{p} = 2p/(\rho \bar{v}_z^2)$.

A.3.3 Numerical Calculation of the Entrance Pressure Drop

For comparison with the analytical solution the entrance pressure drop was calculated using FLOW-3D and FIDAP. A half zone of the tube with radius $R = 0.25$ cm and length $l > z_e = 0.0575\, D_h \mathrm{Re}_h$ was modeled with a constant inlet velocity boundary condition $v_z(z = 0) = \bar{v}_z$ (see A.17). The parameters of the calculations with FIDAP and FLOW-3D are given in Tables A.7 and A.8, respectively. The results of the numerical calculations are nondimension-

Table A.7. Parameters for the FIDAP calculations

Re_h (-)	l (cm)	q (cm^3/s)	$\bar{v}_z$ (cm/s)	ρ (g/cm^3)	μ 10^{-2} (g/(cm s))
10	1	0.04	0.2	1.000	1.000
47	2	0.2	1.0	0.824	0.896
100	3	0.4	2.0	1.000	1.000
234	10	1	5.1	0.824	0.896
937	30	4	20.4	0.824	0.896
1000	30	4	20.0	1.000	1.000
1639	50	7	35.7	0.824	0.896
2295	70	9.8	50.0	0.824	0.896

Table A.8. Parameters for the FLOW-3D calculations

Re_h (-)	l (cm)	q (cm^3/s)	$\bar{v}_z$ (cm/s)	ρ (g/cm^3)	μ 10^{-2} (g/(cm s))
100	9.85	0.4	2.2	0.824	0.896
234	11.8	1.0	5.1	0.824	0.896
1000	49.4	4.0	21.8	0.824	0.896
2000	119	8.5	43.5	0.824	0.896

alized in the following way:

$$\hat{p} = \frac{2p}{\rho \bar{v}_z^2}; \hat{v}_z = \frac{v_z}{\bar{v}_z}; \hat{z} = \frac{z}{D_h \mathrm{Re}_h}; \hat{r} = \frac{r}{D_h}; D_h = 2R \, .$$

For simplification the superscripts of the non-dimensional quantities are dropped for the rest of this section.

The center velocity $v_z(0)$ is depicted in Fig. A.18. The criterion for the entrance length ($v_z \geq 0.99\bar{v}_z$) is not fulfilled for $\mathrm{Re}_h < 50$. For $\mathrm{Re}_h > 50$ the increase of the velocity due to the displacement of the boundary layer at the tube wall starts earlier with increasing Re_h and the velocity gradient increases with decreasing Re_h. The pressure gradient was calculated from the original CFD data using a finite difference formulation such that

$$\left(\frac{\mathrm{d}p}{\mathrm{d}z}\right)_j = \frac{1}{4}\left\{\frac{p_j - p_{j-1}}{z_j - z_{j-1}} + \frac{p_{j+1} - p_j}{z_{j+1} - z_j} + 2\left[\frac{p_{j+1} - p_{j-1}}{z_{j+1} - z_{j-1}}\right]\right\} \, .$$

Please remark that the spacing in z is not constant due to the adaptation of the mesh to the entrance flow problem. The pressure gradients for Reynolds numbers $10 < \mathrm{Re}_h < 2300$ from the numerical calculation and the pressure gradient given by [5] (computed from A.76) are shown in Fig. A.19. The pressure gradient was calculated from the center line pressure and it does not

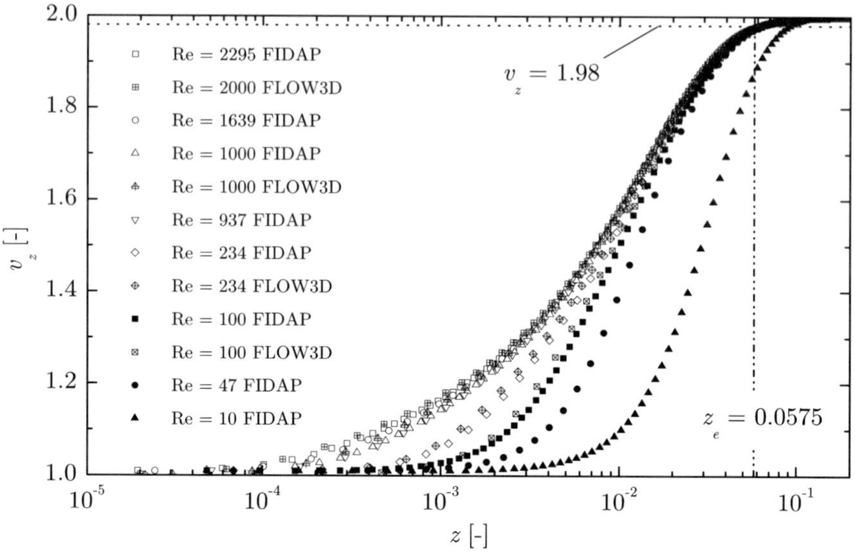

Fig. A.18. Center velocity $\hat{v}_z(\hat{r} = 0)$ in the cylindrical tube entrance region as a function of the REYNOLDS number Re_h

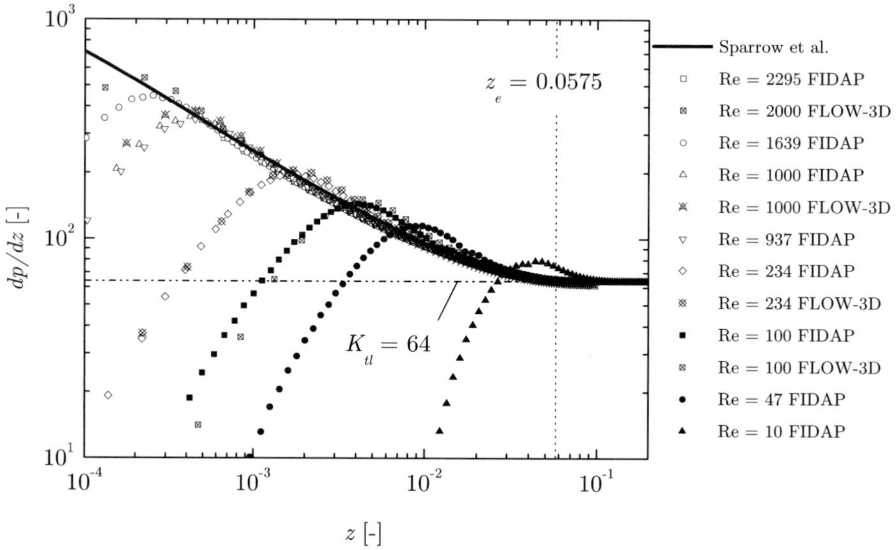

Fig. A.19. Pressure gradient $d\hat{p}/d\hat{z}$ at $\hat{r} = 0$ in the cylindrical tube entrance region as a function of the REYNOLDS number Re_h

represent the cross section averaged pressure. This explains the increase of the pressure at the tube entrance and the following overshoot. The boundary layer develops and the flow area is narrowed by the displacement thickness of the boundary layer. It can be seen on Fig. A.19 that the computed values from FIDAP and FLOW-3D approach the theoretical curve given by [5] better with increasing Reynolds number. Finally it can be concluded that the solution from [5] is useful to calculate the excess pressure drop in the mathematical model for the capillary rise and for the determination of the critical velocities in open capillary channels.

References

1. S. Levine, P. Reed, E.J. Watson, G. Neale: A Theory of the Rate of Rise of a Liquid in a Capillary. In: *Colloid and Interface Science*, Chap. 3, ed by M. Kerker (Academic Press, New York 1976) pp. 403–419
2. F.M. White, F. M.: *Viscous Fluid Flow*, (McGraw-Hill, New York 1974)
3. K. Wieghardt: *Theoretische Strömungslehre*, (Teubner, Stuttgart 1974)
4. K. Stephan: Chemie - Ingenieur - Technik **31.12**, 773 (1959)
5. E.M. Sparrow, S.H. Lin, T.S. Lundgren: Phys. Fluids **7.3**, 338 (1964)

B

List of Symbols

Characteristic values

g_0 gravity acceleration
k characteristic acceleration
L characteristic length
L_c capillary length, $L_c = \sqrt{\beta/k}$
p_c characteristic pressure
t_c characteristic time
t_{cu} convective time scale, $t_{cu} = L/V$
t_{pu} pressure unsteady time scale, $t_{pu} = \sqrt{L^3/\beta}$
t_{pv} pressure viscous time scale, $t_{pv} = \nu L/\beta$
t_{vu} viscous time scale, $t_{vu} = L^2/\nu$
V characteristic velocity
v_c characteristic velocity for tubes, $v_c = \sqrt{2\beta/R}$
v_0 characteristic velocity for channels, $v_0 = \sqrt{2\beta/a}$

Field variables

$\boldsymbol{k}$ acceleration vector
p pressure
T temperature
v mean velocity
U, V velocity
$\boldsymbol{u}$ velocity vector $(u_x, u_y, u_z)^T$

Independent variables

r, φ, z cylindrical coordinates
$\tilde{r}, \theta, \tilde{\varphi}$ spherical coordinates
t time
x, y, z cartesian coordinates

Material properties

c_s	speed of sound
β	kinematic surface tension, $\beta = \sigma/\rho$
γ	contact angle
γ_d	dynamic contact angle
γ_s	static contact angle
γ_{s0}	initial static contact angle
κ	compressibility
μ	dynamic viscosity
ν	kinematic viscosity, μ/ρ
ρ	density
σ	surface tension
n	refractive index

Miscellaneous

a	plate distance
A	amplitude
A	cross-sectional area
b	plate breadth
C	constants
D	damping ratio
D, d	cylinder diameter
D_h	hydraulic diameter
f	recording frequency of video images
F	force
h	coordinate of the free surface
h	meniscus height in the cylindrical tube
h_0	fill height of liquid
H	mean curvature
k	contour of the free surface
k_z	acceleration in z-direction
k_{zi}	initial acceleration in z-direction
K	friction coefficient
l_s	slip length
L	coefficient
L_e	entrance length
m	mass
n	period number of oscillation
Q, q	volume flux
R	cylinder radius
R_1, R_2	principal radii of curvature
S	entropy
T	period of oscillation
T	temperature

Miscellaneous continued

V liquid volume
w_f irreversible pressure loss
α dimensionless segment of spherical surface
α_i eigenvalues
β dimensionless radius of spherical surface
β Sparrow factor
δ exponential decay coefficient
δ_w thickness of Stokes boundary layer
ε stretching coefficient
φ_d zero phase angle
Λ logarithmic decrement
Λ aspect ratio
τ stress tensor
Γ profile factor
Γ free parameter for the contact angle boundary condition
ω_0 undamped natural frequency
ω_d damped natural frequency
Ω dimensionless frequency of undamped oscillation

Operator and Vectors

∇ Nabla operator
∇_s gradient of the normal vector
$\triangle$ difference
$\boldsymbol{n}$ normal vector

Sub- and Superscripts

$\ldots_0$ initial
$\ldots_{cd}$ center disturbance
$\ldots_{ce}$ center equilibrium
$\ldots_{cp}$ center peak
$\ldots_{cr}$ center rise
$\ldots_e$ equilibrium
$\ldots_n$ normal
$\ldots_m$ meniscus
$\ldots_{pu}$ pressure unsteady
$\ldots_{pv}$ pressure viscous
$\ldots_s$ static
$\ldots_{s0}$ static initial
$\ldots_t$ tangential
$\ldots_{vu}$ viscous unsteady

$\ldots_w$	wall
$\ldots_{w0}$	wall initial
$\ldots_{we}$	wall equilibrium
$\ldots_{wf}$	wall fixed
$\ldots_{wm}$	wall maximum
$\ldots_{wp}$	wall peak
$\ldots_x$	in $x-$ direction
$\ldots_y$	in $y-$ direction
$\ldots_z$	in $z-$ direction
$\ldots^*$	dimensionless

Dimensionless numbers

BOND number	$\mathrm{Bo} = \dfrac{kL^2}{\beta}$
Capillary number	$\mathrm{Ca} = \dfrac{\nu V}{\beta}$
FROUDE number	$\mathrm{Fr} = \dfrac{V^2}{kL}$
MACH number	$\mathrm{Ma} = \dfrac{v_g}{c_s}$
MORTON number	$\mathrm{Mo} = \left(\dfrac{\nu^4 k}{\beta^3}\right)^{\frac{1}{4}}$
OHNESORGE number	$\mathrm{Oh} = \left(\dfrac{\nu^2}{\beta L}\right)^{\frac{1}{2}}$
REYNOLDS number	$\mathrm{Re} = \dfrac{VL}{\nu}$
WEBER number	$\mathrm{We} = \dfrac{V^2 L}{\beta}$
GALILEI number	$\mathrm{Ga} = \dfrac{kL^3}{\nu^2}$

Index

Springer Tracts in Modern Physics

179 **Anomalous X-ray Scattering for Materials Characterization**
Atomic-Scale Structure Determination
By Yoshio Waseda 2002. 132 figs. XIV, 214 pages

180 **Coverings of Discrete Quasiperiodic Sets**
Theory and Applications to Quasicrystals
Edited by P. Kramer and Z. Papadopolos 2002. 128 figs., XIV, 274 pages

181 **Emulsion Science**
Basic Principles. An Overview
By J. Bibette, F. Leal-Calderon, V. Schmitt, and P. Poulin 2002. 50 figs., IX, 140 pages

182 **Transmission Electron Microscopy of Semiconductor Nanostructures**
An Analysis of Composition and Strain State
By A. Rosenauer 2003. 136 figs., XII, 238 pages

183 **Transverse Patterns in Nonlinear Optical Resonators**
By K. Staliūnas, V. J. Sánchez-Morcillo 2003. 132 figs., XII, 226 pages

184 **Statistical Physics and Economics**
Concepts, Tools and Applications
By M. Schulz 2003. 54 figs., XII, 244 pages

185 **Electronic Defect States in Alkali Halides**
Effects of Interaction with Molecular Ions
By V. Dierolf 2003. 80 figs., XII, 196 pages

186 **Electron-Beam Interactions with Solids**
Application of the Monte Carlo Method to Electron Scattering Problems
By M. Dapor 2003. 27 figs., X, 110 pages

187 **High-Field Transport in Semiconductor Superlattices**
By K. Leo 2003. 164 figs.,XIV, 240 pages

188 **Transverse Pattern Formation in Photorefractive Optics**
By C. Denz, M. Schwab, and C. Weilnau 2003. 143 figs., XVIII, 331 pages

189 **Spatio-Temporal Dynamics and Quantum Fluctuations in Semiconductor Lasers**
By O. Hess, E. Gehrig 2003. 91 figs., XIV, 232 pages

190 **Neutrino Mass**
Edited by G. Altarelli, K. Winter 2003. 118 figs., XII, 248 pages

191 **Spin-orbit Coupling Effects in Two-dimensional Electron and Hole Systems**
By R. Winkler 2003. 66 figs., XII, 224 pages

192 **Electronic Quantum Transport in Mesoscopic Semiconductor Structures**
By T. Ihn 2003. 90 figs., XII, 280 pages

193 **Spinning Particles – Semiclassics and Spectral Statistics**
By S. Keppeler 2003. 15 figs., X, 190 pages

194 **Light Emitting Silicon for Microphotonics**
By S. Ossicini, L. Pavesi, and F. Priolo 2003. 206 figs., XII, 284 pages

195 **Uncovering *CP* Violation**
Experimental Clarification in the Neutral K Meson and B Meson Systems
By K. Kleinknecht 2003. 67 figs., XII, 144 pages

196 **Ising-type Antiferromagnets**
Model Systems in Statistical Physics and in the Magnetism of Exchange Bias
By C. Binek 2003. 52 figs., X, 120 pages

197 **Electroweak Processes in External Electromagnetic Fields**
By A. Kuznetsov and N. Mikheev 2003. 24 figs., XII, 136 pages

198 **Electroweak Symmetry Breaking**
The Bottom-Up Approach
By W. Kilian 2003. 25 figs., X, 128 pages

Springer Tracts in Modern Physics

199 **X-Ray Diffuse Scattering from Self-Organized Mesoscopic Semiconductor Structures**
By M. Schmidbauer 2003. 102 figs., X, 204 pages

200 **Compton Scattering**
Investigating the Structure of the Nucleon with Real Photons
By F. Wissmann 2003. 68 figs., VIII, 142 pages

201 **Heavy Quark Effective Theory**
By A. Grozin 2004. 72 figs., X, 213 pages

202 **Theory of Unconventional Superconductors**
By D. Manske 2004. 84 figs., XII, 228 pages

203 **Effective Field Theories in Flavour Physics**
By T. Mannel 2004. 29 figs., VIII, 175 pages

204 **Stopping of Heavy Ions**
By P. Sigmund 2004. 43 figs., XIV, 157 pages

205 **Three-Dimensional X-Ray Diffraction Microscopy**
Mapping Polycrystals and Their Dynamics
By H. Poulsen 2004. 49 figs., XI, 154 pages

206 **Ultrathin Metal Films**
Magnetic and Structural Properties
By M. Wuttig and X. Liu 2004. 234 figs., XII, 375 pages

207 **Dynamics of Spatio-Temporal Cellular Structures**
Henri Benard Centenary Review
Edited by I. Mutabazi, J.E. Wesfreid, and E. Guyon 2005. approx. 50 figs., 150 pages

208 **Nuclear Condensed Matter Physics with Synchrotron Radiation**
Basic Principles, Methodology and Applications
By R. Röhlsberger 2004. 152 figs., XVI, 318 pages

209 **Infrared Ellipsometry on Semiconductor Layer Structures**
Phonons, Plasmons, and Polaritons
By M. Schubert 2004. 77 figs., XI, 193 pages

210 **Cosmology**
By D.-E. Liebscher 2005. Approx. 100 figs., 300 pages

211 **Evaluating Feynman Integrals**
By V.A. Smirnov 2004. 48 figs., IX, 247 pages

213 **Parametric X-ray Radiation in Crystals**
By V.G. Baryshevsky, I.D. Feranchuk, and A.P. Ulyanenkov 2006. 63 figs., IX, 172 pages

214 **Unconventional Superconductors**
Experimental Investigation of the Order-Parameter Symmetry
By G. Goll 2006. 67 figs., XII, 172 pages

215 **Control Theory in Physics and other Fields of Science**
Concepts, Tools, and Applications
By M. Schulz 2006. 46 figs., X, 294 pages

216 **Theory of the Muon Anomalous Magnetic Moment**
By K. Melnikov, A. Vainshtein 2006. 33 figs., XII, 176 pages

217 **The Flow Equation Approach to Many-Particle Systems**
By S. Kehrein 2006. 24 figs., XII, 170 pages

219 **Inelastic Light Scattering of Semiconductor Nanostructures**
By Christian Schüller 2007. 105 figs., XII, 178 pages

220 **Precision Electroweak Physics at Electron-Positron Colliders**
By Stefan Roth 2007. 107 figs., X, 174 pages

221 **Free Surface Flows under Compensated Gravity Conditions**
By Michael Dreyer 2007. 128 figs., X, 272 pages

Printing: Krips bv, Meppel
Binding: Stürtz, Würzburg